Energiewende "Made in Germany"

Christian von Hirschhausen • Clemens Gerbaulet •
Claudia Kemfert • Casimir Lorenz • Pao-Yu Oei
Editors

Energiewende "Made in Germany"

Low Carbon Electricity Sector Reform
in the European Context

 Springer

Editors
Christian von Hirschhausen
Economics and Management
University of Technology (TU Berlin)
Berlin, Germany

Clemens Gerbaulet
Economics and Management
University of Technology (TU Berlin)
Berlin, Germany

Claudia Kemfert
DIW, Energy-Transport-Environment
German Institute for Economic Research
Berlin, Germany

Casimir Lorenz
DIW, Energy-Transport-Environment
German Institute for Economic Research
Berlin, Germany

Pao-Yu Oei
Junior Research Group "CoalExit"
University of Technology (TU Berlin)
Berlin, Germany

ISBN 978-3-319-95125-6   ISBN 978-3-319-95126-3   (eBook)
https://doi.org/10.1007/978-3-319-95126-3

Library of Congress Control Number: 2018962959

This Springer imprint is published by the registered company Springer Nature Switzerland AG
The registered company address is: Gewerbestrasse 11, 6330 Cham, Switzerland

Contents

List of Contributors

Hanna Brauers Junior Research Group "CoalExit", Berlin, Germany

TU Berlin, Berlin, Germany

DIW Berlin, Berlin, Germany

Thorsten Burandt Junior Research Group "CoalExit", Berlin, Germany

TU Berlin, Berlin, Germany

DIW Berlin, Berlin, Germany

Jonas Egerer Friedrich-Alexander-Universität (FAU), Erlangen, Nürnberg, Germany

Clemens Gerbaulet TU Berlin, Berlin, Germany

DIW Berlin, Berlin, Germany

Karlo Hainsch TU Berlin, Berlin, Germany

Philipp Herpich TU Berlin, Berlin, Germany

DIW Berlin, Berlin, Germany

Christian von Hirschhausen TU Berlin, Berlin, Germany

DIW Berlin, Berlin, Germany

Franziska Holz DIW Berlin, Berlin, Germany

Norwegian University of Science and Technology (NTNU), Trondheim, Norway

Claudia Kemfert DIW Berlin, Berlin, Germany

Hertie School of Governance, Berlin, Germany

German Advisory Council on the Environment (SRU), Berlin, Germany

Friedrich Kunz TenneT, Berlin, Germany

Konstantin Löffler Junior Research Group "CoalExit", Berlin, Germany

TU Berlin, Berlin, Germany

DIW Berlin, Berlin, Germany

Casimir Lorenz Aurora Energy Research, Berlin, Germany

Roman Mendelevitch HU Berlin, Berlin, Germany

DIW Berlin, Berlin, Germany

Anne Neumann University Potsdam, Potsdam, Germany

DIW Berlin, Berlin, Germany

Pao-Yu Oei Junior Research Group "CoalExit", Berlin, Germany

TU Berlin, Berlin, Germany

DIW Berlin, Berlin, Germany

Petra Opitz DIW Econ, Berlin, Germany

Felix Reitz Europe beyond Coal, Brussels, Belgium

Thure Traber DTU, Kongens Lyngby, Denmark

Ben Wealer TU Berlin, Berlin, Germany

DIW Berlin, Berlin, Germany

Jens Weibezahn TU Berlin, Berlin, Germany

Abbreviations

AC	Alternating Current
ACER	Agency for the Cooperation of Energy Regulators
AEG	Allgemeine Elektricitäts-Gesellschaft
AFRR	Automatic Frequency Restoration Reserve
ATC	Available Transfer Capacity
AtG	Atomgesetz (Law on Nuclear Energy)
BAU	Business as usual
BBPIG	Bundesbedarfsplangesetz (Law on transmission development)
BEMIP	Baltic Energy Market Interconnection Plan
BEWAG	Berliner Elektrizitätswirtschafts-Aktiengesellschaft (Berlin Utility)
BIP	Bruttoinlandsprodukt (gross domestic product)
BMUB	Bundesministerium für Umwelt, Naturschutz, Bau und Reaktorsicherheit (Federal Ministry for the Environment, Nature Conservation, Building and Nuclear Safety)
BMWI	Bundesministerium für Wirtschaft und Energie (Federal Ministry for Economic Affairs and Energy)
bn	Billion
BNetzA	Bundesnetzagentur (German Federal Network Agency)
BWR	Boiling Water Reactor
CA-CM	Capacity Allocation and Congestion Management
CCGT	Combined Cycle Gas Turbine
CCTS	Carbon Capture, Transport and Storage
CDM	Clean Development Mechanism
CDU	Christlich Demokratische Union Deutschlands (Christian Democratic Union)
CH_4	Methane
CHP	Combined Heat and Power
CO_2	Carbon Dioxide
CoBAs	Coordinated Balancing Areas

CPF	Carbon Price Floor
CSU	Christlich-Soziale Union in Bayern (Christian Social Union of Bavaria)
CWE	Central West Europe
DC	Direct Current
dena	Deutsche Energie-Agentur (German Energy Agency)
DIW Berlin	Deutsches Institut für Wirtschaftsforschung (German Institute for Economic Research)
DVG	Deutsche Verbundgesellschaft (German Network Association)
EC	European Commission
ECSC	European Coal and Steel Community
EDF	Électricité de France
EEEP	Economics of Energy & Environmental Policy
EEG	Erneuerbare-Energie-Gesetz (Law on Renewable Energies)
EESS	European Energy Security Strategy
EEX	European Energy Exchange
EIB	European Investment Bank
EMF	Energy Modeling Forum
EN	European Norm
EnBW	Energie Baden-Württemberg
EnEv	Energieeinsparverordnung (Energy Efficiency Ordinance)
EnLAG	Energieleitungsausbaugesetz (Law on Developing Electricity Transmission Infrastructure)
ENTSO-E	European Network of Transmission System Operators for Electricity
ENTSO-G	European Network of Transmission System Operators for Gas
EOR	Enhanced Oil Recovery
EPS	Emissions Performance Standard
ErP	Energy-Related Products
ETS	Emission Trading System
EU	European Union
EU-ETS	European Emission Trading System
EUR	Euro
EURATOM	European Atomic Energy Community
EVS	Energie-Versorgung Schwaben (Energy Supply Schwaben)
FAZ	Frankfurter Allgemeine Zeitung
FBMC	Flow-Based Market Coupling
FDP	Freie Demokratische Partei (Free Democratic Party) Germany
FERC	Federal Energy Regulatory Commission
FIT	Feed-in tariff
FOSG	Friends of the Supergrid
FRG	Federal Republic of Germany
G7	Group of Seven
GBP	British Pound
GDP	Gross Domestic Product

GDR	German Democratic Republic
GGM	Global Gas Model
GHG	Greenhouse Gases
GW	Gigawatt
GWh	Gigawatt hour
GWB	Gesetz gegen Wettbewerbsbeschränkungen (Law against Restraints of Competition)
H_2	Hydrogen
HEW	Hamburgische Elektrizitäts-Werke (Utility of Hamburg)
Hg	Mercury
HVDC	High-Voltage Direct Current
IA	Impact Assessment
IAEA	International Atomic Energy Agency
IAEE	International Association for Energy Economics
IEA	International Energy Agency
IGCC	International Grid Control Cooperation
IPCC	Intergovernmental Panel on Climate Change
ISO	Independent System Operator
ITC	Inter-TSO Compensation Mechanism
JI	Joint Implementation
JRC	Joint Research Centre
kV	Kilovolt
kW	Kilowatt
kWh	Kilowatt hour
KWU	Kraftwerks Union
LCOE	Levelized Cost of Electricity
LNG	Liquefied Natural Gas
MENA	Middle East and North Africa
mn	million
MSR	Market Stability Reserve
Mt	Megaton
MW	Megawatt
MWh	Megawatt hour
NABEG	Netzausbaubeschleunigungsgesetz (Transmission Network Development Acceleration Law)
NAPE	National Action Plan on Energy Efficiency
NC EB	Network Code on Electricity Balancing
NC LFCR	Network Code on Load-Frequency Control and Reserves
NEP	Netzentwicklungsplan (Network Development Plan)
NGO	Non-governmental Organization
NOVA	Netzoptimierung, -verstärkung und -ausbau (Network Optimization, Strengthening and Expansion)
NO_X	Nitrogen Oxide
NPP	Nuclear Power Plant

NPS	New Policies Scenario
NRA	National Regulatory Authority
NRW	Nordrhein-Westfalen (North Rhine-Westphalia)
NSCOGI	North Seas Countries Offshore Grid Initiative
NTC	Net Transfer Capacity
OCGT	Open Cycle Gas Turbine
P2G	Power-to-Gas
P2H	Power-to-Heat
PC	Primary Control Reserve
PCR	Price Coupling of Regions
PLEF	Pentalateral Energy Forum
PS	Horsepower
PSP	Pumped Storage Power Plant
PV	Photovoltaic
PWR	Pressurized Water Reactor
R&D	Research and Development
RBMK	Reaktor Bolshoy Moshchnosty Kanalny (Graphite-Moderated Boiling Water Reactor)
RES	Renewable Energy Sources
RSK	Reaktor-Sicherheitskommission (Commission on the Security of Nuclear Reactors)
RWE	Rheinisch-Westfälisches Elektrizitätswerk
SC	Secondary Control Reserve
SME	Small and Medium-Sized Enterprises
SO_2	Sulphur dioxide
SOAF	Scenario Outlook and Adequacy Forecast
SOC	Social Overhead Capital
SoS	Security of Supply
SPD	Sozialdemokratische Partei Deutschlands (German Social Democratic Party)
StandAG	Standortauswahlgesetz (Site Selection Act)
StrEG	Stromeinspeisegesetz (Law on Renewable Feed-in)
t	Metric Tonne (1000 kg)
TC	Tertiary Control Reserve
tce	Tons of Coal Equivalent
TFEU	Treaty on the Functioning of the European Union
TSO	Transmission System Operator
TWh	Terawatt hour
TYNDP	Ten-Year Network Development Plan
UK	United Kingdom
UKR	Ukraine
UNFCCC	United Nations Framework Convention on Climate Change
US	United States
USA	United States of America

USD	US-Dollar
VEB	Socialist combine ("enterprise owned by the people")
VEW	Vereinigte Elektrizitätswerke Westfalen (Utility of Westphalia)
VVER	Wodo-wodjanoi Energetitscheski Reaktor (Water-Water Energetic Reactor)
WACC	Weighted Average Cost of Capital
WIP	Workgroup for Infrastructure Policy
WW	World War
WWF	World Wide Fund for Nature

Chapter 1
Introduction

Christian von Hirschhausen, Clemens Gerbaulet, Claudia Kemfert, Casimir Lorenz, and Pao-Yu Oei

> *"The second path combines a prompt and serious commitment to efficient use of energy, rapid development of renewable energy sources matched in scale and in energy quality to end-use needs, and special transitional fossil-fuel technologies. This path, a whole greater than the sum of its parts, diverges radically from incremental past practices to pursue long-term goals."*
> Amory B. Lovins (1976). Energy Strategy: The Road Not Taken? Foreign Affairs, 6(20), p. 9.

1.1 Introduction

When Amory Lovins, Director of the Rocky Mountain Institute, set out the conditions for a "soft path" of decentral, renewables-based energy development, in 1976 (see quote), he could not foresee that four decades later, he would receive the highest recognition for public service, the German Federal Cross of Merit

C. von Hirschhausen (✉) · C. Gerbaulet · C. Lorenz
TU Berlin, Berlin, Germany

DIW Berlin, Berlin, Germany
e-mail: cvh@wip.tu-berlin.de

C. Kemfert
DIW Berlin, Berlin, Germany

Hertie School of Governance, Berlin, Germany

German Advisory Council on the Environment (SRU), Berlin, Germany

P.-Y. Oei
TU Berlin, Berlin, Germany

DIW Berlin, Berlin, Germany

Junior Research Group, "CoalExit", Berlin, Germany

© Springer Nature Switzerland AG 2018
C. von Hirschhausen et al. (eds.), *Energiewende "Made in Germany"*,
https://doi.org/10.1007/978-3-319-95126-3_1

("Bundesverdienstkreuz"), for having spearheaded what is now called "energiewende". And in fact, between the publication of the book "Energiewende" in 1980 by a scholar of Lovins', Florentin Krause et al. (1980) to the ground-breaking events in 2010/2011, pushing the energiewende further, many things happened in energy and climate policy, in Germany, Europe, and the world, that may have not been forecast by Lovins, and that have altered energy and climate policy altogether.

Yet, under the impression of the Fukushima nuclear power plant disaster in March 2011, the German government, legal system, civil society, and energy industry again changed course in long-term energy and climate policy, confirming earlier attempts to embark on a "soft path". On a timeline extending to 2050, plans were made to set strict emission caps on greenhouse gas (GHG) emissions, to rapidly decommission all nuclear power plants (NPP), to significantly increase the share of renewables in energy production, and to implement ambitious efficiency targets. The long germination period, since the mid 1970s, leading to the re-orientation of energy and climate policy—at a time when the German electricity sector was still largely reliant on coal and nuclear power—is now commonly referred to as the "energiewende" (*Wende* meaning turn or turnaround, sometimes also called the "energy transformation", "energy transition", etc.) and has attracted substantial attention, both in Germany and internationally. Initially considered a short-lived epiphenomenon by many observers and openly opposed by the incumbent conventional energy industry, the energiewende proved its critics and skeptics wrong, at least partially. Overall, the reforms of the last decade can be considered a success, with some of the targets being accomplished, and it still continues today with widespread public support.

The energiewende (term we will use throughout this book) emerged at a time of increasing debates of global warming and climate change (Houghton et al. 1990; Stern 2007; WBGU 2011). Many countries in Europe and around the globe were considering how to move to lower carbon energy systems, and most of them still are. Thus, the European Union is still pursuing its decarbonization objectives of a 40% reduction of greenhouse gas (GHG) emissions by 2030 (reference: 1990), and a reduction of 80–95% in the longer term. The US, too, launched a program to reduce GHG emissions under the Federal Clean Air Act, even though the current adminis-tration has set out to stop this initiative; nonetheless, the US power sector is constantly retiring coal plants, and is moving towards a coal exit as well (Heal 2017). Even in Asia, the region with the highest energy consumption growth rates worldwide, countries such as China and India have identified the need for more environmentally sustainable energy strategies and reduced coal consumption (IEA 2016). The 2015 Paris Agreement of the UN Convention on Climate Change, to limit the rise of the mean global temperature to 2° and possibly even to 1.5°C has increased pressure on governments and industry to accelerate their low-carbon transformation policies. It comes as no surprise, therefore, that the energiewende is being followed with great interest by observers worldwide, both with high hopes for its potential positive impacts and with skepticism about its costs and financial sustainability.

The objective of this book is to present an in-depth look at the energiewende, from its origins to its concrete implementation in Germany, as well as its impacts within the European context and its medium- and long-term perspectives. Our working hypothesis, based on extensive modeling exercises, policy consulting, personal on-site case studies, and the growing literature, is that the energiewende is a unique political-historical period that will transform the structure of the German energy sector, leading to more decentralized energy production and decision-making and transforming the structure of the energy industry within Germany and beyond. So far, the energiewende has been a success overall, in particular because the foundation for a renewables-based electricity system has been laid. Yet other objectives had to be postponed, though, such as the GHG emission reduction target for 2020 (-40%, relative to 1990). While the lessons of the energiewende do not apply directly to all countries and regions worldwide, they offer insights from the natural experiment of transforming a large-scale, conventional electricity system based on coal and nuclear energy into a renewables-based system. Our analysis focuses on the electricity sector, but we also address other challenges in the transport and heating sectors, as well as the upcoming interconnectedness between the three, called "sector coupling".

The next section of this introductory chapter spells out the key characteristics of the energiewende, which later chapters will analyze in more detail. Section 1.3 looks at the German energiewende in the context of the energy and climate policy literature. Section 1.4 presents a detailed outline of the book, and the last - Section concludes with acknowledgements.

1.2 The Main Ingredients of the Energiewende

The German term "energiewende" is now commonly used throughout the world and is now penetrating the English language the same was as have other German words like kindergarten, bratwurst, wanderlust, or zeitgeist. In this book, we use the term "energiewende" to refer to a political and societal process in the realm of energy and climate policy, that was ongoing for quite some time already, but that accelerated in Germany between September 2010, the German energy concept 2050, and June 2011, moment of the nuclear phase-out law. In the framework of the energiewende, a series of decisions were made to pursue an energy and environmental policy that would shift the German energy sector away from reliance on fossil fuels and nuclear energy and make it more efficient, more decentralized, and more renewables-based. The concrete targets include (see individual chapters for details):

- Reducing greenhouse gas emissions, compared to 1990 levels, by 40% by 2020, 55% by 2030, 70% by 2040, and 80–95% by 2050;
- Closing all nuclear power stations: seven units were taken offline in March 2011 and the remaining nine plants are scheduled to close by 2022;

• Increasing the share of renewables for electricity generation to at least 38% in 2020, 50% in 2030, 67% in 2040, and 80% in 2050, and the share of renewables in final energy consumption to at least 30% by 2030 and at least 60% by 2050;
• Setting ambitious targets for energy efficiency.

These quantitative objectives were designed with the general intention to foster civil society participation in decision-making processes, in the production of energy, and in the distribution of profits and rents. Thus, the energiewende has also introduced a new energy policy paradigm in which a large share of decentralized, individually and cooperatively owned companies generate power alongside "big energy" companies; in 2015, over 67% of the new renewable electricity (wind and sun) was generated outside the traditional utilities, by cooperatives, private producers, etc. Although this objective is not set down in law, it forms the basis of the public consensus on the energiewende (Rosenkranz (2014), Morris and Jungjohann (2016), Davidson (2012)).

1.3   Current State of the Literature

There is a small, but rapidly growing literature on the energiewende. Members of our team have done extensive work on the energiewende, including a first survey by Kemfert (2014), a symposium volume published in the *Journal Economics of Energy & Environmental Policy* (EEEP, Vol. 3, No. 2, Fall 2014, some of which has been updated for the present book), a study on deep decarbonization in Germany (Kemfert et al. 2013), and a collection of papers published by the German Institute for Economic Research (DIW Berlin) in the quarterly *Vierteljahreshefte* ("Quarterly Journal of Economic Research," see Kemfert et al. (2013), in German). Similar research by other scientists include a book by Unnerstall (2017) providing an assessment of the current status of the transformation, with a focus on corporate perspectives. Grubb et al. (2014) textbook on "Planetary Economics" contains many of the discussions around the low-carbon energy transformation. Another book by Schippl et al. (2017) covers virtually all facets of the energy transition but is, however, restricted to German-speaking audiences. Other academic work is extensively cited in the following chapters in this book.

A second, more policy-oriented branch of the literature looks at concrete technical, legal, and institutional aspects of the energiewende, mainly at the level of individual sectors and/or projects. This applied literature stems from formal government entities, public bodies, stakeholder circles, think-tanks, etc. The German government issues the yearly Monitoring Report "Energy of the Future" (see BMWi 2015, 2016), which is accompanied by a detailed assessment from an advisory board to the Ministry of Economics and Energy (see BMWi and BMU

2012; BMWi 2015, 2016; Löschel et al. 2015; Löschel et al. 2014, 2018). Agora Energiewende, a think-tank financed by private foundations, has a 20-expert team dedicated to analyzing the results of technical studies for the use in policy and the public debate (see, for example, Agora's twelve theses on the energiewende and the big study on "Energiewende 2030—The Big Picture", respectively (Agora Energiewende 2013, 2018)).[1] The same foundations also have a series of publications for journalists and the interested public, "Clean Energy Wire", including detailed off-the-shelf material.[2] Another political foundation produces a series of publications on the energiewende in an international context.[3] Morris and Pehnt (2016) and Morris and Jungjohann (2016) and Fechner (2018, in German) provide a detailed account of this policy-oriented research.

A third branch of literature consists of comparative analyses situating the German experience in a broader cross-country context. Early, most influential work on "The Big Transformation" was carried out by the German Advisory Council on Global Change (WBGU 2011), with several updates later on. The 2° target, that has become a benchmark for global climate policy, also originated from this work (see for details Schellnhuber (2015)). The German Section of the World Energy Council (2014) conducted a comparison of six country-specific energy transformation processes (in addition to Germany: USA, Brazil, China, Saudi Arabia, and South Africa). The International Energy Agency (IEA) (2014) published the report "The Power of Transformation—Wind, Sun and the Economics of Flexible Power Systems" on the perspectives of wind- and solar-based electricity systems. The Intergovernmental Panel on Climate Change (IPCC) on renewables provides valuable information at a very detailed level (IPCC 2012), and the IPCC 5th Assessment Report (IPCC 2014) discusses a variety of low-carbon pathways. Another book by Hager and Stefes (2016) provides a comparison of Germany's energy transition with international peers such as the US and Japan. There is also ample literature on other countries undergoing a low-carbon energy transformation, such as Denmark (Danish government (2011)), the UK (DECC (2011), Foxon (2013)), France (Criqui and Hourcade 2015), the USA (Burtraw et al. (2014) and Heal (2017)), China (Li et al. 2018), and India (Bhushan 2017; Singh et al. 2018). Last but not least, the energiewende has even prompted business consultancies to develop specific indicators to put the German experience into international context, such as McKinsey's "Energy Transition Indicator".[4]

---

[1]See https://www.agora-energiewende.de/en/

[2]See: https://www.cleanenergywire.org/

[3]See: https://energytransition.org/

[4]See: http://reports.weforum.org/fostering-effective-energy-transition-2018/?code=wrl23

1.4   Structure of the Book

1.4.1   Part I: Historic Origins: The Energiewende
and the Transformation of the German Coal Sector

This book is divided into four parts. Part I (Chaps. 2 and 3) lays out the historical origins of the energiewende with respect to the energy, environmental, and climate policies that led up to it, as well as the specific transformation of the German coal sector. In Chap. 2, we identify some long-term trends of energy policy in Germany, going back to the turn of the last century; we also retrace subsequent developments such as the Energy Industry Act of 1935, discuss similarities and differences between energy policies in East and West Germany and their consolidation after reunification, and look at more recent European attempts to liberalize national energy sectors and create a single market. The chapter also covers important developments that occurred with the "wind of change" in the early 1990s: the emergence of a European climate policy, the first formal pushes towards renewables targets, and the drafting of unbundling directives for the electricity and the natural gas sectors. We identify elements of the energiewende as we go through time, since 1980, but focus on the developments between September 2010 and summer 2011, which is, from a historical perspective, a crucial period. It took nothing less than the March 2011 nuclear disaster in Fukushima, Japan, to establish the broad consensus on the acceleration of the energiewende, i.e. the shutdown of nuclear power plants, in combination with GHG emission reductions, and renewables targets. This includes the historic decision by Chancellor Angela Merkel to declare a moratorium on the lifetime extension on nuclear power only three days after Fukushima and the closure of the seven oldest reactors in Germany.[5]

Chapter 3 provides an historic account of the German coal industry over the last 70 years, a unique transformation process dominated by steady decline of an industry that previously employed more than 700,000 people. One focus of this historic case study therefore lies on the Ruhr area—Germany's largest hard coal mining area that was particularly hit by this economically driven transformation. Likewise, in East Germany significant efforts were undertaken, after the reunification, to smooth the transformation process, and to rescue the lignite industry from the assault from West German competitors. The analysis is divided into the quantitative consideration of the significance of coal for the energy system and the regional economies, as well as an evaluation of implemented political instruments accompanying the reductions in the coal sector. The political instruments on regional, national and supranational level can be differentiated between measures for the conservation of coal production, the economic reorientation in the regions as well as easing negative social impacts. The good news for the upcoming final phase

---

[5]In fact, a narrow interpretation of the term energiewende would limit the actual "turn" to the 72 h between the Fukushima accident (March 11, 2011), and the "Declaration of the energiewende," including the nuclear moratorium (March 14, 2011).

out of coal in Germany is that the largest part of the transformation process is already achieved. The analysis of past transformation processes of mining areas and energy systems in Germany might provide other countries and regions with valuable lessons of how to structure their upcoming coal phase-out period and therefore provides a useful addition to the existing literature.

1.4.2   Part II: The Energiewende Underway in the Electricity Sector

Part II leads us straight into the "engine room" of the energiewende, a process with very concrete, hands-on technical, institutional, and economic issues. In this part, we shed light on diverse facets of the energiewende based on our own research work, official data and publications, and a survey of the literature. We address the main issues at the heart of the energiewende, with a focus on the electricity sector: decarbonization, the closure of nuclear plants, the focus on renewables, efficiency targets, and infrastructure, and the emergence of coupling between the electricity, transport, and heat sectors.

Decarbonization of the electricity sector and the phase-out of coal is a key element of the energiewende addressed in Chap. 4; the objectives for GHG emission reductions set out in the energy concept were 40% in 2020, 55% in 2030, 70% in 2040, and 80–95% in 2050 (base year: 1990). However, between 1990 and 2017, this reduction was only 30%, and projections for 2020 hovered around a 33% reduction. Due to the collapse of the European Union CO_2-trading price, which had fallen from an average €20/t to around €5/t, there was even a slight increase of CO_2 emissions in 2013 and 2014. Thus, the German government decided to develop more focused national policy instruments to accompany European efforts and to curb national coal consumption. This resulted in the Climate Action Program 2020 and the subsequent longer-term Climate Plan 2050. The chapter discusses different instruments, their economic effects on the electricity market, and concludes that the decarbonization targets imply the phasing-out of coal in Germany in the 2030s.

Chapter 5 is dedicated to nuclear policies in Germany, a particularly controversial field. The chapter focusses on the period between the first phase-out decision, taken in 1998, and the second, final one, in June 2011. Immediately following the March 2011 Fukushima accident, a moratorium was decided on German nuclear power plants, seven of which were shut down immediately; until 2022, all others will follow. Looking back, the effects of the nuclear moratorium on the German and Central-West European electricity markets were small, because ample generation capacity was available to compensate for the loss of capacity. After the March 2011 moratorium, wholesale electricity prices increased slightly by €2–3/megawatt hours (MWh), whereas the German net export surplus declined slightly. Germany turned into a net exporter again in the subsequent year, 2012, when it showed record net exports of 54 terawatt hours (TWh). The chapter also looks forward to 2022, when the last remaining reactors will close, and examines German nuclear policy in a European context as well. While the closure of the nuclear power plants is irreversible from a political perspective, policies to structure and facilitate the process are still needed, in particular with respect to the decommissioning of the plants and the final storage of radioactive waste. Chapter 5 thus also provides an analysis of the post-closure challenges, the uncertainties surrounding this process, and the financial stakes and the expected timelines, which extend over centuries of dealing with the legacies of nuclear power.

Chapter 6 looks in more depth at another focus of the energiewende: renewables targets. As we observe, the targets defined in 2010/2011 have now been translated into concrete policy measures. For example, the "scenario framework"—the planning document that the energy network regulator has to produce every year as a framework for network development—covers renewables targets for a 20-year time period, during which other fuels such as hard coal and lignite will be reduced. In 2017, the share of renewables in electricity production was 37%, with no signs of instability of the system. Given the strong institutional framework, there seems to be no obstacle to reaching the 2030 target of beyond 50%. The chapter also describes the evolution of support schemes for renewables in Germany and the effects of investment strategies in conventional and renewable capacity, as well as the costs associated with this policy.

Energy efficiency is a crucial element of any low-carbon transformation process, as the most cost-effective kilowatt hour (kWh) is an hour that is not used at all ("saved"). Traditionally energy efficiency policies have had a difficult time gaining support from both policy makers and the general public, and the "energy efficiency gap"—the difference between observed consumption and a hypothetical reference case—has been a topic of debate for some time. It comes as no surprise, therefore, that energy efficiency policies are among the most challenging elements of the energiewende. Chapter 7 focuses on energy efficiency, and reports that although some successes have been observed in the reduction of primary and gross electricity consumption, energy productivity improvements still lag behind the targets. A significant gap remains with respect to the 2020 primary energy consumption target (−20%), and further energy productivity increases are necessary to stay on track. The chapter also describes specific approaches to energy efficiency in the construction, industry, and transport sectors, and provides concrete recommendations for how to move this difficult reform process forward.

Chapter 8 focusses on the role of electricity transmission infrastructure in the energiewende. The chapter analyzes approaches to and developments in the electricity network infrastructure, and asks if the glass is half full or half empty. In fact, skeptics of the energiewende see transmission bottlenecks everywhere, whereas optimists insist on the steady progress of the modernization and extension of the high voltage grid in Germany. Our analysis tends to see the glass half full: Although some expansion projects are behind schedule, grid reconstruction is advancing steadily, thanks to the considerable progress made in recent years on several essential lines connecting the states of the former East and West Germany. Congestion management measures, in particular redispatch, have been necessary, but have caused no major problems to the system. We even observe that the current methodology for the long-run grid expansion planning tends to overestimate expansion needs because it assumes a "copper plate" when siting generation. The chapter concludes that while electricity transmission is an important element of any reform process, the debates around network expansion have exaggerated the potential pitfalls, and the focus should be on sustainable electricity generation.

Chapter 9 extends the analysis of electricity sector reform to other important sectors of the energiewende. The first stage of the energy transformation was characterized by the nuclear phase-out and the parallel endeavor to decarbonize the German electricity sector. Yet, in order to reach the climate goal of a 40% reduction in greenhouse gases by 2020, with an additional 80 to 95% reduction in the coming decades until 2050, the second stage needs to focus on all energy usage, especially heat, transportation, and usage as a raw material in the chemical industry. Enhancing energy efficiency, reducing primary energy usage until 2050, and the increased use of renewable power from wind and photovoltaics is the predominant strategy to further decrease greenhouse gas emissions in all energy sectors. However, this strategy requires an increased coupling of energy sectors and is the corner stone for an integrated approach, activating additional degrees of freedom in the energy system facilitating the further integration of renewable energy sources. The chapter sketches out the technical-economic foundations of sector coupling, and then compares different analyses for Germany. While these differ in the level of detail, they all consider ambitious climate targets to be feasible, provided that appropriate institutions and incentives be put in place. Consequently, the distinct energy sectors coalesce and have to be assessed in an integrated way.

1.4.3 Part III: The German Energiewende in the Context of the European Low-Carbon Transformation

The German energiewende is not a national phenomenon: it is taking place within an increasingly integrated European market and in the context of close relationships to Germany's "electrical neighbors". The very nature of the interconnected European electricity system means that the reform process in Germany has effects on the broader European market, including price effects, cross-border flows, and the sharing of backup capacity. In return, the German electricity sector is affected by developments in its neighboring countries, be they EU members or not. Part III of the book therefore addresses important questions concerning the interdependence between the German energiewende and the European low-carbon transformation reform process at large.

Chapter 10 analyzes the electricity mix in the European low-carbon transformation and highlights similarities to and differences from the German energiewende. The European Union, too, has set ambitious decarbonization targets (-40% GHG emissions by 2030, -80 to 95% by 2050), but the EU's current roadmap to attaining them is different from that of the German energiewende, since it still includes high shares of fossil fuels (with carbon capture) and nuclear energy. The chapter first reviews the broad trends in European energy and climate policy since the 1950s, explaining that the European electricity mix was based on coal and nuclear from the beginning, through the 1951 Treaty of the European Coal and Steel Community (ECSC) as well as the 1957 EURATOM Treaty. The chapter then takes a closer look at the 2030 and 2050 targets, at instruments such as the European Emission Trading System (ETS), and at the fuels that will be used to attain these targets: conventional fossil fuels, nuclear, and renewables. In particular, we ask whether the Energy Roadmap 2050, the analytical basis of the strategy, properly reflects recent techno-logical developments in Europe and elsewhere. For instance, the continued use of coal electrification was based on the assumption that a clean and economic technol-ogy known as Carbon Capture, Transport, and Storage (CCTS) would be available soon, and despite considerable research and development (R&D) and demonstration attempts, these projects have not brought about significant progress so far. Nuclear power has high and rising private costs and by far the highest social costs of all energy sources, due not only to high capital costs and unknown costs of long-term storage of nuclear waste, but also to the risks of accidents, which no market has been able to insure. Two of the post-World War II nuclear countries, the UK and France, have expressed serious intentions to build new nuclear power plants, which we explain by the synergy effects they expect to reap from the civil and military use of nuclear power. Renewable energy sources offer not only the cleanest but potentially also the most economical alternative for many Member States, a development that has not been given adequate consideration in the European scenario process to date.

Chapter 11 analyzes the importance of infrastructure for the European low-carbon energy transformation. In this area, "easy" solutions surface frequently in the scientific and policy communities, but implementation on the ground has proven to be much more difficult. Thus, during the first years of the energiewende, a large number of pan-European "supergrid" projects were proposed, including electricity highways stretching from Saudi Arabia to Iceland, and from Morocco to the Arctic Circle. However, none of these mega-projects has materialized, and more realistic, more focused, and less complex solutions need to be found and developed. Using our own modelling results, as well as a large model comparison in the framework of the international Energy Modeling Forum No. 28 ("The Effects of Technology Choices on EU Climate Policy") subgroup on infrastructure, we discuss alternatives to the pan-European infrastructure development plans in three critical sectors: electricity transmission, natural gas, and CO_2-pipeline infrastructure. It turns out that the finding for the German electricity network infrastructure (Chap. 8) can, to a certain extent, be transposed to the European level: energy infrastructure can help the low-carbon transformation, but is not really a critical factor thus far.

Chapter 12 discusses how the German electricity sector and energiewende fits into the regional and European context. The chapter also draws some general conclusions on the role of cross-border and other cooperation in the German energiewende and the European low-carbon transformation. Very different forms of cooperation can be observed, ranging from bilateral mechanisms (e.g., Austrian generators contracting capacity with independent system operators (ISO) in Germany), multilateral coordination (e.g., the Pentalateral Forum including initially five, now seven countries: Germany, France, the Netherlands, Belgium, Luxembourg, Switzerland, and Austria), to more fully-fledged European solutions such as the Inter-TSO-Compensation mechanism (ITC). The chapter provides evidence from our own modelling work on cross-border compensation in balancing markets in the Alpine region and the development of an integrated transmission grid in the North and Baltic Sea region. The chapter covers both European and more nationally oriented strategies, and defines different levels of cross-border cooperation.

Part III closes with an extension of the previous, electricity-focused analysis: Chap. 13 showcases multiple decarbonization pathways for the European energy system with varying carbon dioxide constraints. The Global Energy System Model framework (GENeSYS-MOD), a linear mathematical optimization model, is used to compute low-carbon scenarios for 15 European countries or regions between 2015 and 2050. The traditionally segregated sectors power, low- and high-temperature heating, and passenger and freight transportation are included, with the model endogenously constructing capacities in each period. Emission constraints differ between different scenarios and are either optimized endogenously by the model, or distributed on a per-capita basis, GDP dependent, or based on current emissions. The results show the need for a rapid phase-in of renewable energies, if a carbon budget in line with established climate targets is enforced. The carbon constraints are quite severe (corresponding to the 1,5° and 2° targets), so that new technologies come into play at scale, such as heat pumps for low temperature heat, and electricity overhead infrastructure for low-carbon freight transportation. In fact, it can be shown that the commitment for a 2°C target only comes with a cost increase of about 1–2% (dependent on the emission share) compared to a business-as-usual-pathway, while yielding reduced emissions of about 25%. The different regions and demand sectors experience different decarbonization pathways, depending on their potentials, political settings, and technology options.

1.4.4 Part IV: Assessment, Lessons, and Perspectives

Part IV closes our survey of the energiewende. In Chap. 14, we derive lessons from our research and discuss perspectives for the future. The energiewende has been going on for quite some time now, and results so far are impressive, though not only positive. However, only a small part of the journey is made, and there are at least three decades before us in which further reforms, technical innovation, and political consensus will be required. However, the empirical evidence from the recent past, together with a technical and political assessment of the feasibility of the next reform steps, allows us to formulate hypotheses summarizing the previous chapters and opening up perspectives for the future. Our assessment of the current state of the energiewende is positive in most areas: The renewables-based focus of the German electricity system is on track, and no technical or economic obstacles to the low-carbon transformation are observable, yet this will require more stringent policy instruments, such as phasing out coal, and advancing the technical-economic sector coupling. The longer-term perspective, going back as far as the nineteenth century, is useful in identifying the continuities and discontinuities in the history of energy and climate policy and in understanding the energiewende as a significant historic break with the conventional energy system.

We then draw 15 lessons from the analysis, following the structure of the previous Parts I–III: some lessons are drawn from the long-term analysis of energy and climate policies (Part I); another series of lessons focuses on the lessons from the energiewende in Germany, which is still ongoing (Part II). The chapter also provides lessons from the interplay between the German setting and the low-carbon transformation at the European level (Part III).

Acknowledgements An undertaking like this book is necessarily a collective work, and there are many individuals and organizations involved in it. Special thanks go to our colleagues who participated in or commented on the research projects reported on in this book. We hope that our list of references includes the relevant names, but want to specifically acknowledge Thorsten Beckers, Jochen Dieckmann, Leonard Göke, Christian Hauenstein, Albert Hoffrichter, Mario Kendziorski, Martin Kittel, Philipp Litz, Ann-Katrin Lenz, Jonas Mugge-Durum, Ralf Ott, Catharina Rieve, Wolf-Peter Schill, Julian Schwarzkopf, Daniel Weber, and Alexander Zerrahn; Alexander Weber and Bobby Xiong served as our consultants for editing and bibliography, and Linus Lawrenz and Ben Wealer helped to format and submit the final manuscript. Once more, Deborah Bowen helped with language and style.

We also thank four anonymous reviewers for concise and useful reports, and Patrick Graichen, Hans-Joachim Fell, and Uwe Nestle for suggestions on an earlier manuscript. Special thanks also go to our Publisher Springer, in particular Barbara Fess and Marion Kreisel, for regular and efficient exchange on various issues, and for bearing with us during the publication process. We also thank those who have helped us discover the energiewende during the last decade, researchers, stakeholders, activists, policymakers, etc.!!!

Some of the research presented in this book was developed in two large projects funded by the Stiftung Mercator, in its Center for Climate Change: (1) the project MASMIE ("modelling the energiewende"), led by DIW Berlin (Department of Energy, Transport and the Environment); and (2) the project EE-Netze ("networks for renewables"), led by the Workgroup for Infrastructure Policy (WIP) at Berlin University of Technology, both with various research partners. We thank the Stiftung Mercator for their support, in particular Dr. Lars Grotewold, Head of the Department Climate Change, and Philipp Offergeld, our project manager. In addition, this book includes research from two projects funded by the German Federal Ministry of Education and Research (BMBF): (1) The junior research group CoalExit ("Economics of Coal Phase-Out—Identifying Building Blocks for Future Regional Transition Frameworks", grant no. 01LN1704A), located at TU Berlin; and (2) RESOURCES ("International Energy Resource Markets under Climate Constraints–Strategic Behavior and Carbon Leakage in Coal, Oil, and Natural Gas Markets"; grant no. 01LA1135B) carried out at DIW Berlin. Numerous participants at three Berlin Conferences on Energy Economics (BELEC) from 2015 until 2017 and at other workshops, through own work, discussions, and critique. Last but not least, we thank our administrative support, Dagmar Rauh at DIW Berlin, and Petra Haase at TU Berlin.

References

Agora Energiewende. 2013. 12 Insights on Germany's Energiewende – a discussion paper exploring key challenges for the power sector. Berlin, Germany.

———. 2018. Energiewende 2030: The Big Picture – megatrends, targets, strategies and a 10-point agenda for the second phase of Germany's energy transition. Impulse. Berlin: Agora Energiewende.

Bhushan, Chandra. 2017. *India's Energy Transition – Potential and Prospects*. New Delhi: Heinrich Boell Foundation – India.

BMWi. 2015. The energy of the future, fourth 'energy transition' monitoring report – summary. Berlin, Germany.

———. 2016. The energy of the future, fifth 'energy transition' monitoring report – summary. Berlin.

BMWi, and BMU. 2012. First monitoring report 'energy of the future' – summary. Berlin, Germany.

Burtraw, Dallas, Josh Linn, Karen Palmer, and Anthony Paul. 2014. The costs and consequences of clean air act regulation of CO_2 from power plants. *American Economic Review* 104 (5): 557–562.

Criqui, Patrick, and Jean-Charles Hourcade. 2015. *Pathways to Deep Decarbonization in France*. Paris: SDSN - IDDRI.

Danish Government. 2011. Our future energy. Copenhagen, Denmark.

Davidson, Osha Gray. 2012. In *Clean Break: The Story of Germany's Energy Transformation and What Americans Can Learn from It*, ed. Susan White, Catherine Mann, and Christopher Flavin. Brooklyn, NY: InsideClimate News.

DECC. 2011. *Planning Our Electric Future: A White Paper for Secure, Affordable and Low-Carbon Electricity*. London: Department of Energy and Climate Change.

Fechner, Carl-A. 2018. *Power to Change: Die Energierevolution ist möglich!* 1. Auflage. Gütersloh: Gütersloher Verlagshaus.

Foxon, Timothy J. 2013. Transition pathways for a UK low carbon electricity future. *Energy Policy*, Special Section: Transition Pathways to a Low Carbon Economy 52 (January): 10–24.

Grubb, Michael, Jean-Charles Hourcade, and Karsten Neuhoff. 2014. *Planetary Economics – Energy, Climate Change and the Three Domains of Sustainable Development*. London: Routledge.

Hager, Carol, and Christoph H. Stefes, eds. 2016. *Germany's Energy Transition*. New York: Palgrave Macmillan US.

Heal, Geoffrey. 2017. What would it take to reduce U.S. Greenhouse Gas emissions 80 percent by 2050? *Review of Environmental Economics and Policy* 11 (2): 1–18.

Houghton, John Theodore, G.J. Jenkins, and J.J. Ephraums. 1990. *Climate Change: The IPCC Scientific Assessment*. Cambridge: Cambridge University Press.

IEA. 2014. *The Power of Transformation – Wind, Sun and the Economics of Flexible Power Systems*. Paris: International Energy Agency (IEA).

———. 2016. *World Energy Outlook 2016*. Paris: OECD.

IPCC. 2012. Renewable energy sources and climate change mitigation: special report of the Intergovernmental Panel on Climate Change. In *United Nations Environment Programme, World Meteorological Organization, Intergovernmental Panel on Climate Change, and Potsdam-Institut für Klimafolgenforschung*, ed. Ottmar Edenhofer, Ramón Pichs Madruga, and Y. Sokona. Cambridge: Cambridge University Press.

———. 2014. Climate change 2014: impacts, adaptation, and vulnerability. Part A: global and sectoral aspects. Contribution of Working Group II to the Fifth Assessment Report of the Intergovernmental Panel on Climate Change. Cambridge, UK: Cambridge University Press.

Kemfert, Claudia. 2014. *The Battle About Electricity*. Hamburg: Murmann.

Kemfert, Claudia, Wolf-Peter Schill, and Thure Traber, eds. 2013. Energiewende in Deutschland – Chancen und Herausforderungen. *Vierteljahrshefte Zur Wirtschaftsforschung* 82 (3): 1–206.

Krause, Florentin, Hartmut Bossel, and Karl-Friedrich Müller-Reissmann. 1980. *Energie-Wende: Wachstum und Wohlstand ohne Erdöl und Uran.* Edited by Öko-Institut Freiburg. Frankfurt am Main, Germany: S. Fischer.

Li, Mingwei, Da Zhang, Chiao-Ting Li, Kathleen M. Mulvaney, Noelle E. Selin, and Valerie J. Karplus. 2018. Air quality co-benefits of carbon pricing in China. *Nature Climate Change* 8 (5): 398–403.

Löschel, Andreas, Georg Erdmann, Frithjof Staiß, and Hans-Joachim Ziesing. 2014. Statement on the first progress report by the German Government for 2013. Summary. Berlin, Münster and Stuttgart, Germany.

———. 2015. Statement on the fourth monitoring report of the Federal Government for 2014. Summary. Berlin, Münster and Stuttgart, Germany: Federal Ministry for Economic Affairs and Energy.

———. 2018. Statement on the sixth monitoring report of the Federal German Government for 2016 – expert commission on the 'energy of the future' monitoring process. Berlin, Münster and Stuttgart, Germany: Federal Ministry for Economic Affairs and Energy.

Lovins, Amory B. 1976. Energy strategy: the road not taken? *Foreign Affairs* 6 (20): 9–19.

Morris, Craig, and Arne Jungjohann. 2016. *Energy Democracy – Germany's Energiewende to Renewables.* London: Palgrave Macmillan.

Morris, Craig, and Martin Pehnt. 2016. Energy transition: the German Energiewende. An initiative of the Heinrich Böll Foundation. First Released in November 2012, Revised in July 2016. Berlin, Germany.

Rosenkranz, Gerd. 2014. *Energiewende 2.0: aus der Nische zum Mainstream,* Schriften zur Ökologie. Vol. 36. Berlin: Heinrich-Böll-Stiftung.

Schellnhuber, Hans Joachim. 2015. *Selbstverbrennung: Die fatale Dreiecksbeziehung zwischen Klima, Mensch und Kohlenstoff.* Munich: C. Bertelsmann Verlag.

Schippl, Jens, Armin Grunwald, and Ortwin Renn. 2017. *Die Energiewende verstehen – orientieren – gestalten.* Stuttgart: Helmholtz-Allianz ENERGY-TRANS.

Singh, Arun, Valerie J. Karplus, and Niven Winchester. 2018. Evaluating India's climate targets: the implications of economy-wide and sector specific policies. Report 327. MIT Joint Program. Cambridge, MA.

Stern, Nicholas Herbert. 2007. *The Economics of Climate Change: Stern Review on the Economics of Climate Change.* New-York: Cambridge University Press.

Unnerstall, Thomas. 2017. *The German Energy Transition: Design, Implementation, Cost and Lessons.* Berlin: Springer.

WBGU. 2011. *World in Transition: A Social Contract for Sustainability – Flagship Report.* Berlin: German Advisory Council on Global Change (WBGU).

World Energy Council. 2014. *Global Energy Transitions: A Comparative Analysis of Key Countries and Implications for the International Energy Debate.* Berlin: World Energy Council – A.T. Kearney.

Part I
The Historical Origins and Emergence of the Energiewende

Chapter 2
German Energy and Climate Policies: A Historical Overview

Christian von Hirschhausen

> *The events in Japan teach us that things we consider*
> *impossible according to scientific criteria can nonetheless*
> *become reality. (...) We will suspend the recent decision to*
> *extend the lifetime of the German nuclear power plants. This*
> *is a moratorium that will last 3 months. ... The situation after*
> *the moratorium will be different than before. (...) We speak*
> *about nuclear energy as a "bridge technology," which means*
> *nothing other than that we are discontinuing the use of*
> *nuclear energy and want to ensure the German energy supply*
> *through the use of renewables as quickly as possible. ... The*
> *only honest response is to accelerate the path towards the age*
> *of renewable energies.*
> *Chancellor Angela Merkel, televised press conference,*
> *March, 14, 2011*
> *Bundesregierung (2011), authors' translation.*

2.1 Introduction

The energiewende marked a major turn in German energy and climate policy in two main respects. First, with respect to the energy mix, the energiewende aims at replacing coal and nuclear power with renewable energies. Second, with respect to governance structures, the energiewende aims at restructuring the traditional energy

---

The first part of this chapter draws on historical sources, such as Becker (2011), Hughes (1993), Stier (1999), and Zängl (1989). Thanks to Alexander Weber for comments and suggestions, the usual disclaimer applies.

C. von Hirschhausen (✉)
TU Berlin, Berlin, Germany

DIW Berlin, Berlin, Germany
e-mail: cvh@wip.tu-berlin.de

© Springer Nature Switzerland AG 2018   17
C. von Hirschhausen et al. (eds.), *Energiewende "Made in Germany"*,
https://doi.org/10.1007/978-3-319-95126-3_2

oligopolists and actively involving other stakeholders that were previously not involved in the policy process, such as citizen cooperatives, non-governmental organizations (NGOs), and others. Thus, the energiewende is not "just another policy change" with only short-reaching consequences; rather, it constitutes a break between the past and the future energy system. What are the major features of this break? Where do the ambitious goals of the energiewende have their roots, and who formulated these goals? Finally, what was the process that led Germany to embark on this voyage and to state ambitious sustainability and renewable objectives in the midst of worldwide recession?

This chapter provides a survey of German energy and climate policies leading up to the important decisions on the energy mix, climate objectives, efficiency, etc., in first decade of this century. This allows us not only to look at the specifics of the energiewende from a longer-term energy policy perspective ("la longue durée," in the words of Fernand Braudel), but also to examine the fuel mix prevalent in each period and the political discussions surrounding it. The energiewende constitutes a break between two systems, in which the incumbent electricity system—dominated by four oligopolists based on fossil fuels and nuclear power—was abandoned, giving rise to a renewables-based electricity system with a significantly higher share of distributed generation. The chapter describes the main trends and characteristics of German energy and climate policies from their inception in the late nineteenth century up to the present energiewende. An understanding of both the technical and institutional idiosyncrasies of the system is useful to assess the specifics of the energiewende.

The chapter is divided into two main parts. The next Sect. 2.2 looks broadly at over a century of German energy policy, examining the governance structures and energy mix dominant in three key periods: (1) 1880s–1945: a period based almost entirely on coal electrification by large, monopolistic industrial-financial trusts, (2) 1950s–1980s: a period in which Germany was divided and attempts were made in both East and West to complement coal by a new generation technology: nuclear fission, and (3) the 1980s–2010s: a period in which renewable energies were introduced gradually alongside the still-dominant coal, accompanied by some natural gas and nuclear production, also driven by external events like the Chornobyl nuclear accident (1986) and the emerging debate on climate policies, especially after the 1992 climate conference in Rio. This period was also marked by discussions about a "soft path" of energy transformation, since the 1970s [following Amory and Hunter Lovins' (1976) concept], in West Germany, the process of German reunification, and political wrangling over whether or not to close nuclear power plants in the 2000s. In this period, nuclear energy was declared as having no future in the German energy mix (a position strongly opposed by the energy industry), whereas renewables had initially low but gradually increasing rates of penetration.

The second main part of this chapter, Sect. 2.3 looks in more detail at the period between the fall of 2010 and the spring and summer of 2011. A focus is on the year 2010 and the Energy Concept 2050, which was voted into law by parliament in September 2010. The concept represents a curious combination of lifetime extensions for nuclear power plants and coal-based generation technologies on the one hand, and ambitious decarbonization objectives and a strong role for renewables (over 80% by 2050) on the other. The section then focuses on Chancellor Angela Merkel's decision on the nuclear moratorium, and the subsequent passage of legislation by parliament to rapidly close down nuclear power plants following the Fukushima-Daichi accident. Section 2.3.2 describes what took place between the nuclear disaster in Japan on March 11, 2011, and Chancellor Merkel's March 14, 2011, declaration of a nuclear moratorium in Germany, continues through June 2011, a period of major public policy decisions, and provides an account of the decisions to close all remaining nuclear power plants by December 2022. Another subsection provides a summary of the key objectives of the energiewende in both the electricity sector and the energy sector as a whole, including a list of policy objectives and concrete quantitative targets of the German energiewende to 2050. Section 2.4 concludes.

2.2 Historic Periods and Fuel Choices: Fossil, Nuclear, and Renewables

2.2.1 1880s–1945: Coal Is King

2.2.1.1 Emergence of the First Concessions and Industrial Cooperation in the Late Nineteenth Century

The dynamo, an electrical generator used to transform mechanical energy into electrical power, was invented in England in 1866 by Wheatstone and Varley and introduced in Germany by Werner von Siemens. Applications of this technology developed rapidly during the 1870s in the areas of street lighting, electric trains, electric refrigerators, and a variety of military uses. The advent of "big electricity" in Germany was in 1884, the year of the first large concession awarded to supply electricity to the nation's capital, Berlin. On January 24, 1884, Emil Rathenau, an industrialist who had purchased a license to develop Thomas Edison's lightbulbs in Germany, obtained the first concession to distribute electricity in the center of Berlin, around Werdersche Markt, just a few yards away from the main boulevard "Unter den Linden".[1] This concession kick started a boom in the sales of light bulbs by the holding company, Rathenau Deutscher Edison, from 90,000 bulbs sold in 1886 to over a million in 1887 (Becker 2011, 10). In addition to technology, large sums of capital were required for the expansion of the business. In 1887, Emil Rathenau, in a quest for capital to finance the expansion of electricity provision to the city of Berlin, obtained large loans from Deutsche Bank, then headed by Georg Siemens (a cousin

---

[1]This concession, granted by the city of Berlin, included a monopoly on electricity generation, transport, and sales in this district against payment of 6% of turnover. The concessionaire, a company called Actiengesellschaft Städtische Elektrizitätswerke, was also obliged to connect all citizens to the grid and was subjected to price control.

of the electrical inventor Werner Siemens), to develop large-scale generators (up to 100 PS). The first technical-financial trust in energy was created, which would become known worldwide under the name of AEG (Allgemeine Electricitäts-Gesellschaft, the successor company to German Edison).

The turn of the nineteenth to the twentieth century was the time of trusts, first, in the United States, and then in Europe: Large, vertically and often horizontally integrated corporations, not yet subject to competition law. These trusts fulfilled many functions of what one would consider to be public policy, with little oversight from state agencies, be it in capacity planning, pricing, fuel choice, or horizontal and vertical coordination. Different state, city, and municipal levels were often active players in this process, but private capital was generally the major driving force. In most instances, these corporations were granted power by public policies, e.g., through monopoly concessions, price strategies, or direct parliamentary intervention and legislation. Three representative examples of such trusts are AEG (Allgemeine Electricitäts-Gesellschaft), the Siemens-Deutsche Bank group, and the RWE-Thyssen-Lahmeyer group, all of which emerged in the late nineteenth century.[2]

Following the blueprint of the North American trusts, Rathenau started to negotiate contracts with the other large equipment producer, Siemens, to control the market and establish what would become a very lucrative cartel (Zängl 1989, 21). Cooperation took place in the areas of both technology and financing. The Siemens equipment trust obtained 1 mn. reichsmark of Walter Rathenau's capital in AEG, strengthening the industrial-financial cartel (Becker 2011, 12). Later on, AEG also established a cartel with the other large manufacturer, Siemens & Halske, allowing it to pay its shareholders a comfortable dividend of 15% up to the outbreak of World War I. Along similar lines, Siemens purchased other independent equipment producers (Schukert, Helios, etc.) in what was to become a duopoly in the electricity industry, similar to that of General Electric and Westinghouse in the USA. Deutsche Bank remained both financier and chief controller of the corporations.

These large energy trusts favored the use of coal, which emerged as the predominant fuel in electricity generation after a brief but very close contest with natural gas, the incumbent for street lighting and industrial heat. This trend became particularly evident in the West German Rhine-Ruhr region, where another large industrial-financial conglomerate was founded under the auspices of the coal industry, banks, and the (monopolistic) municipal utilities of large and medium-sized cities. In 1900, Hugo Stinnes, coal producer and trader, joined forces with RWE, then a conglomerate of municipalities in the Rhine region, to obtain monopoly concessions for electricity in each of the municipalities. Financing for this expansion was provided by August Thyssen, who would later become a major steel producer and who also brought a large manufacturer, Lahmeyer, into the trust. Together, Stinnes and

---

[2]Some federal states also opted for more decentralized electricity generation structures, e.g., in southwest Germany (Baden-Württtemberg). In hindsight, there was a striking continuity in both the choices of fuel (coal) and governance structures (trusts) of the German electricity industry from 1884, the year of the first private concession in Berlin, up to 1945.

Thyssen proceeded with the electrification of the Ruhr region, including large cities such as Essen, Mühlheim, and Gelsenkirchen. There was close cooperation if not complete synergy between the RWE trust and local policymakers in the cities, the latter obtaining high dividends, a voting majority, as well as lucrative jobs in the administration of both the RWE holding and its subsidiaries (Becker 2011).

After World War I, lignite started to compete with (more expensive) hard coal. RWE started to develop large lignite areas and power plants themselves, including the Roddergrube, the second largest lignite production site after Rheinbraun. Stinnes' successor as Chief Executive of RWE, Arthur Kopechen, further extended the trust into southern Germany, Switzerland, and Austria by connecting hydro resources in the South with the coal resources in the Ruhr area. Up to 1930, RWE laid 4100 km of high-voltage cables, including a 800 km corridor between Bavaria (Walchenseewerk) and Essen. The RWE trust had become the largest electricity producer in Europe and number four worldwide behind three US trusts.[3]

2.2.1.2   Energy Corporations Resist Socialization After World War I

The dominance of energy trusts in which public entities such as regional and municipal governments held partial ownership was maintained and protected against repeated attempts to streamline the industry, even throughout the chaotic postwar period and into the Nazi regime of the 1930s. The first attack on the trusts was launched after the end of World War I by the newly elected German parliament, which was dominated by social democrats and leftist parties. In 1919, parliament passed the "law on socializing the electricity industry" ("Gesetz zur Sozialisierung der Elektrizitätswirtschaft"). However, parliamentary majorities changed over the years that followed, and some of the most influential industrialists became members of parliament themselves, among them Hugo Stinnes and the presidents of Siemens, AEG, and the German Association of Electrical Industries. Although the law on socializing the industry remained in place, it was not enforced and largely ignored (Becker 2011, 26).

Another attempt to streamline the sector was undertaken in the 1920s by Oskar von Miller, previously a civil servant in Bavaria who had successfully introduced electrification there through use of the state's rich hydropower resources. Von Miller had already considered a prototype of a nationwide development plan in the late 1890s, and successfully developed a long-term development plan for Bavaria in 1908. After his appointment as Minister of Economy of the Reich, von Miller became the driving force in an industry-wide development plan to coordinate investments in generation and transmission, the so-called Miller Plan (1928), whose objective was to increase interregional coordination under improved

---

[3] Already at that time final consumers of electricity, paying a high surcharge on the true costs, cross-subsidized industrial users, who had higher negotiating power thanks to political pressures, their capacity to self-generate, etc. see Becker (2011, 23).

centralized control.[4] The German Reich was to be split into 13 regional electricity systems that would be centrally managed after nationalization of the trusts. With the 1929 proposal of the "electricity law" before parliament, von Miller and the central administration of the Reich found themselves on one side of a battle for centralization, as proponents of the law, and the eight large incumbent, regionally based energy trusts on the other as its opponents. The large trusts[5] vigorously were able to torpedo the Miller plan, and the proposed electricity law.

Even the Nazi regime that came into power in 1933 did not break the private electricity trusts, although it introduced some control over the heavy industry and the military complex. On December 13, 1935, an energy law was implemented that included supervision of suppliers, price controls, and investment control. However, the law was still pervaded by elements of German cartelism: it allowed the energy industry to resolve internal issues between corporations and limited public oversight to issues that "the industry cannot resolve it by itself".[6] Hjalmar Schacht, then Minister of Economy, banker, and self-declared "friend" of the electricity industry, suggested that the monopoly of electricity supply be maintained for "private initiative, private capital, and private risk."[7] In particular, smaller consumers should not, he argued, be allowed to develop their own energy supply unless they could prove that the trusts were unable to do so (which was seldom the case). Given the high and increasing demand, the electricity business remained highly profitable. As the major generators of foreign exchange, energy trusts like Siemens and AEG played an important role in the economy and lobbied successfully to maintain their monopolistic rents.[8]

2.2.1.3   Energy Mix Dominated by Coal

Throughout the nineteenth century and up to 1945, the German Reich's electricity supply relied on abundant (although relatively expensive and dirty) coal resources. Coal contributed significantly to the rapid industrial and economic growth: The

---

[4]For details, see Boll (1969, 58 sq.).

[5]RWE (Rhine-Ruhr region), Preussen-Elektra (Central Germany), Badenwerke, Energieversorgung Schwaben (EVS), "Interessengebiet Elektrowerke" (assembly of East German utilities), Bayrische Elektrizitätswerke (South), etc. see Stier (1999, 294 sq.).

[6]Preambel to the 1935 law, quoted by Becker (2011, 38).

[7]Cited in Becker (2011, 31). Note that the monopolization of the electricity market was by no means limited to Germany; on the contrary, cartelization had spread all over Europe and was coordinated by the "International Electric Association" to fix prices and quantities for power plants and other equipment (2011, 39).

[8]One finds a long list of German energy providers, industrial energy consumers, and representatives of the energy equipment industry among the donors to the "Museum der Deutschen Kunst," the first prestigious project of the National Socialist government under Chancellor Adolf Hitler in May 1933: Karl-Friedrich von Siemens, son of Werner von Siemens and then head of the Siemens trust, was on the governing council of the "Museum der Deutschen Kunst," as was Fritz Thyssen (see Brantl, Sabine 2007: Haus der Kunst, München. Munich, edition monacensia).

industrialization of Germany in the Rhine-Ruhr region and Central Germany (Leipzig/Halle/Bitterfeld) in the nineteenth century occurred thanks to their abundant coal resources; the same was true of Silesia, part of the German Reich until 1945.

The share of hard coal and lignite differed among the country's large coal basins. Silesia and the Saar, two of Germany's most important coal regions, relied on hard coal. The Rhine-Ruhr region also traditionally relied on hard coal, with lignite only being exploited at large scale starting in the 1950s.[9] Central Germany, by contrast, was fueled almost exclusively by lignite. In fact, the GDR would later become the world's largest lignite producer (up to 300 mn. tons yearly).

2.2.2 1950s–1980s: East and West Germany Enter the Nuclear Age

2.2.2.1 Monopolistic, Corporate Governance Structures Maintained

On May 8, 1945, Germany surrendered to the Allied Forces and was subsequently split into two countries, the Federal Republic of Germany (FRG) and the German Democratic Republic (GDR). And on July 16, 1945, the US tested its first atomic bomb, a weapon that would soon be used to destroy Hiroshima and Nagasaki on August 6 and 9, 1945, respectively. The nuclear era had arrived. Curiously, though the two Germanies developed very differently politically, there are strong parallels in the corporate governance structures, large integrated corporatist structures, and the energy mix, where "big coal" was now complemented by ("small") nuclear energy.

Monopolistic structures in the electricity industry were maintained after World War II on both sides of what was later to become the Iron Curtain. In the German Democratic Republic (GDR, "East" Germany), the Socialist Party established large socialist combines for equipment supply, mining, and electricity supply. These combines worked to implement centrally designed plans of the Party, and took over roles of social policy and civic control. They were maintained up to the end of the GDR, in 1990.

In the Federal Republic of German (FRG, "West" Germany) as well, the monopolistic structures established after the end of World War II were maintained. Following US antitrust movements half a century earlier, the government had begun work on a "law against restraints of competition" (Gesetz gegen Wettbewerbsbeschränkungen, GWB), the first draft of which was published in 1952. However, the electricity industry as a whole was not on board. Bowing to pressure from the industry and many communities as well, the energy industry was

---

[9]Today the situation is quite the opposite: while the last hard coal mines are closing down (the last to be closed in 2018), the two large lignite pits (Garzweiler and Hambach) will still supply the electricity industry for about another decade or so.

ultimately exempted from the competition law [§103 of the Antitrust Law (GWB)].[10]

Therefore, industry structures in the electricity sector changed little in West Germany between the 1950s and 1980s: eight vertically integrated monopolistic suppliers dominated their respective concessions.[11] Their cooperation with other partners such as equipment suppliers (Siemens, AEG) remained very high as well. With respect to electricity transmission, a joint German Network Company (DVG—Deutsche Verbundgesellschaft) was created to coordinate investments in transmission across the country, the costs being rolled over to electricity customers.

2.2.2.2 Energy Mix: Big Coal Still Dominant...

Strong reliance on coal in both East and West Germany also required governance structures throughout the political system that protected coal as an energy source and opposed changes—in particular to the use of other, cleaner fuels such as natural gas. In the GDR, lignite resources were developed rapidly, making the country the largest lignite producer worldwide. The use of domestic lignite offered the country a high degree of self-sufficiency. As a result, GDR lignite production was pushed to the very limit, with up to 300 mn. t of annual production in the 1980s. Abundant but dirty lignite resources assured the survival of heavy industry, electrification, and heating in the GDR over decades.

In West Germany and Western Europe, the dominant role of coal was maintained through agreements between the heavy industries of six countries that would later emerge as the European Commission for Steel and Coal ECSC in 1953. Mechanisms to protect domestic coal suppliers against cheaper imported coal were already broadly applied before World War II. After the war, one of the founding principles on which the European Community was built was the idea that a cartelized energy and heavy industry would maintain jobs and supply security. In West Germany, hard coal was mined with federal subsidies from the 1950s until ... 2018.

Whereas natural gas was used to a very limited extent in the GDR (which benefited to some extent from Soviet natural gas imports), it expanded into the West German power sector against strong opposition from the coal-dominated energy industry, which sought to prevent unwanted competition. The industry's preferred wording was that "natural gas was too valuable" to be burned for electrification; instead it should be used for heating. However, the subsequent electricity legislation ("Verstromungsgesetze") prevented large-scale deployment of natural

---

[10]Formally, a price control clause was introduced in § 104 GWB, but it little effects: All that (understaffed) price control bodies could do was to compare monopoly prices with prices of other monpolists: the Monopoly Commission, advising the Minstry of Economy, concluded in a 1958 report, that "price control has virtually remained without any effect," cited in (Becker 2011, 46).

[11]Badenwerk, Bayernwerk, BEWAG Berlin, EVS Schwaben, HEW Hamburg, Preussen-Elektra, RWE, and VEW.

gas de facto into the mid-1990s.[12] It was only with the liberalization and completion of the European Single Market, spearheaded by the UK, in the 1990s that natural gas gained a more significant share of the German (and European) electricity market.

2.2.2.3 . . . And Supplemented by Nuclear Power

The history of nuclear power in East and West Germany is short, but intense. After World War II, Germany was demilitarized and prevented from developing nuclear technology itself. Civil nuclear activities were only allowed after 1954, and military nuclear applications were forbidden until the end of the Allied government (1990). Contrary, to earlier nuclear countries, like France or the UK, neither East nor West Germany therefore developed a domestic nuclear industry. Once these capacities were imported, from the U.S. and the Soviet Union, respectively, the governments of East and West Germany had to effectively impose the use of nuclear energy on the energy industry, which was conservative and feared competition with its coal-based activities. In the GDR, resistance was overcome with the help of the Soviet Union, which compelled its "brother" nation to adopt Soviet-style nuclear technology under relatively favorable conditions of technology and knowledge transfer. In East Germany, the VEB Kernenergie was established as the main driver of nuclear energy development.

In West Germany, the conflict between nuclear elites aspiring to develop the technology and the incumbent, coal-dominated energy trusts resisting their efforts was particularly pronounced. On the one hand, West German nuclear elites wanted to make Germany a nuclear power and catch up with the US, the UK, France, and the Soviet Union. A "military-scientific" alliance (Lévêque 2014, von Hirschhausen 2017) sought support from the new Adenauer government, which was itself interested in following other countries' lead in the pursuit of nuclear energy. On the other hand, nuclear power was considered a particularly unwelcome competitor by the incumbent, coal-dominated electricity industry in the 1950s. Attempts by the government to spur the industry's interest in commercial nuclear power failed or were even actively opposed.[13] Radkau (1983), Radkau and Hahn (2013), Müller (1990, 1996), and Matthes (2000) describe multiple maneuvers by energy and fiscal policy makers to encourage the energy industry to embark on the adventure of nuclear power, including generous financial support, coverage of construction risks, and limitation of financial liability for accidents (to DM 2.5 bn. or around 1 euro bn.). Structural changes such as the merger of nuclear equipment suppliers into one

---

[12]The third German law on electritic generation (Verstromungsgesetz, 1974) made the expansion of existing or the construction of new natural gas (and mineral fuel oil) power plants de facto impossible (for details, see Matthes (2000, 126).

[13]Radkau (1983) and Radkau and Hahn (2013) provide a full account of both the remarkable fascination with nuclear power in the post-war period—Willy Brandt, then Mayor of West Berlin, wanted to construct a nuclear power plant in the western part of the divided city—but also the resistance to a non-competitive energy technology by the incumbent energy industry.

national champion, Kraftwerk Union (KWU, a subsidiary of Siemens and AEG), in 1969 provided additional favorable conditions. Thus, there emerged a consensus among energy utilities to embrace a "coal and some nuclear" strategy.

In terms of technologies, East and West Germany both relied on the nuclear powers to develop their own industries. The GDR adopted Soviet technology developed after World War II and tried to build on technology transfer to push its own technologies further. West Germany underwent an effort to develop its own nuclear value-added chain under the leadership of Physics Nobel-prize winner (1932) Werner Heisenberg. Having come to the realization that the "German solution," based on heavy water, would not work, West German industry and politics took quite some time to begin adopting an "American" technological solution in the 1960s.[14] Broadly speaking, a choice had to be made between the pressurized water reactor (PWR) and the boiling water reactor, both light water reactors. For each of these technologies there was only one US supplier available: General Electric for the boiling water reactor and Westinghouse for the pressurized water reactor. In the early years, more potential was attributed to the boiling water reactor, which had an easier construction and, apparently, lower costs. With increasing experience, and also more stringent security regulations, the pressurized water reactor took over as the default technology.[15]

2.2.2.4   Fuel Mix

By the 1980s, nuclear power had gained a significant market share, in particular in West Germany, but the dominant role of coal and lignite had not been challenged. In the GDR, one large nuclear complex had been developed in Greifswald/Lubmin, featuring six blocks of the Soviet-type VVER reactors (a total of 2.6 GW). A second site was under development at the time of German reunification, in 1990, in Stendal, but did not come online. Lignite remained by far the largest provider of electricity. In West Germany, coal also remained the dominant fuel into the 1980s, when generation from nuclear power plant developments began to increase substantially. In West Germany, the development of nuclear power capacities in West Germany was quite steep growth in the 1980s, though no new developments followed thereafter.

---

[14]The history of nuclear power in Japan follows the same pattern.

[15]After Germany succeeded in building a nuclear industry from scratch, a complex process of technology transfer began, with German engineering companies being given more and more autonomy in the construction of nuclear plants. AEG obtained a license from GE to build boiling water reactors, as did Siemens for the Westinghouse pressurized water reactor (similarly to Toshiba in Japan). The merger of AEG and Siemens into one unified, monopolistic nuclear engineering firm, Kraftwerk Union (KWU), then led to the internalization of this technology competition into one firm.

2.2.3 1980s–2010: Wind of Change and the Rise of Renewables

2.2.3.1 1980s: Wind of Change

During the three decades from the early 1980s to 2010, some of the founding principles of German energy policy were increasingly challenged, but without calling the system into question as such. As early as the 1970s, Amory Lovins, founder of the independent think-tank Rocky Mountain Institute (RMI), called for a transformation of energy strategy from a "hard" to a "soft" path (Lovins 1976). The shift he was suggesting was not only from nuclear and fossil fuels to renewables (mainly solar) but also from centralized energy production to more decentralized, democratically controllable structures.

Lovin's article "Energy Strategy: The Road Not Taken?" (Lovins 1976) inspired many environmentally oriented politicians, researchers, and intellectuals in Germany as well. In 1975, Erhard Eppler, a left-wing SPD leader and Federal Minster for Economic Cooperation, established the concept of the "Wende" (turnaround) in a book called "End or Turnaround: On the necessity of doing the feasible".[16] In it, Eppler argued that there existed a close connection between unsustainable development and autocratic governance structures in West Germany: "Never has the mismatch been more untenable between the vision of the scientists, or what they consider to be necessary, and what politicians say and do." (Eppler 1975, 3, own translation).

Then, in 1980, Florentin Krause, a German native researcher working with Lovins at the RMI, together with Hartmut Bossel and Karl-Friedrich Müller-Reißmann (1980) published the first study on the "energiewende" proper: "Energiewende—Growth and Wealth without Oil and Uranium".[17] In it, they presented the concept of an "energiewende" that would hinge on a decided rejection of mineral oil and nuclear power. This concept had its roots in crucial events of the preceding decade. The two oil crises of 1973 had 1979 had increased awareness of the world economy's dependence on oil imports, particularly from countries in the Arab Gulf region, and had sparked movements for oil independence. In addition, the 1979 Harrisburg (PA) nuclear accident, where the core almost melted down, had proven beyond all shadow of doubt the dangers of nuclear energy, even in the world's most technologically advanced country. Inspired by Amory Lovins, Krause et al. outlined a "soft" scenario of energy turnaround involving a massive reduction of energy consumption by 2030 and an energy mix relying on coal (!) and solar.[18]

---

[16]Eppler (1975): Ende oder Wende. Von der Notwendigkeit des Machbaren (authors' translation from German original).

[17]"Energie-Wende: Wachstum und Wohlstand ohne Erdöl und Uran" (authors' translation from German original).

[18]In fact, Florentin Krause, the main author of "Energiewende," had been a research associate with the Rocky Mountain Institute and was hired to develop a "soft path" concept for Germany, as a guest researcher with the Institute for Applied Ecology (Öko-Institut).

They developed a mathematical model leading up to 2030 with a "hard" policy pathway corresponding to the status quo and a "soft" pathway that would reduce primary energy consumption form 400 mn. tce (tons of coal equivalent) to 216 mn. tce. Although their scenarios envisaged the phase-out of nuclear energy and oil, coal continued to play a role: 125 mn. Tce came from coal, 26 mn. tce from solar, 12 mn. tce from wind, 3 mn. tce hydro, and 50 mn. tce biofuels (Krause et al. 1980, 40). In summary, the authors conclude "The often cited 'energy gap' is nothing but the gap between the imagination and political will of those in charge of energy policy. The increasingly tight situation on the oil and uranium markets requires an energiewende, for which the current policy, based on large-scale technology, is not only too risky, but also too inflexible and costly." (p. 40).[19]

In general, the 1980s can also be considered a period of turnaround in public opinion because the environmental limits to growth of the conventional energy system had become patently clear. The study by Krause et al. reflected the growing disenchantment of a large part of the population in general with the existing energy policy models—or lack thereof—in the energy industry. Energiewende became a political buzzword among those who were opposed to the existing large-scale electricity industry, who advocated more ambitious environmental goals and democratic control over industry, and who were fundamentally critical of nuclear energy. The history of energiewende thinking can thus be traced back to the post-1968 movements to "work through" Germany's Nazi past, to social opposition to nuclear weapons projects of the German Bundeswehr after World War II, and more broadly, to public frustration with the German energy sector's longstanding failure to address sustainability issues in any substantive way.

The 1980s also destroyed hopes that nuclear energy, once predicted to become "too cheap to meter," would soon become a viable source of cheap energy.[20] Climate change and the unsustainability of fossil fuels had already become topics of public discussion in the preceding decades. But it took the Chornobyl nuclear accident to finally create consensus in the population on the closure of Germany's nuclear power plants. On April 26, 1986, the core of the nuclear reactor No. 4 in the Soviet Ukrainian nuclear power plant "Vladimir I. Lenin" in Chornobyl exploded and melted down due to a human error in restarting the reactor in combination with a faulty design. After a period of silence from the Soviet authorities, it became clear from measurements in countries neighboring Ukraine that the nuclear cloud originating in Chornobyl was moving northwest across Belarus and would reach Germany 2 days later. The vast scale of the catastrophe, its irresponsible handling by Soviet authorities and politicians, and the German population's first experience of

[19]In a follow-up study, Hennicke et al. (1990) provided a detailled analysis of what needed to take place at the institutional level for the Energiewende to happen. In particular, they argued for the necessity of overcoming big corporate structures that were resistant to change.

[20]See Hirschhausen (2017) with further references; the term "too cheap to meter" comes from a speech by Lewis Strauss, Chairman of the United States Atomic Energy Commission to the National Association of Science Writers in (1954): "our children will enjoy in their homes electrical energy too cheap to meter".

being directly affected by radiation reinforced anti-nuclear sentiments and established the view that nuclear power was not an option for a safe and economic energy supply. The idea of closing nuclear power plants gained ground, and preparatory work on a reprocessing factory in Wackersdorf (Bavaria) was stopped following massive public protest.

Two more events in the 1980s added elements of a "softer" energy policy path: the first was the move in the US and UK towards unbundling of vertically integrated governance structures in an attempt to overcome the monopolistic legacy. In 1983, Paul Joskow and Richard Schmalensee (1983) published the first reform proposals for unbundling, and these would inspire US FERC Orders 436 and 636 on restructuring and pave the way for other countries as well, including the UK. It was hoped that separating transmission assets, a natural monopoly, from generation, a potentially competitive activity, and privatizing a large share of both would increase the efficiency of electricity supply could be increased and overcome resistance from the incumbent coal industry to technical change, in particular to new natural gas plants. Although it took another 10 years for this to be implemented at the European level (Directive 96/92) and even more time to be implemented in Germany (energy law of 1998), the foundation had been laid for unbundling, without which the energiewende (and many other developments) would never have taken place.[21]

In addition, German reunification—commonly referred to as *die Wende*, a political turnaround—brought about an important additional step towards more decentralized decision-making in energy policy. In 1990, Germany's Supreme Court (*Bundesverfassungsgericht*) issued a decision preventing the big West German utilities from taking over the East German electricity industry. Initially, the three largest West German utilities (Preussen-Elektra, Bayernwerk, RWE) had prepared the takeover not only of generation but also of transmission lines as well as the 15 regional distributors at the district (*Kreis*) level in East Germany. They were joined later by the remaining West German utilities (VEW, Hamburg, Berlin, and EnBW (a merger of Badenwerk and EVS Schwaben). On August 22, 1990, the electricity contracts (*Stromverträge*) sealing the takeover were signed. Then a Berlin lawyer, Peter Becker, and a handful of colleagues launched a legal battle over the rights of the East German municipalities to set up their own local utilities based on the fundamental right to municipal self-determination, pursuant to Article 28 of the German constitution (*Grundgesetz*). The Supreme Court agreed and rejected the takeover agreements, thus paving the way to a decentralized energy industry in East Germany and stronger local utilities (*Stadtwerke*) in the West as well. This also laid

---

[21]Unbundling in Germany was advocated by the Ministry for Environment (then headed by SPD Minister Siegmar Gabriel), citing the slow progress of the European Directives. The idea was to impose a more sustainable energy mix on the unbundled energy companies, see: Theobald and Theobald (2013) and Federal Ministry of the Environment (2007): "Gabriel welcomes European Commission's legislative package for the EU electricity and gas markets," press release.

the groundwork for a fundamental reshaping of the East German energy mix, reducing the share of lignite and giving natural gas a more significant role.[22]

2.2.3.2 The 1990s: Emergence of Climate Policies and Renewables

The most notable trend in the 1990s was the emergence of climate policies in Germany, Europe, and worldwide. Although the greenhouse effect and other adverse implications of CO_2 and other pollutants had been apparent long before, it was only in the 1990s that policies to combat these effects were debated seriously. Political debate began to emerge in Germany over the coal-based electricity sector following similar discussions in other countries, in particular the US. Germany joined the countries with what can be considered progressive climate and environmental policies in the 1980s. Whether it was because of the realization that the "Deutsche Wälder" that formed the core of nineteenth century romanticist notions of the German landscape could not withstand the ravages of acid rain, or the innovative tactics of the environmental engineering lobby used to strengthen its position on international markets, Germany took an increasingly active role in international negotiations. In the run-up to the Rio 1992 summit on sustainability and thereafter, Germany developed a climate policy aimed at balancing environmental goals with the interests of the incumbent energy industry. Bowing to corporate interests, Germany abandoned the initial idea of a European-wide energy or CO_2 tax and replaced it with an emissions trading scheme with generous grandfathering policies (Corbach 2007).

At the European level, Germany also played an active role in spearheading the 1997 Kyoto Protocol, the first global climate agreement (but one from which important countries were missing, such as the US, China, and India). As a signatory of the 1997 Kyoto Protocol, Germany, as part of the European Union, implemented the three flexible mechanisms foreseen in the agreement: emissions trading, Joint Implementation (JI), and the Clean Development Mechanism (CDM). The Cap-and-Trade European Emission Trading System (ETS), modeled after the US SO_2 trading scheme (Ellerman 2000) was to become a tool for seeking lower-carbon energy technologies, and after a test phase (2005–2006) it became fully operational in 2007. About 1660 German "installations" participated in the ETS in Germany, representing about half of the countries' CO_2 emissions (see Chap. 4 for more details).

The 1990s also witnessed an important breakthrough for intermittent (or: variable) renewable energies, which would later become a pillar of the energiewende. In fact, renewable energy sources had played a certain role since as early as 1990, when the first law was passed on renewable feed-in tariffs, responding to some extent to the rising discussion about the sustainability of nuclear and fossil fuel-based

---

[22]Becker (2011, Chap. 2), one of the main advocates of municipal interests, provides a detailed account of this battle.

electricity generation. Consequently, all the ideas about an energy turnaround or a "soft path" included plans to increase the development and use of renewable energies. In response to this widespread sentiment as well as the aftermath of the Chornobyl disaster (1986), the conservative administration of Helmut Kohl, under whom Klaus Töpfer served as Environmental Minister, passed the first "law on feeding in renewables" in 1990.[23] This law followed a similar suggestion formulated by the European Commission favoring the emergence of renewable energies (Theobald and Theobald 2013, 495). The law adopted a classical feed-in approach and led to a modest but steady rise in RES towards an initial cap of 5%. The renewables law was based on the idea of decentralized production, and the big utilities fought against the introduction of renewables from the outset. They launched a major public relations campaign in the early 1990s, arguing that "technical constraints prevent any more than a 4% share of renewables in the electricity system."[24]

The election of the Green Party to federal government in 1998 favored the growth of renewables. In the run-up to the first European Directives on the Common electricity market (96/92/EC) and natural gas market (98/30/EC), there was some debate within Germany over whether to integrate renewables legislation into the more general energy law (of 1998) or to keep it separate. The latter solution was adopted for a simple political reason: With the Green Party's entry into parliament (Schroeder-Fischer), Juergen Trittin, a leading Green activist, became Minister for the Environment and obtained the government dossiers on renewable energy from the Economics Ministry, which was traditionally responsible for energy policy.

2.2.3.3 The 2000s: Controversies Battles Over the Electricity Mix

The first decade of the twenty-first century will be remembered as a time of struggle between the incumbent structures and technologies—coal and nuclear—and the emerging political, technical, and economic ideas of a turnaround in energy and climate policies. It was a decade in which no clear tendency emerged on the preferred electricity mix, but where the three power sources—coal (and some natural gas), nuclear, and renewables—more or less co-existed. In fact, the September 2010 Energy Concept 2050 sought to strike a compromise and to include all three: renewables, nuclear (lifetime extensions with the option of new builds later on), and coal (with carbon capture). Policies on each of the three elements of this triad are independent in some respects, but also intertwined in others:

- On the coal and climate policy front, conflicts intensified. The incumbent coal industry benefitted from the anticipated closure of nuclear plants and launched a

[23]Gesetz über die Einspeisung von Strom aus erneuerbaren Energien in das öffentliche Netz (Stromeinspeisungsgesetz) v. 7.12.1990 (StrEG), BGBl. I S. 2633).

[24]Published as an advertisement by the big eight utilities in the Süddeutsche Zeitung, 1993, Nr. 152 [authors' translation].

major investment program. However, towards the end of the decade, climate policy objectives were tightened in the wake of EU activities led by Germany as EU Council Head in 2007 and the G8 (also headed by Germany in 2008). Germany was among the proactive forces behind the European Energy and Climate Package of 2008/2009 as well as efforts towards a global agreement on climate. Whereas pressure increased on the incumbent coal-burning industry to reduce emissions, the energy industry insisted that its investment program could fill the alleged "electricity gap" created by the shortfall of nuclear capacity, and that without this program, resource adequacy and supply security would be endangered. Policy makers argued that the renewal of the fleet of lignite plants was necessary to ensure the survival of the East German energy industry. Thus, the government funded a power plant investment program in the amount of 20 GW of new fossil-fuel power plants, 80% of which based on coal in the first decade of the new century.

- Debates on <u>nuclear</u> power were the most controversial over the course of the decade. In fact, the phase-out of nuclear power in Germany had been declared a political priority by the coalition of Social Democrats and Greens that came into power in 1998 (under Chancellor Gerhard Schroeder and Vice-Chancellor Joschka Fischer). The agreement with the four energy utilities (now E.on, RWE, Vattenfall, and EnBW) negotiated by State Secretary of the Environment and former member of the parliament of Wetzlar, Rainer Baake, included a cap on total electricity generation from nuclear power plants, calibrated such that the last plant would close sometime in 2025. Given the fact that the generation quota could be shifted from older to younger plants, the scheme introduced some incentives for efficient use of the remaining capacities [see Matthes (2012) for details]. Naturally, the agreement became the subject of political wrangling, and the energy utilities soon announced that they were working on a revision. In the meantime, they negotiated favorable conditions for the expansion of fossil-fuel power plants including free allocation of CO_2 emission allowances. The opposition, an alliance of conservative parties (CDU/CSU) and liberals (FDP), swore "revenge" and included reversal of the closure agreement in their official election programs. The 2009 federal elections, which propelled these two groups to power, marked the beginning of a new nuclear policy in Germany and the renewal of efforts towards the lifetime extension of all NPPs. The idea of developing an "Energy Concept 2050" in the context of longer-term European energy and climate goals was seen as a means to bring the nuclear industry back into business.

- Last but not least, <u>renewables</u> gained ground as the third pillar of this "troika," clearly revealing that a rising share of intermittent renewables posed no obvious risk to supply security. The feed-in law of 1991 was replaced by a more comprehensive Renewable Energy Act (*Erneuerbare-Energien-Gesetz*, EEG, of 2000), also intended to place renewables on equal footing with other energy sources and bring them into the mainstream of electricity generation. The 2000 EEG continued the tradition of technology-specific, feed-in-focused, long-term (20-year) support mechanisms for renewables, mainly wind, solar, and geothermal. This was later complemented by regulations on bioenergy. The basic approach was maintained throughout the decade and has become a model for many other countries introducing renewable energies. In contrast to the feed-in law of 1990, which had support from all the parties, the EEG faced opposition mainly because the incumbent energy industry had realized that it might become a danger for conventional generation. In fact, from 4.8% of electricity generation in 1998 (just below the cap of 5%), the share of renewables rose to 16% (~ 85 TWh) by 2009.

With respect to corporate governance and policymaking, the major utilities continued to lead most initiatives and remained actively involved in the political process. European liberalization in 1998 and the unbundling prescribed in the Second Energy Package of 2003 on liberalizing electricity and natural gas (the "acceleration" directives) did not change the balance of power significantly.

2.2.3.4 Energy Mix: An Emerging "Troika" of Coal, Nuclear, and Renewables

Figure 2.1 presents our interpretation of the electricity "troika" of the 1980s–2010. During this period, coal was still king, nuclear power gained a significant share, and renewables left the realm of experimental technologies to become one of the pillars of the electricity supply. Hydropower remained the largest contributor of renewable energy, but variable renewables—in particular wind—increased their share significantly.

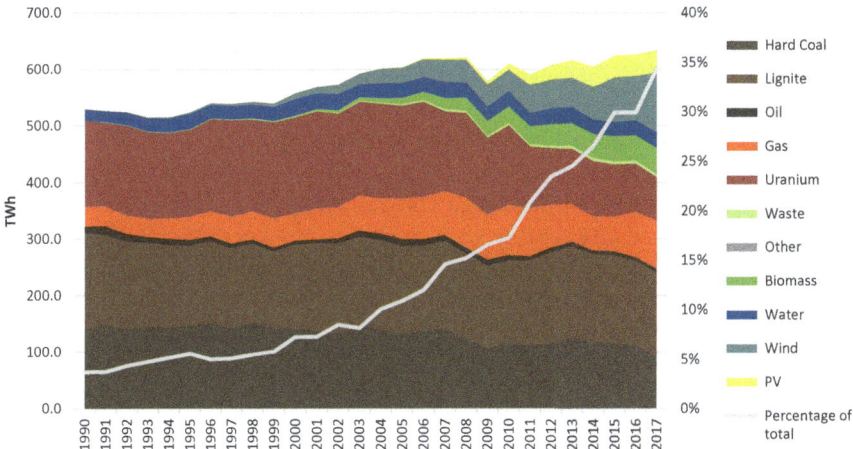

Fig. 2.1 Electricity mix in Germany, 1990–2010 (in TWh and %)

2.3   2010–2011: A Critical Moment of the Energiewende

The 12 months between the summer of 2010 and the summer of 2011 will go down
in history as an eventful period in which the energiewende accelerated, to become a
truly low-carbon, nuclear-free, and renewables-based transformation process. However, the story has to be told in two acts: the first is the ambivalent Energy Concept
2050 of September 2010, which included ambitious climate and renewables targets,
but also put nuclear power back into the longer-term electricity mix; and the second
is Angela Merkel's March 2011 "Declaration of the energiewende," a renewables-
based energy system with no nuclear energy and little coal.

2.3.1   September 2010: An Ambivalent "Energy Concept 2050"

After ambitious energy and climate targets were set at the European level in the
period 2007–2009, EU Member States began translating these aggregate targets into
national and climate programs. An early example was the UK (DECC 2011); another
one was Denmark[25]; a survey of the early 2050 debates is provided by Meeus et al.
(2011, 2012) and Förster et al. (2012). In Germany, too, work had begun on a longer-
term energy program that was to become the basis of the Energy Concept 2050.
However, this concept still relied on the troika of energy sources that had co-existed
for two decades: coal, nuclear, and renewables. In fact, although the Energy Concept
2050 included ambitious climate targets (80–95%, reduction of GHG emissions by
2050 from the base year 1990), an increasing share of renewables, and ambitious
efficiency targets, it also contained provisions to keep the two traditional pillars of
the previous system firmly in place:

- The future use of coal was justified by the anticipation of a "clean coal" technology known as carbon capture, transport, and storage (CCTS). This technology
  had gained popularity through the Special Report of the IPCC (2005) and had
  spurred worldwide hopes of reconciling fossil fuels with the vision of a
  low-carbon future (Jaccard 2006). Thus, although the technology was inexistent
  at the time (and even a decade later), the scenario calculations underlying the
  Energy Concept 2050 all included CCTS to attain the GHG emission targets
  (EWI, Prognos, and GWS 2010).
- Discussions over the second pillar, nuclear energy, delayed the publication of the
  Energy Concept 2050 through the summer of 2010. Negotiations between the
  utilities and the government were intense, the former insisting on an "electricity

---

[25]Danish Government. 2011. "Our Future Energy." Copenhagen, Denmark.

gap" and the need to continue the use of nuclear energy as a cheap source of electricity. A compromise was finally struck between the government and the utilities in the very early hours of September 6, 2010. The utilities were granted a lifetime extension for nuclear power plants by 12 years on average (older plants: 8 years, younger plants: 14 years); in return, they had to pay a tax on the uranium fuel used and provide some support for the development of renewable energies. In the morning of September 6, Chancellor Merkel presented the Energy Concept 2050 to the public as a breakthrough for German energy supply and the basis for a low-carbon future that would include nuclear power.

What Chancellor Merkel failed to mention, however, was the part of the agreement that had been negotiated in secret the night before and signed by the participants at precisely 5:23 am on September 6, 2010.[26] This side agreement (called a "term sheet") was designed such that the utilities would maintain their economic benefits from the lifetime extension even after a subsequent change of government; it also assured the utilities a reduction of payments for renewables and capped future expenses for safety upgrades at 500 million €.[27] It seemed that the incumbent energy industry had won again. The agreement might never even have become public had it not been challenged in a public discussion on Tuesday, September 7, 2011, by environmental activist Tobias Muenchmeyer, who demanded information about how the additional income from the lifetime extension was to be transferred to the federal government. In responding to this question, the spokesman of the RWE executive committee, Rolf-Martin Schmitz, accidentally mentioned the secret deal between the energy industry and the government. Facing rising public pressure after this slip of the tongue, the government published the secret deal on the Internet in the afternoon of Thursday, September 9, 2010.[28]

Thus, when the Energy Concept 2050, including nuclear lifetime extensions, was adopted in a parliamentary vote on September 26, 2010, the differences of opinion between the ruling conservative coalition and the opposition, both in parliament and in the public at large, could not have been wider. The window for mobilizing support for the energiewende across the political spectrum seemed to have closed.

---

[26]For details, see the comment by Bank (2010): Der Atomdeal—Eine kleine Chronologie undemokratischer Politik.; accessed on March 14, 2014, see www.lobbycontrol.de, as well as the daily press, e.g., Süddeutsche Zeitung, September 09, 2011; a personal account of the events is provided by Tobias Münchmeyer himself, in a conference presentation given at DIW Berlin: http://www.diw.de/de/diw_01.c.409266.de/forschung_beratung/projekte/projekt_homepages/masmie_modellieren_fuer_die_energiewende/nbsp_nbsp_veranstaltungen/nbsp_nbsp_veranstaltungen.html, last accessed 20 September 2016.

[27]Förderfondsvertrag: Term Sheet aus Besprechung Bund und EVUs (energy utilities).

[28]Source   https://www.lobbycontrol.de/2010/09/der-atomdeal-eine-kleine-chronologie-undemokratischer-politik/. What seems like an anecdote is important because it was the last time that the traditional nuclear lobby was able to impose its preferences on the policymaking process.

2.3.2   The Fukushima Nuclear Accident and the Events
of Spring 2011

2.3.2.1   Chancellor Merkel Changes Course: The March 14, 2011,
Moratorium on Seven Nuclear Power Plants

Tragically, it took the nuclear catastrophe in Fukushima to bring about the end of
nuclear power generation in Germany. On March 11, 2011, when an earthquake
followed by a tsunami hit the East coast of Japan, water flooded the Fukushima
Daichi power plant, cutting off the emergency electricity supply to the entire plant
and leading to the meltdown of three reactor cores and the release of significant
amounts of radioactivity. It was one of the worst accidents in the history of nuclear
power (Radkau and Hahn (2013). Shortly thereafter, Japan shut down all its nuclear
reactors (56) due to the high risk, and the country embarked on its own path towards
energy transformation with a critical assessment of the role of nuclear power in its
energy mix.[29]

The political fallout in Germany was unprecedented. The day the tsunami hit
Fukushima, the true dimensions of the unfolding disaster were still unknown. But
the meltdown of the cores of reactors 1–3 and several explosions in the buildings
heightened pressures on Chancellor Angela Merkel, who declared on public televi-
sion on March 14 that "a new situation" had occurred (see Box 1): If a highly
industrialized country like Japan, with its high security precautions, was not able to
avoid nuclear accidents, this had to have consequences for Europe—Germany
included (Bundesregierung 2011). The German government declared an initial
3-month moratorium on the lifetime extension of the existing nuclear power plants.
During this period, options would be assessed for how to move forward quickly
toward an age of renewable energies: "the only honest answer [to the Japanese
nuclear accidents] is a forced and accelerated path towards the age of renewable
energies." (Bundesregierung 2011).

Concretely, Merkel announced a "moratorium" on the operation of the seven
oldest nuclear power plants in Germany, which had been identified as such in the
first phase-out process in the early part of the decade.[30] The federal government
enacted provisions allowing the state governments, already empowered with issuing
operating licenses for nuclear power plants, to halt the operation of these seven
plants for 3 months. The plants were never reopened.

There are two interpretations of the decision Chancellor Merkel made relatively
independently, in consultation with just a few close advisors but not with her cabinet
nor the representatives of the ruling parties (CDU/CSU and FDP):

---

[29] See the Symposium of Economics of Energy and Environmental Policy (EEEP), Vol. 5, No. 1 on
"The Japanese Energy Policy after Fukushima".

[30] From North to South: Brunsbüttel, Unterweser Biblis A, Biblis B, Philippsburg 1, Neckarwestheim
1, Isar/Ohu 1. In addition, the plant at Krümmel is sometimes mentioned as part of the moratorium,
although it had already been shut down in 2007 due to technical problems.

- The first is that the decision was made, quite rationally, looking towards the upcoming state election, only 2 weeks later, on March, 27, 2011, in Baden-Wuerttemberg, a traditional CDU stronghold and strong supporter of lifetime extension for nuclear facilities. Voters in that state appeared at that point to be leaning strongly towards the opposition, led by the anti-nuclear Green Party.
- The other interpretation is that Merkel, a GDR-trained doctor in physics, had already been convinced that nuclear power was not safe, and her "turnaround" was a reaction to the intensive lobbying she and the entire German government had been subject to in the run-up to the lifetime extension decision in the fall of 2010.

Thus, the question of whether Merkel's decision was based on her true conviction that nuclear power was not safe or whether it was a purely tactical maneuver to save her party (CDU) in the upcoming state elections in Baden Wuertemberg—or something in between—should be left to historians. The fact of the matter is that with her decision of March 14, 2011, Merkel made the crucial final step toward a full turnaround in Germany's energy policy: the end of nuclear electricity generation in Germany, combined with the move towards the renewable age.

Box 1 The "Declaration of the Nuclear Moratorium": Chancellor Angela Merkel at a Government Press Conference, Monday, March 14, 2011
"It was and is not cheap talk when I say: We must not simply go back to business as usual and pretend as if the current and up to now undisputed safety of our nuclear operations is enough to guide our future actions, without stopping for serious reflection in the light of recent events. The events in Japan teach us that things we consider impossible according to scientific criteria can nonetheless become reality. These events also teach us that risks that were considered highly improbable are possible after all. And if this is the case, and when even a highly developed country such as Japan, a country with the highest safety standards and security requirements, cannot avoid the nuclear consequences of an earthquake and a tsunami, this must have consequences for the entire world. It has consequences for Europe as well, and it has consequences for us in Germany. It changes the situation in Germany. We have to analyze this situation uncompromisingly, comprehensively, and without hesitation. Only then can decisions be made.

Last Saturday, we therefore decided that in the light of the findings we have from Japan, all German nuclear power plants will be subject to a comprehensive security check. I state this very clearly: in these security checks nothing will be off limits. For the same reason, we will suspend the recent decision to extend the lifetime of German nuclear power plants. This is a moratorium that will last 3 months. We are in discussions with the operators of the nuclear power plants about the concrete implications of this moratorium.

To leave no doubt: the situation after the moratorium will be another one than before the moratorium. We will seize the occasion of the moratorium to

(continued)

Box 1 (continued)

explore pathways, how to accelerate our path towards the age of renewable energies, and how to reach this goal even faster. Because if we are speaking about nuclear energy as a "bridge technology," this means nothing other than that we are discontinuing the use of nuclear energy and want to ensure the German energy supply through the use of renewables as quickly as possible. However, disconnecting German nuclear power plants and accepting the impact of nuclear energy from other countries—this I also state very clearly—cannot and must not be our only answer. The only honest response is to accelerate our path towards the age of renewable energies."

Source: Authors' translation of official press statement (Bundesregierung 2011): "Pressestatement von Bundeskanzlerin Angela Merkel zu den Folgen der Naturkatastrophen in Japan sowie den Auswirkungen auf die deutschen Kernkraftwerke," March 14. Accessed on August, 16, 2013: http://www. bundesregierung.de/Content/DE/Mitschrift/Pressekonferenzen/2011/03/ 2011-03-14-bkin-lage-japan-atomkraftwerke.html

2.3.2.2 Implementation: The Chain of Events up to June 2011

A speech by a government leader does not suffice to fundamentally change the (energy) policies of an entire country. What followed Merkel's declaration were 3 months of intensive political work to translate the energiewende nuclear moratorium into legislation. In order to build broad consensus on these decisions, Merkel established a non-partisan "Ethics Commission for a Safe Energy Supply" only a week after, on March 22, 2011. The commission was to study the technical and ethical aspects of nuclear energy, to prepare the foundation for discussions that could build consensus on the phasing-out of nuclear power, and to develop concrete suggestions for the turnaround towards renewable energies.[31] The Ethics Commission was given 3 months, until June 15, 2001, to deliver findings.[32] Its conclusions were clear: nuclear energy was neither secure nor economical and should be phased out rapidly, and the path towards a renewables-based system could be accelerated

---

[31]The commission, which worked from April 4 to May 28, 2011, was headed by Klaus Toepfer, former environmental minister under the Kohl administration, which had implemented the first renewables law in 1990, and by Matthias Kleiner, then head of the German Research Foundation. Other members of the ethics commission included former high-ranking politicians, scientists, representatives of the civil organizations (churches, labor unions), and one industrial enterprise (BASF). Source: http://www.bmbf.de/pubRD/2011_05_30_abschlussbericht_ethikkommission_ property_publicationFile.pdf, accessed March 15, 2015.

[32]In parallel, a "Commission on the Security of Nuclear Reactors" (Reaktor-Sicherheitskommission, RSK), was to investigate the safety of German nuclear power plants according to updated security standards and was to deliver its report by June 15, 2011, as well.

without putting the energy system at risk (Ethics Commission for a Safe Energy Supply 2011).

Based on this report as well as additional documents provided by the government, administration, stakeholders, and other participants in the process, a legislative package was prepared in the summer of 2011 including the law on the closure of nuclear power plants. The 13th Amendment of the Law on Nuclear Energy (*Atomgesetz*, AtG) was accepted by the government cabinet on June 6, passed by the lower house of parliament (Bundestag) on June 30, 2011, and confirmed by the upper house (Bundesrat) on July 8, 2011. It was signed by the German Federal President on July 11 and entered into force a day later, on July 12, 2011.

Contrary to previous decisions on nuclear energy, the final phase-out decision of June 2011 had the broadest conceivable support from the government, parliament, and the German public. A look at the parliamentary votes on nuclear policies makes this very clear: both the votes on phasing out nuclear energy on December 14, 2001, and on the lifetime extension of nuclear facilities on September 28, 2010, were very tight, passing by a majority of only a few votes each.[33] Yet in its vote on June 30, 2011, parliament agreed with an overwhelming majority on the nuclear phase-out by 2022. The 13th Amendment of the Nuclear Energy Law received 510 yes votes and only 86 no votes, with about 80 of the latter coming from The Left party, which was in favor of an even faster nuclear phase-out.

2.3.3 The Objectives of the Energiewende

There are different interpretations what precisely constitute the energiewende, yet there is a broad consensus on its core objectives, summarized in Table 2.1, based on the Energy Concept 2050 and the 13th revision of the Law on Nuclear Energy:

- Greenhouse gases (GHG) are to be reduced by 40% (2020), 55% (2030), 70% (2040), and up to 80–95% by 2050. This longer-term perspective, in conjunction with a low probability of "clean coal," implies the full decarbonization of the power sector, and the phase-out of coal. Although it looked like a mere detail, the question of whether 80% or 95% GHG emission reductions would be required by 2050 is already emerging as an important issue (for details, see Chap. 4).
- Nuclear power plants are to be shut down between 2015 and 2022. The roadmap to the nuclear phase-out was presented in detail in the Law on Nuclear Energy of June 2011 (*Atomgesetz*, AtG): the seven oldest remaining nuclear power plants shut down temporarily following the moratorium were not to restart operations,

---

[33]The law on the first nuclear phase-out of December 14, 2001 was voted with support by the governing parites (SPD, Greens) but against the votes of all the opposition (CDU/CSU, FDP); the October 2010 decision on the lifetime extension received 309 "yes" votes versus 280 "no" votes (with 2 abstentions) in favor of the 11th Ammendment of the Law on nuclear power. Source: http://www.bundestag.de/bundestag/plenum/abstimmung/20101028_energie1.pdf download March 2013, see for a survey Matthes (2012).

Table 2.1 Main objectives of the energiewende

	Reduction of nuclear energy	Share of renewable energy		Reduction GHG-emissions	Reduction of energy demand			
		Gross final energy	Electricity production		Primary energy	Domestic heat	Final energy transport	Electricity demand
2015	−47%							
2017	−56%							
2019	−60%							
2020		18%	35%	−40%	−20%	−20%	−10%	−10%
2021	−80%		40–45%					
2022	−100%							
2025								
2030		30%	50%	−55%				
2035			55–60%					
2040		45%	65%	−70%				
2050		60%	80%	−80% to 95%	−50%	−80%	−40%	−25%
Base	2010	–	–	1990	2008	2008	2005	2008

Source: Energy Concept 2050 (BReg 2010), 13th Amendment of the Law on Nuclear Energy (*Atomgesetz*, AtG)

and a concrete timetable for the closure of the remaining nuclear power plants was defined, starting with Grafenrheinfeld in 2015, and going up to the last closure of Isar 2 by the end of 2022 (for details, see Chap. 5).

- Two different roadmaps were drawn up for renewables and subsequently entered into energy law as well: the share of renewables in electricity generation is to increase to at least 35% (2020), 50% (2030), 65% (2040), and 80% (2050); the share of renewables in gross final energy consumption is set to be at least 18% (2020), 30% (2030), 45% (2040), and 60% (2050, for details, see Chap. 6).
- Last but not least, ambitious targets for energy efficiency have also been defined (for details, see Chap. 7).

The specifics of the German energiewende, that differentiate it from other forms of low-carbon transformation processes, are the nature of the energy mix. As requested by proponents of the soft path since the 1970s, and confirmed in Chancellor Merkel's declaration on the nuclear moratorium, the energiewende targets renewable energies. The secret hope of those behind the Energy Concept 2050 had thus not materialized: nuclear power had been eliminated from the list of so-called "low-carbon technologies." With respect to the ambitions on GHG emission reductions, the energiewende also implied the end of coal-fired electricity. The key feature of the low-carbon transformation in Germany is clearly the strong focus on renewables, in particular in the electricity sector. The ambitious objectives first formulated in the Energy Concept 2050 were confirmed by subsequent policy documents and also entered into the Amendment of the Law on Renewables (EEG) in 2012: renewables would comprise at least 60% of the energy system as a whole and at least 80% of the electricity sector.

2.4   Conclusions

The energiewende in Germany has to be analyzed in the historical context of energy and climate policy trends in Germany, in Europe, and worldwide. In this chapter, we have provided an overview of energy policies in Germany since the 1880s, looking at these in the context of governance structures, important corporate and policy decisions, and the overall energy mix. This historical background is useful in understanding key features of the German low-carbon transformation: the energiewende marks a path towards a largely renewables-based electricity and energy system, and it explicitly renounces to the use of nuclear power (after 2022) and of coal and other fossil fuels in the medium term. With respect to governance structures, the energiewende constitutes the end of the traditional, utilities-led business model based on coal electrification (and some nuclear), and opens the way toward more decentralized supply structures. Understood in this way, the energiewende constitutes a "soft path" of energy policy, in contrast to the "hard path" that predominated in the 1970s (Lovins 1976).

The chapter has established that the energiewende is the German answer to the challenge of a low-carbon energy transformation, the key feature of which is its focus on the phase-out of coal and nuclear, and the path towards a renewable electricity and energy system. In a broader context, one can understand the energiewende as a political and societal process of negotiation that took began in the 1970s, and culminated between the summer of 2010 and the summer of 2011, resulting in the decision to pursue an energy and environmental policy that would make it possible to achieve a reduction of greenhouse gas emissions by 80–95% by 2050 (compared to 1990), to close down the nuclear power plants by 2022, achieve at least 80% of renewable electricity and 60% overall renewable energy by 2050, and to achieve a reduction of electricity consumption of 25% and of total primary energy of 50% (by 2050, respectively, compared to 2008). The energiewende decision enjoyed majority support from all of the participating bodies and the population at large. Its scope is long-term but it is based on a shared understanding that the conditions conducive to reaching the targets must be laid in the short term.

Since an understanding of the present requires an understanding of the past, it was important for us to identify the main characteristics of German energy and climate policies. In particular, since we consider the energiewende a major break between a centralized and a monopolistic energy system, and as a move toward more decentralized and transparent governance structures, we first looked at the preceding system, which had its roots in the late nineteenth century. In order to understand the origins of the energiewende in political, technical, and economic terms, and to trace the steps that led to broad parliamentary support for several laws on energy passed in the summer of 2011, we provided a brief history of German energy and climate policies. We distinguished three key periods preceding the energiewende: (1) the establishment of large industrial-financial energy corporations from the late nineteenth century through 1945 based on coal; (2) the nuclear period from the 1950s to the 1980s, when both Germanies tried to supplement coal with nuclear energy; and (3) the 1980s–2010, characterized by growing opposition to conventional governance structures and the fossil-nuclear energy mix, when renewables gradually entered the troika of the electricity mix. Our historical analysis also revealed interdependencies between the corporate structures and the policymaking process.

There are several ways to approach the low-carbon transformation, and the energiewende is a specifically German approach. As laid out in more detail in Chap. 9, most of the 195 signatory countries of the Paris climate convention are currently pondering different approaches to the low-carbon transformation. Some of them still propose to pursue the low-carbon transformation with a troika of coal (and other fossil fuels), nuclear, and renewables, as outlined, e.g., in the European Energy Roadmap 2050. With hindsight, it is noteworthy that the German Energy Concept 2050 presented in 2010 was also designed in an effort to keep coal and nuclear energy alive: although it contained ambitious targets, it still upheld the troika of "clean" coal, nuclear, and renewables. On the contrary, the true energiewende, in the spirit of Amory Lovins' "soft path", is based on distributed renewables, a choice confirmed by German energy policy in 2011, without "clean coal" neither nuclear power. The fact that Amory Lovins, who originated the concept of the "soft path," was awarded the Federal Cross of Merit (Bundesverdienstkreuz) by the German government in 2016—a very rare event for a foreign citizen—is evidence of the high recognition accorded to his early ideas and their importance for German energy policy. The remainder of this book will examine key aspects of the energiewende in the electricity sector and their implications in more detail.

References

Becker, Peter. 2011. *Aufstieg und Krise der deutschen Stromkonzerne – Zugleich ein Beitrag zur Entwicklung des Energierechts*. 2nd ed. Bochum: Ponte Press.

Boll, Georg. 1969. *Geschichte des Verbundbetriebs: Entstehung und Entwicklung des Verbundbetriebs in der deutschen Elektrizitätswirtschaft bis zum europäischen Verbund*. Frankfurt: Verlags- u. Wirtschaftsgesellschaft der Elektrizitätswerke.

Bundesregierung. 2011. Bundesregierung setzt Laufzeitverlängerung für drei Monate aus. https://www.bundeskanzlerin.de/ContentArchiv/DE/Archiv17/Artikel/2011/03/2011-03-14-moratrium-kernkraft-deutschland.html.

Corbach, Matthias. 2007. *Die deutsche Stromwirtschaft und der Emissionshandel. Ecological Energy Policy (EEP)*. Vol. 5. Stuttgart: ibidem-Verlag.

DECC. 2011. *Planning Our Electric Future: A White Paper For Secure, Affordable and Low-carbon Electricity*. London: Department of Energy and Climate Change.

Ellerman, A. Denny, ed. 2000. *Markets for Clean Air: The U.S. Acid Rain Program*. Cambridge: Cambridge University Press.

Eppler, Erhard. 1975. *Ende oder Wende – Von der Machbarkeit des Notwendigen*. Stuttgart: W. Kohlhammer.

Ethics Commission for a Safe Energy Supply. 2011. *Germany's Energy Transition – A Collective Project for the Future*. Report to the German Government.

EWI, Prognos, and GWS. 2010. *Energieszenarien für ein Energiekonzept der Bundesregierung*. Studie für das Bundesministerium für Wirtschaft und Technologie Projekt Nr. 12/10. Basel; Köln/Osnabrück

Förster, Hannah, Sean Healy, Charlotte Loreck, Felix Matthes, Manfred Fischedick, Stefan Lechtenböhmer, Sascha Samadi, and Johannes Venjakob. 2012 Metastudy Analysis on 2050 Energy Scenarios – Policy Briefing. SEFEP working paper 2012-05. SEFEP, Berlin

Hennicke, Peter, Jeffrey P. Johnson, and Stephan Kohler. 1990. *Die Energiewende ist möglich*. Frankfurt am Main: Fischer S. Verlag GmbH.

Hughes, Thomas Parke. 1993. *Networks of Power: Electrification in Western Society, 1880–1930*. Baltimore: Johns Hopkins University Press.

IPCC. 2005. *IPCC Special Report on Carbon Dioxide Capture and Storage*. Prepared by working group III of the Intergovernmental Panel on Climate Change. Cambridge University Press, Cambridge

Jaccard, Mark. 2006. *Sustainable Fossil Fuels – The Unusual Suspect in the Quest for Clean and Enduring Energy*. Cambridge: Cambridge University Press.

Joskow, Paul L., and Richard Schmalensee. 1983. *Markets for Power: An Analysis of Electric Utility Deregulation*. Cambridge: MIT Press.

Krause, Florentin, Hartmut Bossel, and Karl-Friedrich Müller-Reissmann. 1980. In *Energie-Wende: Wachstum und Wohlstand ohne Erdöl und Uran*, ed. Öko-Institut Freiburg. Frankfurt: S. Fischer.

Lévêque, François. 2014. *The Economics and Uncertainties of Nuclear Power*. Cambridge: Cambridge University Press.

Lovins, Amory B. 1976. Energy strategy: the road not taken? *Foreign Affairs* 6 (20): 9–19.

Matthes, Felix Christian. 2000. *Stromwirtschaft und deutsche Einheit: Eine Fallstudie zur Trans-formation der Elektrizitätswirtschaft in Ost-Deutschland*. Berlin: Germany.
———. 2012. Exit economics: the relatively low cost of germany's nuclear phase-out. *The Bulletin of the Atomic Scientists* 68 (6): 42–54.
Meeus, Leonardo, Isabel Azevedo, Claudio Marcantonini, Jean-Michel Glachant, and Manfred Hafner. 2012. EU 2050 low-carbon energy future: visions and strategies. *The Electricity Journal* 25 (5): 57–63.
Meeus, Leonardo, Manfred Hafner, Isabel Azevedo, Claudio Marcantonini, and Jean-Michel Glachant (2011) *Transition Towards a Low Carbon Energy System by 2050: What Role for the EU?* Final report 3. THINK project, European University Institute
Müller, Wolfgang D. 1990. *Geschichte der Kernenergie in der Bundesrepublik Deutschland: Anfänge und Weichenstellungen*. Stuttgart: Schäffer Verlag für Wirtschaft und Steuern.
———. 1996. *Geschichte der Kernenergie in der Bundesrepublik Deutschland: Auf der Suche nach dem Erfolg, die sechziger Jahre*. Stuttgart: Schäffer Verlag für Wirtschaft und Steuern.
Radkau, Joachim. 1983. *Aufstieg und Krise der deutschen Atomwirtschaft 1945–1975: Verdrängte Alternativen in der Kerntechnik und der Ursprung der nuklearen Kontroverse*. Hamburg: Rowohlt.
Radkau, Joachim, and Lothar Hahn. 2013. *Aufstieg und Fall der deutschen Atomwirtschaft*. Munich: Oekom Verlag.
Stier, Bernhard. 1999. *Die Politische Steuerung des Elektrizitätssystems in Deutschland 1890–1950. Staat und Strom*. Technik und Arbeit, 10. Ubstadt-Weiher: Verlag Regionalkultur.
Theobald, Christian, and Christiane Theobald. 2013. *Grundzüge Des Energiewirtschafts-rechts*. 3rd ed. Munich: Verlag C.H. Beck.
von Hirschhausen, Christian. 2017. *Nuclear Power in the 21st Century – An Assessment* (Part I). DIW Discussion Paper 1700, Berlin
Zängl, Wolfgang. 1989. *Deutschlands Strom: Die Politik der Elektrifizierung von 1866 bis Heute*. Frankfurt: Campus.

Chapter 3
The Transformation of the German Coal Sector from 1950 to 2017: An Historical Overview

Hanna Brauers, Philipp Herpich, and Pao-Yu Oei

"The mission of the European Coal and Steel Community is to contribute to economic expansion, the development of employment and the improvement of the standard of living in the participating countries through the institution, in harmony with the general economy of the member States, of a common market [. . .].
The Community must progressively establish conditions which will in themselves assure the most rational distribution of production at the highest possible level of productivity, while safeguarding the continuity of employment and avoiding the creation of fundamental and persistent disturbances in the economies of the member States."
Treaty constituting the European Coal and Steel Community (1951) Title 1, Article 2

This chapter is an updated version of the background report "An historical case study on previous coal transitions in Germany" which is part of the project "Coal Transitions: Research and Dialogue on the Future of Coal", led by IDDRI and Climate Strategies in co-operation with DIW Berlin. We thank especially Jonas Mugge, Jörn Richstein, Oliver Sartor, and Christian von Hirschhausen and the participants of the project workshops for helpful feedback; the usual disclaimer applies.

H. Brauers (✉) · P.-Y. Oei
Junior Research Group "CoalExit", Berlin, Germany

TU Berlin, Berlin, Germany

DIW Berlin, Berlin, Germany
e-mail: hbr@wip.tu-berlin.de

P. Herpich
TU Berlin, Berlin, Germany

DIW Berlin, Berlin, Germany

© Springer Nature Switzerland AG 2018
C. von Hirschhausen et al. (eds.), *Energiewende "Made in Germany"*,
https://doi.org/10.1007/978-3-319-95126-3_3

3.1 Introduction

The German economic and industrial development in the nineteenth and twentieth century was based (among other things) on coal. After World War II, the reconstruction of both German states, too, was largely organized around the coal and steel industry. Therefore, it is a particular challenge, that the objectives of the energiewende require a complete phase-out of coal in only about two decades.

Whereas the previous chapter has provided a survey of energy and climate policies in Germany and Europe over the past decades, this chapter focusses on the past transformation of the coal sector in Germany. It provides lessons to be learned for other countries undergoing similar transformation processes. Our main working hypothesis is that the coal industry was reduced gradually in all large industrial basins, both in East and West Germany, in a rather structured and orderly manner. What is left over today, in the middle of the energiewende, is but a marginal share of previous activity and employment. Conditions are different, though, between the rather comfortable situation in the Rhine and Ruhr areas of prosperous West Germany, compared to the East German coal basin Lusatia, which was hit particularly hard.

The chapter is structured in the following way: The next two Sects. 3.2 and 3.3 report the history of hard coal and lignite, respectively, between 1950 and 2017, including the time of the separation between East and West. Section 3.2 describes the role hard coal played in the energy system and economy of the mining areas in Western Germany from the 1950s until 2017. Section 3.3 describes the role of lignite in Germany, focusing on the drastic decline of lignite in East Germany after reunification. It is shown that both in terms of production and employment, the largest part of the transformation process has already taken place, with a particularly rapid speed in East German lignite between 1990 and 2000. The following Sect. 3.4 analyzes the implemented political measures which accompanied the decline in hard coal and lignite production. Section 3.5 then derives some lessons learned from the transformation process for other regions, and Sect. 3.6 concludes.

3.2 History of Hard Coal in Germany 1950–2017

After the Second World War, Germany was divided into West and East Germany. The entire production of underground hard coal was based in West Germany. For West Germany, the domestic hard coal reserves were more than just an energy carrier, since it helped to rebuild its industry and enabled its "economic miracle". Furthermore, coal helped to reintegrate Germany into an international union: The European Coal and Steel Community (ECSC), the predecessor of the European Union, was founded in 1951 together with Italy, Belgium, France, Luxemburg and the Netherlands.

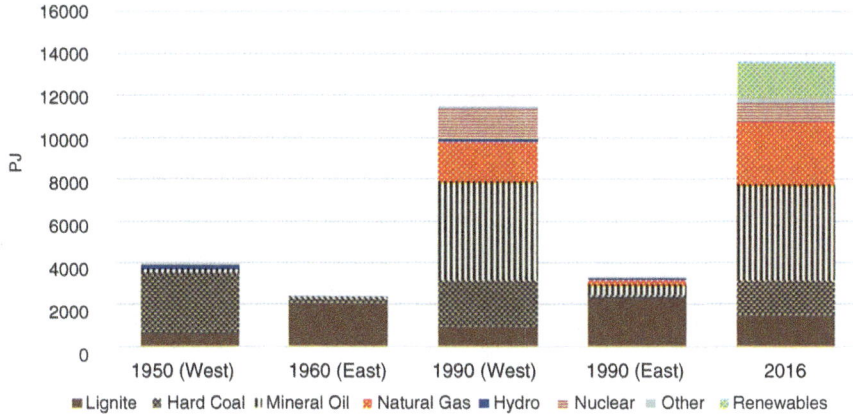

Fig. 3.1 Primary energy consumption in West and East Germany in 1950 and 1990 and in Germany in 2016. Source: AG Energiebilanzen e.V. (2017a, b) (AG Energiebilanzen e.V. 2017a. 'Zeitreihen bis 1989'. 2017. https://ag-energiebilanzen.de/12-0-Zeitreihen-bis-1989.html; AG Energiebilanzen e.V. 2017b. 'Primärenergieverbrauch'. 2 March 2017. https://www.ag-energiebilanzen.de/6-0-Primaerenergieverbrauch.html) and Kahlert (1988, 10)

3.2.1   Hard Coal as Energy Carrier: Primary Energy Consumption and Electricity Generation

Hard coal was the backbone of West Germany's energy supply after the war. Its importance can be illustrated by the fact that Germany introduced the so-called Hard Coal Units (HCU) to measure energy, analogue to the oil equivalent (OE).[1] In 1950, hard coal provided 98.7 million t HCU (in the following in the international abbreviation: tce) (2893 PJ), or more than 70% of primary energy consumption (PEC).[2] Hard coal was eventually substituted, mainly with imported mineral oil, and its share dropped to 19% in 1990 and 12% in 2016. Absolute consumption of hard coal declined in the same period from 74 million tce (2169 PJ) in 1990 to 55 million tce (1612 PJ) in 2016 (see also Fig. 3.1).

From 1950 to 1990, the PEC almost trippled to 392.2 million tce (11,494 PJ). After Germany's reunification, PEC increased further to 455 million tce (13,335 PJ) in 2016. The decrease in coal consumption was covered mainly by increasing imports of oil and natural gas. In 1950, mineral oil provided only 5% of PEC; which increased to a share of 41% in 1990. Gas had a negligible share in 1950, by 1990 it contributed 18% and in 2016 22% of PEC.

Figure 3.1 displays the primary energy consumption for West and East Germany in 1950 and 1990 and for the reunified Germany in 2016, illustrating the increasing

---

[1] 1 kg HCU = 0.7 kg OE = 29.3076 MJ = 10^{-3} tonnes of coal equivalent (tce).

[2] AG Energiebilanzen e.V. 2017a. 'Zeitreihen bis 1989'. 2017. https://ag-energiebilanzen.de/12-0-Zeitreihen-bis-1989.html

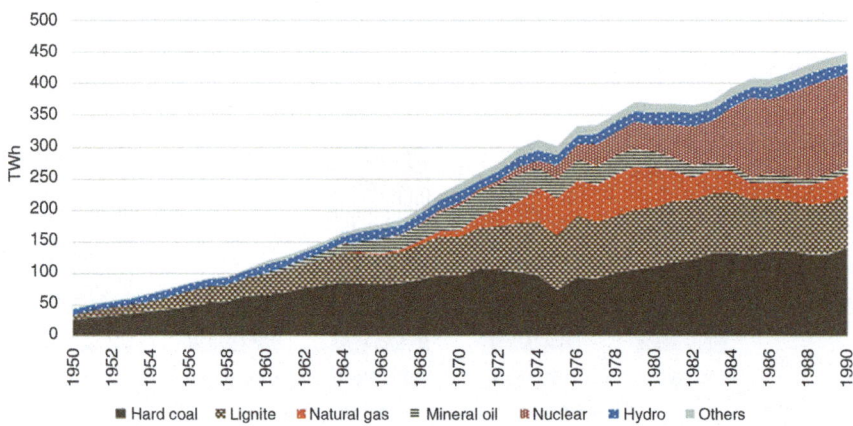

Fig. 3.2 Gross electricity generation in West Germany 1950–1990. Source: Statistik der Kohlenwirtschaft e.V. (2017a) (Statistik der Kohlenwirtschaft e.V. 2017a. 'Bruttostromerzeugung'. Download. 2017. https://kohlenstatistik.de/17-0-Deutschland.html)

diversification of the energy system: Before the reunification, mineral oil, natural gas and nuclear energy consumption increased drastically while hard coal consumption decreased. The main change after 1990 is the increased usage of renewable energy, while hard coal consumption decreased and lignite consumption (as well as total PEC) continued to grow.

In contrast to the development in the energy sector, during the German separation hard coal consumption increased in the electricity sector of West Germany (with the exception of the years shortly before the first oil crisis in 1973). Hard coal reached its highest share with over 60% at the end of the 1950s and has been fluctuating around 30% since the reunification.

Gross electricity generation (GEG) increased tenfold from 44 TWh in 1950 to 450 TWh by 1990.[3] After 1960, mineral oil and natural gas gained in importance in the electricity sector, but after the two oil crises the electricity sector started to shift away from mineral oil. During that time, nuclear energy gained importance and covered 30% of gross electricity generation since the 1980s. After the accident in Fukushima in 2011, Germany decided to phase-out nuclear power by 2022. Figure 3.2 displays West Germany's electricity generation from 1950 to 1990 and Fig. 3.3 for the reunified Germany from 1990 to 2016. Coal consumption for electricity generation has increased until the 1990s and has been on a gradual decline since the 2000s. However, coal's share in PEC has declined more strongly (see Fig. 3.1), due to varying competition over time from oil, natural gas, nuclear power and renewable energies.

---

[3]Statistik der Kohlenwirtschaft e.V. 2017a. 'Bruttostromerzeugung'. Download. 2017. https://kohlenstatistik.de/17-0-Deutschland.html

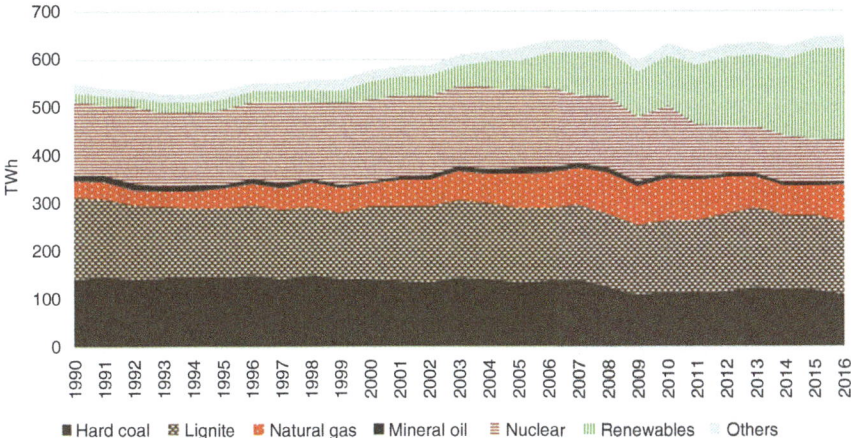

Fig. 3.3 Gross electricity generation in Germany 1990–2016. Source: Statistik der Kohlenwirtschaft e.V. (2017a) (Statistik der Kohlenwirtschaft e.V. 2017a. 'Bruttostromerzeugung'. Download. 2017. https://kohlenstatistik.de/17-0-Deutschland.html)

3.2.2 Hard Coal Production and Employment

After the end of the war, underground hard coal production rose until 1958—the first year of the coal crisis. After the Suez crisis and the first attempts to liberalize the energy sector in the mid-1950s, cheap import oil gained in significance; hard coal consumption began to decline. Before, the government had set the price for coal on a low level for the reconstruction of Germany, but in 1956, the ECSC demanded a market based price in Germany.[4] The sales and production figures started to strongly decline in the mining as well as the steel industry, one of the biggest consumers of German hard coal. Germany did not only import oil but also comparably cheap foreign hard coal, which additionally decreased the demand for domestic coal. In order to protect domestic production, the hard coal industry received subsidies to level-out the price difference between domestic and imported coal since 1968. Since 1964, the prices for domestic coal exceeded the ones of imported coal (Fig. 3.4).

The coal and steel industry formed a powerful network together with influential unions and politicians (especially the social democratic party), protecting domestic coal production. Besides its regional significance, hard coal was considered important for other political and strategic reasons: Coal guaranteed a certain level of supply security, making Germany less dependent on foreign oil, coal and later natural gas imports. Additionally, hard coal was a gateway for Germany into international affairs: Being a member of the ECSC was beneficial in forming strong relations with other European nations. Therefore, and to prevent structural disruptions at the

---

[4]Nonn, Christoph. 2009. 'Der Höhepunkt der Bergbaukrise (1958–1969)'. In Kumpel und Kohle—Der Landtag NRW und die Ruhrkohle 1946 bis 2008, edited by Die Präsidentin des Landtages Nordrhein-Westfalen, 19:96–124. Schriften des Landtags Nordrhein-Westfalen. Düsseldorf, Germany.

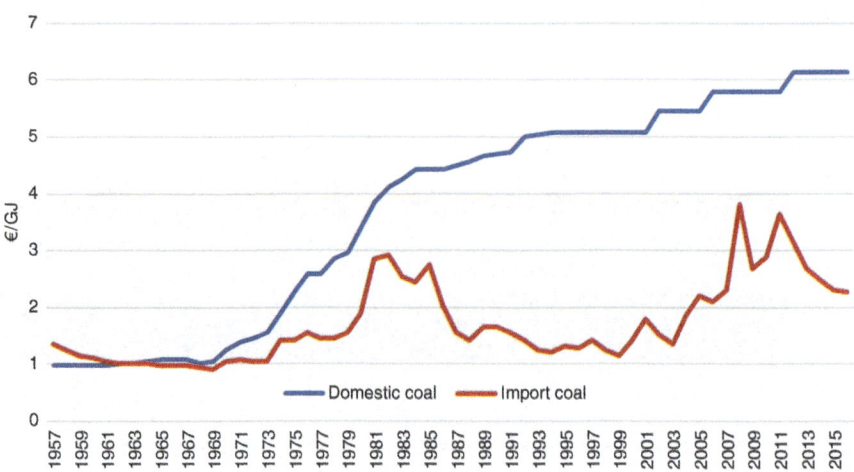

Fig. 3.4 Prices for domestic and imported hard coal. *Source:* Verein der Kohleimporteure e.V. (2017, 111) (Verein der Kohleimporteure e.V. 2017. 'Jahresbericht 2017—Fakten und Trends 16/17'. Hamburg, Germany. http://www.kohlenimporteure.de/publikationen/jahresbericht-2017.html)

regional level, the government provided the hard coal sector with various subsidies for more than 60 years.

Besides the increasing amount of imported energy carriers, the ongoing mechanization of the mining sector led to a lower employment which had an additional impact on the total number of employees in the Ruhr area. At the peak of production, right before the coal crisis in 1958, over 600,000 people were employed. Within ten years, 320,000 people had lost their jobs. Figure 3.5 shows the development of employment in hard coal mining as well as produced and imported hard coal.

The majority of German hard coal production came from the Ruhr area, which is located between Dortmund and Düsseldorf. Therefore, this study focuses on the development in the Ruhr area. The coal mining and steel industry (in German 'Montanindustrie') made this area the most densely populated area in Germany— until today. The region depended strongly on the economic circumstances of this industry and suffered repeatedly of high unemployment rates over the years. Figure 3.6 displays the development of the unemployment rate for (West-) Germany, North Rhine-Westphalia (NRW) and the Ruhr area from 1960 to 2015. The figure does not display the years before 1960 for NRW and West Germany and before 1967 for the Ruhr area due to a lack of yearly data. However, total unemployment in West Germany fell from around 1.9 million in 1950 to just 150,000 in 1962.[5] The rising

---

[5]Bundesagentur für Arbeit. 2018. 'Arbeitslose und Unterbeschäftigung—Deutschland und West/ Ost (Zeitreihe Monats- und Jahreszahlen ab 1950)'. Bundesagentur für Arbeit—Statistik. 19 February 2018. https://statistik.arbeitsagentur.de/nn_31892/SiteGlobals/Forms/Rubrikensuche/Rubrikensuche_Form.html?view=processForm&resourceId=210368&input_=&pageLocale=de&topicId=17722&year_month=aktuell&year_month.GROUP=1&search=Suchen

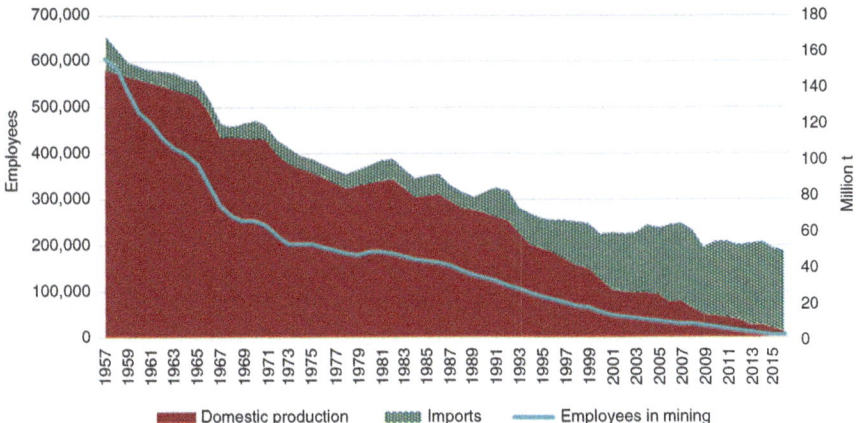

Fig. 3.5 Domestic hard coal production, imports and employees (mining only) of West Germany. Source: Own calculations based on Statistik der Kohlenwirtschaft (2017b, c) and Verein der Kohleimporteure (2017) (Statistik der Kohlenwirtschaft e.V. 2017b. 'Steinkohle'. Statistik der Kohlenwirtschaft. 2017. http://www.kohlenstatistik.de/18-0-Steinkohle.html; Statistik der Kohlenwirtschaft e.V. 2017c. 'Steinkohle—Belegschaft im Steinkohlebergbau'. Steinkohle. 2017. https://kohlenstatistik.de/18-0-Steinkohle.html; Verein der Kohleimporteure e.V. 2017. 'Jahresbericht 2017—Fakten und Trends 16/17'. Hamburg, Germany. http://www. kohlenimporteure.de/publikationen/jahresbericht-2017.html)

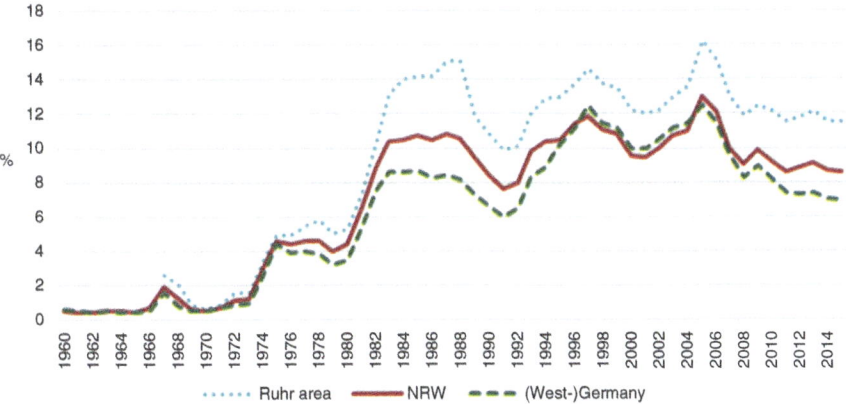

Fig. 3.6 Development of unemployment rates in the Ruhr area, North Rhine-Westphalia and (West-)Germany from 1960 to 2015. *Source:* Regionalverband Ruhr (2017c) (Regionalverband Ruhr. 2017c. 'Arbeitsmarkt'. 2 October 2017. http://www.metropoleruhr.de/fileadmin/user_upload/metropoleruhr.de/01_PDFs/Regionalverband/Regionalstatistik/Arbeit_und_Soziales/Arbeitsmarkt/2015_Zeitreihe_Arbeitsmarkt.pdf). Note: The depicted (West-)Germany values are only for West Germany from 1960 until 1990 and from then onwards for the reunified Germany; data for the Ruhr area was available only for the years after 1966

development of the metal industry was able to cover most of the job losses in the mining sector.[6] Yet, in the following years, the "economic miracle" ended in Germany and the Ruhr area as well as Germany as a whole suffered from global macroeconomic changes like the oil crises in 1973 and 1979. The induced economic recessions led to a doubling of the unemployment rates in the Ruhr area within only a few years (1973: 1.6% → 1974: 3.3% and 1979: 5% → 1982: 10.3%). A first peak in unemployment was reached with 15.1% in 1987, which dropped to around 10% in the early 1990s. In 2005, a new maximum was reached with 16.4%. In the past 10 years, the region's average unemployment rate equates to approximately 11%. The development of the unemployment figures corresponded with the trends of NRW and the rest of Germany; however, unemployment rates of the Ruhr area were always higher. The gap widened especially in the 1980s, where they were between 5% and 7% higher than the West German average. The gap was reduced to only about 2% in 2002 and remains at a level between 4 and 5% since 2010.[7]

3.3   History of Lignite in Germany 1950–2017

Besides hard coal, lignite is the only energy carrier mined in a significant amount in Germany. Unlike hard coal, lignite was available in both parts of Germany during the separation. Lignite contains a higher share of water than hard coal, which makes the transportation over large distances uneconomic. Therefore, lignite production and power plants are clustered in the mining regions. A phase-out of lignite production, would thus also lead to a phase-out of lignite-fired power plants. In 2017, lignite is produced in Germany in the open pits of the Rhineland (West Germany, close to the Ruhr area), Lusatia (East Germany) and Central Germany (East Germany) region. The eastern coal mining regions are mostly rural areas with low population figures, unlike the lignite and hard coal mining areas in West Germany.[8]

---

[6]Nonn, Christoph 2001. Die Ruhrbergbaukrise: Entindustrialisierung und Politik 1958–1969. Kritische Studien zur Geschichtswissenschaft, Bd. 149. Göttingen: Vandenhoeck & Ruprecht.

[7]Regionalverband Ruhr. 2017c. 'Arbeitsmarkt'. 2 October 2017. http://www.metropoleruhr.de/fileadmin/user_upload/metropoleruhr.de/01_PDFs/Regionalverband/Regionalstatistik/Arbeit_und_Soziales/Arbeitsmarkt/2015_Zeitreihe_Arbeitsmarkt.pdf

[8]During the last 150 years, coal production was also located in other regions but with an accumulated production of 4% of total production, the contribution is rather insignificant. Öko-Institut. 2017. 'Die deutsche Braunkohlenwirtschaft—Historische Entwicklungen, Ressourcen, Technik, wirtschaftliche Strukturen und Umweltauswirkungen'. Studie im Auftrag von Agora Energiewende und der European Climate Foundation. Berlin. https://www.agora-energiewende.de/fileadmin/Projekte/2017/Deutsche_Braunkohlenwirtschaft/Agora_Die-deutsche-Braunkohlenwirtschaft_WEB.pdf

3.3.1 Lignite in East Germany's Energy System

East Germany covered around 90% of its primary energy consumption via domestically produced lignite in the years right after the second world war (Kahlert 1988, 10). Its PEC almost tripled in the years from 1950 to 1990 from 51 million tce (1495 PJ) to closely 130 million tce (3810 PJ). In 1960, lignite contributed 88%, in 1970 it had dropped to 75% of the primary energy consumption and continued to decrease until the end of the 1970s. The reasons were increasing imports of mineral oil and the rising share of nuclear power. The decline of lignite production stopped when the oil crises from 1973 and 1979 raised the prices of oil, and the ambitious plans for the deployment of nuclear power plants could not be realized (Matthes 2000, 53). East Germany started to increase the share of lignite in its energy system again in the 1980s. On the one hand, this increase led to a state where East Germany was able to cover 70% of its PEC by domestic energy carriers in 1986. On the other hand, East Germany's economic stability was threatened by high consumption of expensive and uncompetitive lignite (Kahlert 1988, 10). From the mid-1980s, annual investments into lignite and energy summed up to GDR-Mark 9–10 billion[9] (East Germany's currency, equivalent to ~1.15–1.28 billion €),[10] which corresponded to approximately one quarter of total industrial investments (Matthes 2000, 54). Figure 3.1 displays East Germany's primary energy consumption in 1960 and 1990. In 1950, the amount of hard coal and lignite combined exceeded 99% of the PEC[11] (Kahlert 1988, 10) (Fig. 3.7).

Figure 3.8 displays the development of gross electricity generation (GEG) in East Germany from 1979 to 1990 (natural gas is listed in "others"). From 1955 until 1977, lignite contributed around 90% of GEG. After 1978, it declined to around 80–85% due to the deployment of nuclear power. The absolute amount of lignite increased from 25 TWh in 1955 to around 100 TWh in 1990 (Matthes 2000, 67). Shortly before the reunification, East Germany's power plant fleet consisted of two thirds lignite-fired power plants with a capacity of 15 GW (Kahlert 1988, 13). The share of lignite peaked in East Germany right after the reunification with 91%, when East Germany phased out nuclear power. The total GEG increased to 118 TWh in 1990.[12] After the reunification, the lignite sector broke down, because it was less productive and more expensive compared to the West German lignite sector. Additionally,

---

[9]The exchange rate from Deutsche Mark (DM) to GDR-Mark in 1987 was 1:4. Exchange rate for € to DM = 1:1.95583. Baltensperger, Ernst, and Deutsche Bundesbank. 1998. Fünfzig Jahre Deutsche Mark: Notenbank und Währung in Deutschland seit 1948. München: Beck. P. 648.

[10]If not expressed otherwise, the values in this chapter are nominal.

[11]The available data does not differentiate between both energy carriers.

[12]Statistik der Kohlenwirtschaft e.V. 2017a. 'Bruttostromerzeugung'. Download. 2017. https://kohlenstatistik.de/17-0-Deutschland.html

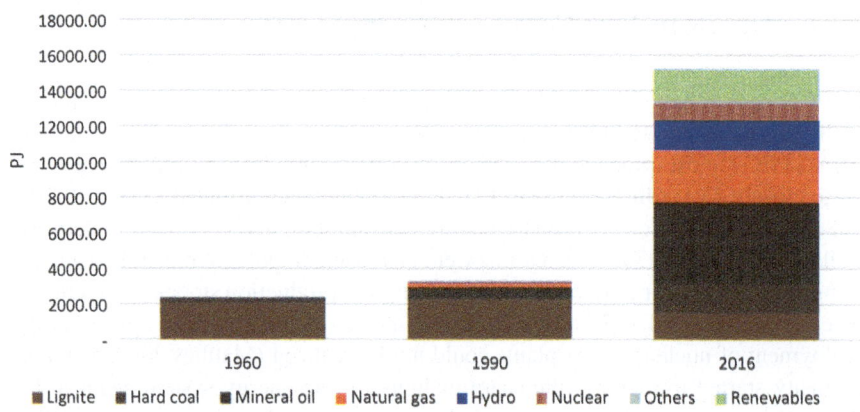

Fig. 3.7 Primary energy consumption in East Germany 1960, 1990 and in Germany 2016. Source:
AG Energiebilanzen e.V. (2017b) and Kahlert (1988, 10)

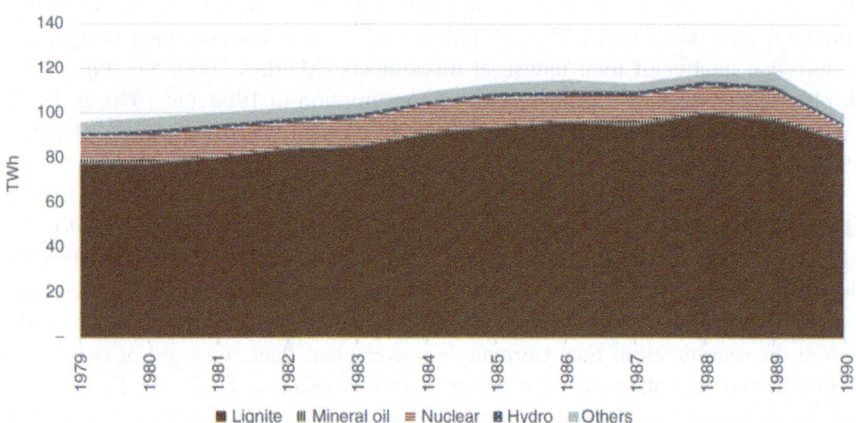

Fig. 3.8 Gross electricity generation in East Germany (The contribution of hard coal is only
between 0.68 and 0.2 TWh for the time period between 1979 and 1990. Due to its limited share,
it is not displayed in the figure). Source: Statistik der Kohlenwirtschaft e.V (2017a) (Statistik der
Kohlenwirtschaft e.V. 2017a. 'Bruttostromerzeugung'. Download. 2017. https://kohlenstatistik.de/
17-0-Deutschland.html)

citizens and regions started to develop ecological concerns (Matthes 2000, 238).
Both figures on PEC and GEG show a strong dominance of lignite in East Germany
until 1990 (Figs. 3.1 and 3.8). However, in 2016 lignite plays a much smaller role in
Germany's PEC, both in absolute and in relative terms.

3.3.2 Lignite's Contribution to the Energy Systems of West Germany and Germany After the Reunification

Compared to East Germany, lignite was not as substantial in the energy system of West Germany. At the beginning of the 1950s, lignite contributed only 15% to the PEC.[13] The share of lignite declined to only 8% by 1990—in absolute figures it rose from 21 to 32 million tce (607—938 PJ). The amount of lignite used in the generation of electricity increased eightfold from 11 to 83 TWh in the same period of time. In the years after the first oil crises, when hard coal contribution was at a low, lignite reached its highest share of one third. By the time of the reunification, the share was only 18.1% due to the high shares of hard coal (32%) and nuclear power (34%).[14]

Since the reunification, the share and the total amount of the three technologies decreased, while simultaneously the electricity production rose. In 2016 lignite-fired power plants generated around 23% of GEG, with 150 TWh, making it the largest producer in the system before hard coal with only 110 TWh (compare Fig. 3.3).

In 2017, Germany produced 648 TWh of electricity of which more than 50 TWh were exported. The increase in the renewable energy consumption has not yet led to a decrease of fossil fuel-fired generation, but instead has turned Germany into a large electricity exporter.

3.3.3 Lignite Production and Employment

Until the reunification, the total production of lignite continuously increased, mainly in East Germany, in order to cover the increasing PEC. The supply rose almost throughout the time of separation. The peak was reached in 1985 when around 140,000 employees produced 312 million tons of lignite. The total production of Germany in 1985 equaled 430 million tons with almost 160,000 employees of which 90% worked in the mines of East Germany. As a reaction to the oil crises in 1973 and 1979, East Germany tried to use political means to increase production, even though its economic condition was not able to sustain it (Matthes 2000, 56). Its maximum level of 1985 slowly decreased in the following years. East Germany was the biggest lignite producer globally, extracting one quarter of the total amount of lignite and doubling the production of the second largest producer—the Soviet Union (Kahlert 1988, 10). 70% of the production in Germany was concentrated in East Germany.[15] The lignite production in East Germany was characterized by high overcapacities

---

[13]AG Energiebilanzen e.V. 2017b. 'Primärenergieverbrauch'. 2 March 2017. https://www.ag-energiebilanzen.de/6-0-Primaerenergieverbrauch.html

[14]Statistik der Kohlenwirtschaft e.V. 2017a. 'Bruttostromerzeugung'. Download. 2017. https://kohlenstatistik.de/17-0-Deutschland.html

[15]Meyer, Bettina, Swantje Küchle, and Oliver Hölzinger. 2010. 'Staatliche Förderungen der Stein- und Braunkohle im Zeitraum 1950–2008'. Berlin: Forum ökologisch-soziale Marktwirtschaft eV. http://www.greenpeace.de/fileadmin/gpd/user_upload/themen/energie/Kohlesubventionen_1950-2008.pdf

(Kahlert 1988, 15). Right before the reunification the average production in tons per worker in West Germany was three times higher than in East Germany. This resulted in a drastic reduction of the lignite production in East Germany after reunification when all mines were forced into inner German competition. In East Germany, between 1989 and 1994 over 100,000 employees lost their job and production decreased by about 200 million tons. Unlike the hard coal decline, lignite broke down within just a few years, leading to a structural disruption in some regions. Since the mid-1990s, lignite production and employment has stayed almost constant, however, at only a fraction of the pre-reunification time. Despite this, Germany is still the largest lignite producing country in the world. Figure 3.9 displays the lignite production and employees in Germany from 1950 to 2016.

3.4   Political Instruments Since the 1950s Until Today

This following section points to the social consequences of the hard coal phase-out and lignite reduction especially in affected mining areas. It highlights policy instruments on regional, national and supranational level which accompanied the decline of both energy carriers. The description starts chronologically with the hard coal decline in the Ruhr area after the 1950s and covers the reduction in lignite production of East Germany, focusing on the Lusatian region.

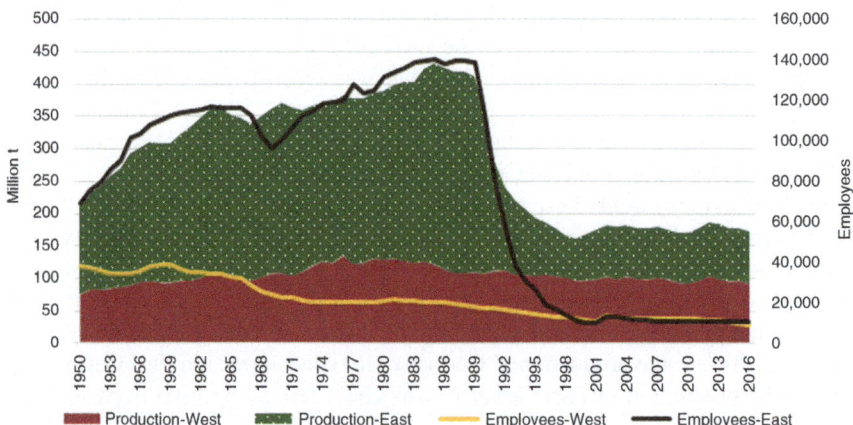

Fig. 3.9 Lignite production and employees in Germany 1950–2016. Source: Statistik der Kohlenwirtschaft e.V. (2017) (Statistik der Kohlenwirtschaft e.V. 2017. 'Braunkohle'. Statistik der Kohlenwirtschaft. 9 September 2017. https://kohlenstatistik.de/19-0-Braunkohle.html). Note: The values for lignite production are displayed as stacked areas for East and West Germany, while employment figures are depicted as individual lines. Since 2002, the employees of lignite-fired power plants are included

3.4.1 The Coal Crisis in 1958 and the First Structural Policy Program of North Rhine-Westphalia (NRW)

German hard coal production and consumption has been declining since the coal crisis in 1958 (see Sect. 3.2). Since cheap oil was one of the main causes of the crisis, some politicians, especially in the Ruhr area, urged to implement a protectionist import tax on oil. The income of that tax was partly used to compensate around 16,000 workers for a shortening of their shifts, which were implemented as reaction to the quickly lowering coal demand. Additionally, early retirement in the mining industry was financially supported by the state.[16] In the short-term, these measures were able to alleviate negative consequences for affected workers, however, the measures did not succeed in addressing the structural problems of the hard coal sector. Between 1957 and 1967, over 300,000 out of 600,000 workers lost their job in hard coal production—most of them in the Ruhr area. The first years of the reduction in coal production due to the oil crisis overlapped with the last years of the "economic miracle" in Germany. The majority of the workers were therefore able to transfer into other jobs, mainly in the metal industry (see Fig. 3.10).[17]

Thus, unemployment payments were only necessary for workers close to their retirement. In 1962, the economic situation changed again, especially in the steel sector. It had become apparent that the mining industry would not be able to recover. Hence, the law for rationalization and decommissioning was implemented in 1963, causing 51 out of 141 coal mines to be shut down by 1967. In order to be able to initiate a controlled decline of coal production, mining companies were forced to combine their production in a newly founded company called RAG AG (Goch 2009, 128). Additionally, in 1968, the coal sector concluded sale contracts with the energy and steel sector which included state subsidies for domestic coal, paying the price difference between domestic and imported hard coal. This framework enabled a structured and slowed down decline in coal production and employment.

As the decline in domestic coal production and related employment accelerated, the government of NRW started to address the need for an economic reorientation in a more strategic way: It launched its first structural policy program called "Development Program Ruhr" in 1968 with a volume of Deutsche Mark (DM) 17 billion (8.7 billion €) (Goch 2009, 146), which bundled hitherto individual and isolated measures. The program intended to attract new enterprises from other sectors. In order to achieve that, the government needed to convince the mining companies to sell the land they owned to the new competition. The fear of losing qualified workers to the potential newcomers made them hold onto the land, so only few enterprises

---

[16]Farrenkopf, Michael. 2009. 'Wirtschaftswunder und erste Kohlekrisen'. In Kumpel und Kohle— Der Landtag in NRW und die Ruhrkohle 1946 bis 2008, edited by Die Präsidentin des Landtages Nordrhein-Westfalen, 49–95. Schriften des Landtags Nordrhein-Westfalen 19. Düsseldorf, Germany.

[17]Nonn, Christoph. 2001. Die Ruhrbergbaukrise: Entindustrialisierung und Politik 1958–1969. Kritische Studien zur Geschichtswissenschaft, Bd. 149. Göttingen: Vandenhoeck & Ruprecht.

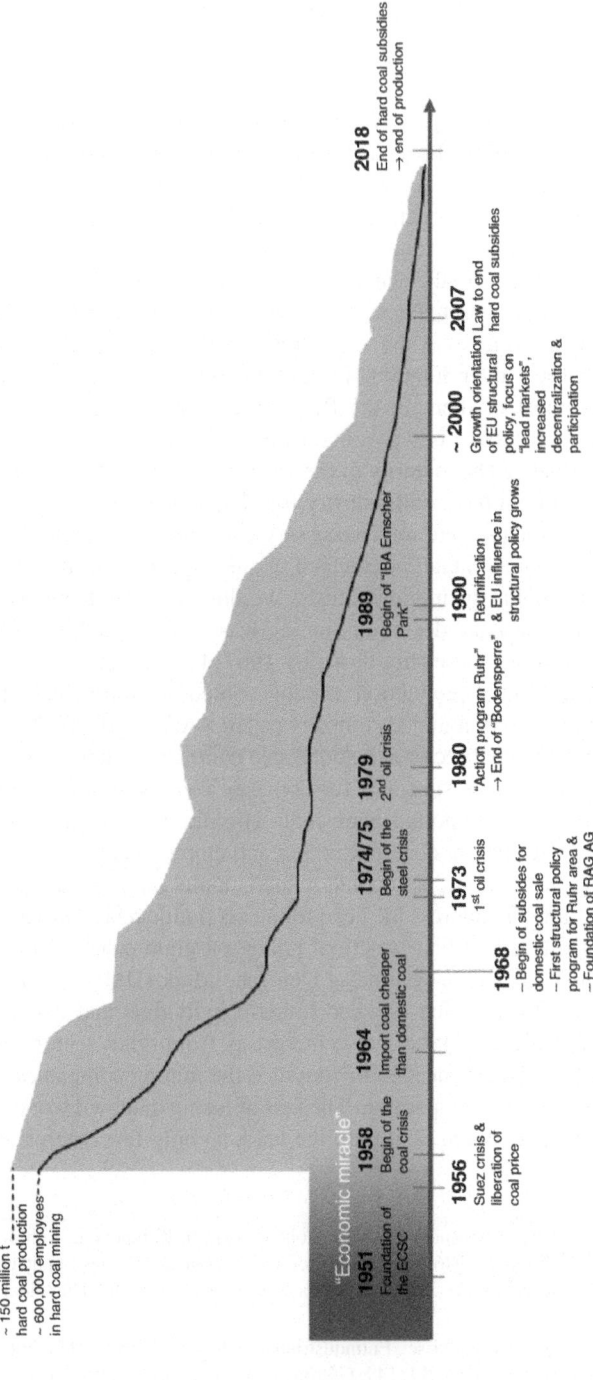

Fig. 3.10 History of hard coal and structural policy programs in the Ruhr area since 1951 (Farrenkopf, Michael. 2009. 'Wirtschaftswunder und erste Kohlekrisen'. In Kumpel und Kohle—Der Landtag in NRW und die Ruhrkohle 1946 bis 2008, edited by Die Präsidentin des Landtages Nordrhein-Westfalen, 49–95. Schriften des Landtags Nordrhein-Westfalen 19. Düsseldorf, Germany. Bundesregierung. 2007. 'Drucksache 557/07 des Deutschen Bundestages—Gesetzesentwurf zur Finanzierung der Beendigung des subventionierten Steinkohlenbergbaus zum Jahr 2018'. Statistik der Kohlenwirtschaft e.V. 2017b. 'Steinkohle'. Statistik der Kohlenwirtschaft. 2017. http://www.kohlenstatistik.de/18-0-Steinkohle.html, Statistik der Kohlenwirtschaft e.V. 2017c. 'Steinkohle—Belegschaft im Steinkohlebergbau'. Steinkohle. 2017. https://kohlenstatistik.de/18-0-Steinkohle.html. Verein der Kohleimporteure e.V. 2017. 'Jahresbericht 2017—Fakten und Trends 16/17'. Hamburg, Germany. http://www.kohlenimporteure.de/publikationen/jahresbericht-2017.html and Goch (2009))

were able to settle in the Ruhr area. The behavior of the mining companies was later referred to as "ground lock" ("Bodensperre").

Besides the economic reorientation, which mostly failed, the program improved both education and traffic infrastructure to accompany the economic changes (Goch 2009, 146). Before, there existed no university in the area and the cities within the Ruhr area were not sufficiently connected by transport routes. The economic reorientation needed a higher mobility of workers since the distances between their homes and jobs were likely to increase. Homes had previously been in close distance to the work places and therefore the need for an infrastructural connection between the cities was neglected.[18]

3.4.2 The Oil Crisis 1973 and the Structural Policy Programs of the Ruhr Area

In the Ruhr area, unemployment figures rose from 12,000 in 1970 to almost 100,000 in 1976 (see also Fig. 3.6).[19] The strategy of settling new industries in the Ruhr area of the previous structural policy program had failed due to the "ground lock". Therefore, the new strategy intended to exploit the endogenous potentials of *existing* industries via investments of DM2 billion (1.0 billion €) into the modernization of the coal mining, energy and steel sectors (Goch 2009, 150). This re-industrialization was partly driven by the hopes of a renaissance of coal as an energy carrier due to the oil crisis. In 1974, steel production, which was one of the biggest consumers of domestic hard coal, reached its peak. The following steel crisis further aggravated the situation for the Ruhr area one year later. Policy makers realized that this development was not due to the economic cycle but rather a structural problem, which required adjustments on the production level.

The rising unemployment figures in the 1970s and the development in the steel sector revealed the problems associated with the high sectoral specialization of the Ruhr area. The structural policy program "Action Program Ruhr" from 1980 until 1984, therefore, focused on an economic reorientation of the Ruhr area and the establishment of new industries. Furthermore, the program intended to improve the technology transfer between universities and companies as well as to increase the tertiarization of the Ruhr area. The program used a new approach by including elements of participation of regional stakeholders, since previous programs had faced their resistance (Goch 2009, 152). The "Action Program Ruhr" with a volume of DM6.9 billion (3.5 billion €) combined several individual measures for technology and innovation support, ecology,

---

[18]Bogumil, Jörg, Rolf G. Heinze, Franz Lehner, and Klaus Peter Strohmeier. 2012. Viel erreicht—wenig gewonnen: ein realistischer Blick auf das Ruhrgebiet. 1. Auflage. Essen: Klartext.

[19]Regionalverband Ruhr. 2017c. 'Arbeitsmarkt'. 2 October 2017. http://www.metropoleruhr.de/fileadmin/user_upload/metropoleruhr.de/01_PDFs/Regionalverband/Regionalstatistik/Arbeit_und_Soziales/Arbeitsmarkt/2015_Zeitreihe_Arbeitsmarkt.pdf

culture and the labor market. One goal of the program was the better coordination of the various measures by the federal government, the state and municipalities. Although a majority of the measures was still implemented in an isolated way, the result was a more dialogue oriented policy making. The program improved the Ruhr area's situation in terms of soft location factors (e.g. improving the regional image, more cultural activities, etc.). Although it led to the creation of several new technology centers, it was not able to substantially diversify the economy, as large part of subsidies still went to the coal and steel industry.

The implementation of the property fund Ruhr and the "State development society" ("Landesentwicklungsgesellschaft"), which bought and restored former industrial sites, led to an end of the "ground lock".[20] Hence, the action program Ruhr was able to remove one of the barriers that prevented reorientation in the area. Nevertheless, the program's focus on slowing down the decline of the coal industry impeded a more rapid establishment of new industries.

3.4.3   Regionalization of the Structural Policy Since the Mid-1980s

Policy makers had realized that there was no single industry likely to replace the steel and coal sector in a way so that it could stabilize the Ruhr area's economy. Therefore, each city within the Ruhr area needed its own strategy of economic reorientation. Previous programs did not take the individual strengths and weaknesses of the cities into account. The new approach regionalized the structural policy, mainly via regionally planned development strategies including individual strengths and weaknesses analyses (Goch 2009, 156). The need for a new structural policy program increased after the second oil crisis in 1979 (see Fig. 3.10). The unemployment rate almost tripled within 6 years to 14.2% in 1985—significantly above the rate of 8.7% in the rest of the country.[21] The government of NRW implemented the so-called "Komission Montanregionen" ("Commission for Coal and Steel Regions"), which elaborated strategies with regional stakeholders. In 1987, the program "Zukunftsinitiative Montanregionen" ("Future Initiative Coal and Steel Regions") with a volume of DM2 billion (1.0 billion €) was launched. Hereby, the state declared fields of development, namely innovation and technology funding, education of workers, infrastructure and improvement of the environment as well as energy matters. It further granted financial resources to regional decision makers: In

---

[20]metropoleruhr. 2010. 'Bodensperre'. Regionalkunde Ruhr. 2010. http://www.ruhrgebiet-regionalkunde.de/html/aufstieg_und_rueckzug_der_montanindustrie/huerden_des_strukturellen_wandels/bodensperre.php%3Fp=4,1.html

[21]Regionalverband Ruhr. 2017c. 'Arbeitsmarkt'. 2 October 2017. http://www.metropoleruhr.de/fileadmin/user_upload/metropoleruhr.de/01_PDFs/Regionalverband/Regionalstatistik/Arbeit_und_Soziales/Arbeitsmarkt/2015_Zeitreihe_Arbeitsmarkt.pdf

order to receive funding, the regions had to submit projects that had been developed together with regional stakeholders such as the chamber of crafts, unions or environmental organizations. The program itself did not introduce new measures but marked the shift to a more regionalized structural policy approach (Goch 2009, 159).

Another example for the consent-based regionalized policy is the so-called "International Building Exhibition Emscher Park". Between 1989 and 1999 the program had a volume of DM5 billion (2.6 billion €), of which two thirds came from the public budget. It combined over 120 small projects aimed at improving soft location factors in order to create a new identity of the Ruhr area. These projects included measures to implement an underground sewage system, improving water quality and opening up new areas for both citizens and nature alike. The cultural and touristic attractiveness of the region was increased by transforming former industrial sites into touristic landmarks, preserving the regions coal history and increasing tourism. Furthermore, 17 technology centers were created while mining damages were remediated as far as possible (Goch 2009, 162).

The structural policy programs caused an image change of the Ruhr area beyond the mining and steel industry, helped to create several universities and research institutions and improved the attractiveness of the region. However, only a limited number of new companies, and hence employment opportunities, was attracted into the Ruhr area, as financial support was focused on preserving the old industries and a powerful network of the coal and steel companies, unions and politicians resisted more rapid changes.

3.4.4   East Germany's Reduction in Lignite Production Due to the Reunification

In 1990 West and East Germany were reunified. As a result, the political and economic system of East Germany broke down and with it the majority of its lignite production (see also Sect. 3.3). The main reason was the comparably low development in labor productivity in East Germany as Fig. 3.11 displays. The Figure shows the labor productivity development for lignite mining in East and West Germany from 1957 to 2016, relative to labor productivity in 1957. Note that the drop in labor productivity in 2002 is due to a statistical change: From 2002 onwards employees in lignite-fired power plants are included in the statistics. Labor productivity in East Germany remained nearly constant from 1957 until 1990, while it increased 2.5-fold in West Germany. Only after the Reunification did labor productivity in East Germany catch up with the levels of West Germany. The labor productivity for hard coal does not include workers in the power plants.

Within five years after Reunification, 86,000 workers lost their jobs in the former East German lignite sector. The policy in the following years was not exclusively for the lignite sector and regions but for the whole former East Germany. East Germany needed to be integrated in the common currency union with West Germany, to open

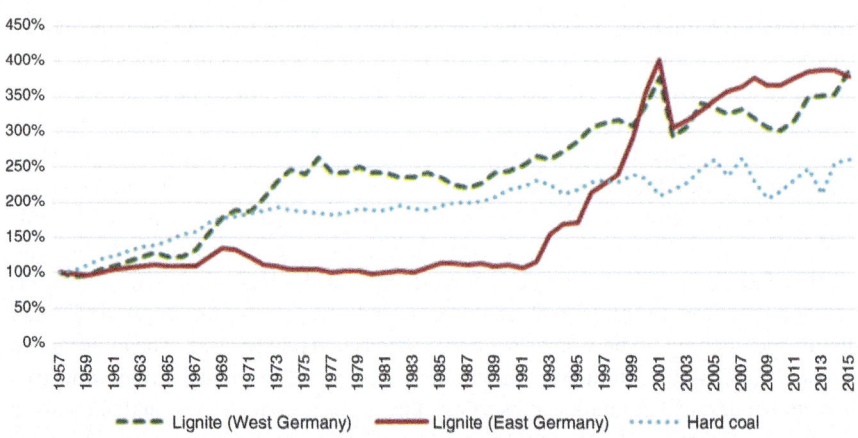

Fig. 3.11 Development of the standardized labor productivity in the German mining industry. *Source:* Own calculation and depiction based on Statistik der Kohlenwirtschaft (2017b, c, d) (Statistik der Kohlenwirtschaft e.V. 2017b. 'Steinkohle'. Statistik der Kohlenwirtschaft. 2017. http://www.kohlenstatistik.de/18-0-Steinkohle.html; Statistik der Kohlenwirtschaft e.V. 2017c. 'Steinkohle—Belegschaft im Steinkohlebergbau'. Steinkohle. 2017. https://kohlenstatistik.de/18-0-Steinkohle.html; Statistik der Kohlenwirtschaft e.V. 2017d. 'Braunkohle'. Statistik der Kohlenwirtschaft. 9 September 2017. https://kohlenstatistik.de/19-0-Braunkohle.html). Note: The drop in labor productivity for lignite in 2002 is due to a statistical change, as from 2002 onwards employees in lignite-fired power plants are included

itself to the market economy and to initiate a process of privatization of large parts of its economy. When creating the common currency union, the exchange rate was 1:1. This secured the interests of the population, but caused a difficult competition situation for companies in East Germany.[22] Additionally, the internal market broke down, partly because people preferred buying western products. The living conditions in East and West Germany were very different, which caused many people to migrate from East to West Germany after the opening of the boarder. The structural policy focused on creating the same standards of living in both parts of Germany to stop the migration. Due to the weak economic situation of East Germany before the reunification, the government neglected investments in infrastructure, education, buildings as well as an efficient production especially in the lignite sector. The political measures in the first years intended to erase those deficits. As some kind of "first aid kit", the counties of East Germany received DM12 billion (6.1 billion €) via the so-called "joint effort upturn east" ("Gemeinschaftswerk Aufschwung Ost") in 1991.[23] In the following years, many programs followed which were later gathered under the term "Reconstruction East" ("Aufbau Ost"). The implemented measures for East Germany focused on investment support, infrastructure and labor market interventions.

---

[22]Schroeder, Klaus. 2000. Der Preis der Einheit: Eine Bilanz. München: Hanser.

[23]Bundespresseamt. 1991. Gemeinschaftswerk Aufschwung-Ost—Eine Dokumentation der wichtigsten Beschlüsse und Vorhaben. Berichte und Dokumentationen. Berlin, Germany.

Depending on the sector and size, companies were granted investment support of up to 50% of total investments. Furthermore, companies could receive loans with low interest rates. These measures had a positive effect on investments and employment in East Germany but created a capital intensive production, which prevented further job creation.[24] In some sectors it even led to overcapacities. The investment support strategy intended an assimilation within the regions, and therefore funded companies in the peripheral areas. Often the regions did not implement sufficient measures to supply educated worker and research facilities, so that companies had few reasons to stay in the periphery.

The programs for infrastructure connected East and West Germany as well as the regions within. Often the projects were implemented on a regional level and due to a lack of a sufficient planning phase; some projects did not take the demographic and economic changes of the region into account. The result was that many infrastructure projects have a low utilization rate due to a lack of traffic.[25] The infrastructure projects and restorations in the private and public housing sector (schools, hospitals, etc.) in combination with the investment support measures created a boom in the building sector, which created jobs but also resulted in the already mentioned overcapacities.

Due to the transformation of the economic system in East Germany many people lost their jobs. This created high tensions on the labor market, especially in the mining areas. The government intervened with different policy measures, one of them being an option of early retirement. At an age of 55, people could receive a so-called "age transitioning payment" ("Altersübergangsgeld") of 65% of their last net income, if they became unemployed.[26] At the age of 60 they could receive a pre-pension payment until they entered the pension fund. Furthermore, the government implemented measures of active labor market policy such as programs for retraining which 400,000 people used in the first half of the 1990s. The high demand for those programs arose due to their included social security aspects: During the retraining program, former coal workers were granted special unemployment payments and after completing the program, those who were still unemployed were able to claim further unemployment payments. Additionally, the government financed so-called "job creation measures" ("Arbeitsbeschaffungsmaßnahmen") for 360,000 unemployed people during the same period of time. Job creation measures were low-paid jobs in order to prepare people for a regular follow-up employment. The social security aspect of the program can be evaluated as positive. However, the measures that were supposed to reintegrate the unemployed into the regular labor

[24]Brenke, Karl, Udo Ludwig, and Joachim Ragnitz. 2011. 'Analyse der Schlüsselentscheidungen im Bereich Wirtschaftspolitik und ihre Wirkung auf die ökonomische Entwicklung der vergangenen zwei Jahrzehnte im Land Brandenburg'. Berlin, Dresden, Halle/Saale, Germany.

[25]Bundespresseamt. 1991. Gemeinschaftswerk Aufschwung-Ost—Eine Dokumentation der wichtigsten Beschlüsse und Vorhaben. Berichte und Dokumentationen. Berlin, Germany.

[26]Buchholz, Sandra. 2008. Die Flexibilisierung des Erwerbsverlaufs: Eine Analyse von Einstiegs- und Ausstiegsprozessen in Ost- und Westdeutschland. Wiesbaden: VS Verlag für Sozialwissenschaften/GWV Fachverlage, Wiesbaden.

market mostly failed. Reasons for that were insufficient amount of available work as well as the perceived stigmatization for people who participated in these kind of measures and the related reduced willingness to apply actively for a regular employment.

In East Germany, the "polluter pays" principal for the renaturation of old mines was not applied after the closure of several lignite mines. Germany—as the legal successor of East Germany—privatized the lignite production, while the responsibility for the renaturation of the former mining sites stayed with the state. Germany created the so-called "Lusatian and Central German Mining Administration Company", which is responsible for renaturation and reuse.

For a future phase-out of the rest of the lignite production, the renaturation process is supposed to be financed by the mining companies themselves. Studies, however, have contested the ability of companies to cover all upcoming costs due to too low existing provisions. Most mining companies in Germany are facing economic problems as the value of their assets as well as possible future income flows experience a downward trend. As soon as these companies generate too little income, e.g. due to the upcoming coal phase-out, it will be very difficult if not impossible for them to pay all costs associated with the renaturation. Measures on how to safeguard sufficient provisions as long as the mining companies are solvent are currently being discussed but not yet in place. They include the introduction of a public fund, foundation or laws to protect the provisions from insolvency (Oei et al. 2017).

3.4.5 Directional Shift in the Structural Policy and Growing Influence of the EU Since the Turn of the Century

In the 1990s, the German structural policy aimed at equalizing regional disparities. The main focus was on funding projects in rural areas and the periphery, neglecting an emphasis on strengthening specific sectors. Especially in East Germany, this funding strategy was not sustainable since a plan on how to develop future oriented sectoral economic structures was missing. Around the turn of the millennium, structural policy became more growth oriented, due to growing influence of the EU.

Cohesion and structural funds are central instruments of EU policy and represent around a third of the EU budget.[27] Cohesion policies target the reduction of disparities between various EU regions. The European Regional Development Fund (ERDF) was central for the support both in Eastern and Western mining regions. The ERDF provides the financial resources for cohesion projects in the EU, focusing on increasing the competitiveness of regions, on developing and distributing technologies and products, and the creation and preservation of jobs.

---

[27]Kambeck, R., and C. M. Schmidt. 2011. 'Den Strukturwandel richtig begleiten – moderne Strukturpolitik statt Erhaltungssubventionen'. In Phönix flieg! Das Ruhrgebiet entdeckt sich neu, edited by Klaus Engel, Jürgen Großmann, and Bodo Hombach, 1. Aufl, 367–87. Essen: Klartext.

Attention was shifted from independent large scale industrial projects to improving competencies in the networks of promising sectors or clusters. Cluster policy addresses the fact that there is no one-size-fits-all solution, focusing on particular strengths of the regions. Clusters are networks of enterprises, associations, research facilities and other institutions within a region.[28] They are meant to strengthen the vertical and horizontal link of companies, suppliers and universities/research institutions. The Ruhr area, for example, declared eight so-called "lead markets", namely health, resource efficiency, mobility, urban building & living, sustainable consumption, digital communication, education & knowledge, leisure time & events as its competencies. Money obtained though ERDF is distributed by the regions themselves to individual projects. After 2007, ERDF shifted its focus towards increasing competition and innovation. Financial support changed from being divided between regions to state-wide support schemes, which meant for the Ruhr Area that funds are now allocated by its federal state NRW.

Progress has been made in the aforementioned lead markets, and especially the southern part of the Ruhr area has experienced a positive economic development. In general, structural policy has led to new employment, increased investments, improved competitiveness and innovativeness, and investments in renaturation that have improved both living standards and environmental quality.[29] However, a clear empirical evaluation of the successes of a single cluster and policy measure is difficult. (Rehfeld 2013) For some regions it might be more helpful to break with old pathways and strengths more rapidly, building up expertise in new sectors to enable an economic system fit for the future.[30] As a general lesson, cluster oriented structural policy needs to be embedded in a broader, more coherent strategy for the development of a region to be successful in the long term.[31]

---

[28]Weingarten, Jörg. 2010. Antizipation des Wandels: Herausforderungen und Handlungsansätze für Kommunen, Unternehmen und Beschäftigte im Rahmen der kohlepolitischen Vereinbarungen in NRW. Dissertation eingereicht an der Ruhr Universität Bochum. http://nbn-resolving.de/urn:nbn:de:hbz:294-30406

[29]Untiedt, Gerhard, Michael Ridder, Stefan Meyer, and Nils Biermann. 2010. 'Zukunft der Europäischen Strukturfonds in Nordrhein-Westfalen—Gutachten im Auftrag der Ministerin für Bundesangelegenheiten, Europa und Medien und des Ministeriums für Wirtschaft, Energie, Bauen, Wohnen und Verkehr des Landes Nordrhein-Westfalen—Endbericht'. Münster/Bremen. http://www2.efre.nrw.de/1_NRW-EU_Ziel_2_Programm_2007-2013/3_Ergebnisse/Gutachten_GEFRA_MR_ZukunftSF_NRW-2010-0801_Final.pdf

[30]Rehfeld, Dieter. 2013. 'Clusterpolitik, intelligente Spezialisierung, soziale Innovationen—neue Impulse in der Innovationspolitik'. Research Report No. 04/2013. Gelsenkirchen, Germany: Institut Arbeit und Technik (IAT), Forschung Aktuell. http://nbn-resolving.de/urn:nbn:de:0176-201304012

[31]Rehfeld, Dieter. 2005. 'Perspektiven des Clusteransatzes—Zur Neujustierung der Strukturpolitik zwischen Wachstum und Ausgleich'. In Theorie und Strategie, 52–55. Dortmund, Germany. http://spw.de/data/rehfeld_spw145.pdf

3.4.6   End of Subsidies for Domestic Hard Coal Production

In 2007, after 40 years of hard coal production subsidies with a total volume of between 289 and 331 billion € (real value for 2008 money) from 1950 to 2008,[32] the federal government passed a law to end the subsidies by 2018.[33] They were no longer in accordance with EU law that forbids such kind of distortion of competition.

The subsidies, which initially were spent to secure the supply with domestic coal and later to prevent an economic disruption in the Ruhr area, rose from 13,500 € per employee in the mining industry in 1980 to 75,000 € per employee in 2005. These annual costs exceeded the average yearly salary of an employee.[34] In the negotiations regarding the end date of the subsidies, the "social compatibility" stood in the center. This term paraphrased, that the exit pathway had to secure that every person working in hard coal production either entered retirement or got a new job. None of the workers should be threatened by unemployment through the law that ended the hard coal subsidies and implicitly the hard coal production in Germany ("Steinkohlefinanzierungsgesetz"). Even after the massive decrease in the hard coal production since the 1950s, around 5–10% of the regularly employed in Ruhr area were still working in the mining sector. Therefore, stakeholders like the IG BCE union emphasized the disruptive effects of a too early end date. In order to decide on an appropriate end date for coal subsidies, hearings were held in 2007. The mining industry, unions and social democrats pleaded for 2018 in the debate, referring to the "social compatibility" and the time needed in the Ruhr area to adjust to the changes. The IG BCE union stated that 11,000 employees would lose their job, if the end date was 2012. On the contrary, research facilities like University Duisburg-Essen and the Leibniz Institute for Economic Research (RWI) stated that the 2012 end date could have saved between 4 and 10 billion € due to lower mining damages and less years of hard coal subsidies. The RWI proposed that those savings could be used to reeducate former employees and give them a new job in the decommissioning of the mining infrastructure. With an end in 2012 there could have been 1 million € per worker to create a "socially compatible" phase-out.

Nevertheless, the powerful network of unions, the mining industry and the social democrats achieved that the end date of coal mining subsidies was deferred to 2018. Every worker with the age of 42 or older was secured by law against unemployment. After the end of their employment in coal mines, workers would work three years in

---

[32]In 2008 €, upper value includes financial support (direct and tax breaks), benefits in the emissions trade system and the costs of higher electricity prices through incomplete competition in the electricity sector.

[33]Meyer, Bettina, Swantje Küchle, and Oliver Hölzinger. 2010. 'Staatliche Förderungen der Stein- und Braunkohle im Zeitraum 1950–2008'. Berlin: Forum ökologisch-soziale Marktwirtschaft eV. http://www.foes.de/pdf/Kohlesubventionen_1950_2008.pdf

[34]Frigelj, Kristian. 2009. 'Der lange Weg zum Ausstiegsbeschluss—das Ende der Steinkohle-Subventionen für Nordrhein-Westfalen'. In Kumpel und Kohle—Der Landtag NRW und die Ruhrkohle 1946 bis 2008, edited by Die Präsidentin des Landtages Nordrhein-Westfalen. Vol. 19. Schriften des Landtags Nordrhein-Westfalen 19. Düsseldorf, Germany.

decommissioning and then receive payments for 5 years to bridge the time until they enter the regular pension fund at age 62 in 2027.[35] The federal parliament estimated the total costs for the phase-out period from 2006 to 2018 at around 38 billion €.[36] The parliament estimated around 2 billion € for pensions and mining damages and additional 7 billion € for the so-called eternity costs.[37]

Eternity costs ("Ewigkeitskosten") are the follow-up costs of the mining activities, especially resulting from water management in the mines. As it is still unclear how many decades these costs will occur, they are called eternity costs. According to German mining law, the polluter-pays principle must be applied. Since the last mining company (RAG AG) will most likely have difficulties to pay for the eternity costs after the end of public subsidies, a foundation was established to assume the task. Therefore, the RAG AG transferred its promising chemical industry—namely the Evonik AG—into the RAG foundation. The revenues of Evonik AG and the sale of its shares are supposed to generate sufficient funds to cover the costs for the eternity burdens of coal mining. This solution is connected with a high risk, since it strongly depends on the economic liability of the Evonik AG. Furthermore, the data to calculate the amount of eternity burdens of coal mining was supplied by the RAG AG itself, without the possibility to verify the assumed costs independently. In case that the foundation does not manage to generate sufficient funds, the government guarantees to pay the costs.

The Ruhr area experienced a long process of decline of its coal and steel industry, beginning with the coal crisis in 1958 and the following steel crisis in the mid-1970s. Over the past 50 years, many different structural and societal policy measures were implemented in order to control the rate of necessary structural changes. At first, the small, locally concentrated structures around the mines needed to be cracked up. This meant, amongst other factors, investments in modern transport infrastructure to interconnect the cities within the Ruhr area. These investments were meant to increase citizens' mobility, to enable them to travel between their homes and potential new workplaces outside the mining industry. Additionally, the opening of the first university in the Ruhr area was an important part for the region's reorientation. However, the intended economic reorientation done by attracting new companies to the area was slowed down substantially by the resistance of the network between mining companies, politicians and unions.

The inability of industrial regions to enable an economic reorientation has been termed "lock-in" (Hospers 2004, 151; Campbell and Coenen 2017, 6f). The institutional lock-in (network of companies, politicians and unions), the economic lock-in (high dependency on the mining and steel industry) as well as a cognitive lock-in

[35]Ibid.

[36]Bundesregierung. 2007. 'Drucksache 557/07 des Deutschen Bundestages—Gesetzesentwurf zur Finanzierung der Beendigung des subventionierten Steinkohlenbergbaus zum Jahr 2018'.

[37]Frigelj, Kristian. 2009. 'Der lange Weg zum Ausstiegsbeschluss—das Ende der Steinkohle-Subventionen für Nordrhein-Westfalen'. In Kumpel und Kohle—Der Landtag NRW und die Ruhrkohle 1946 bis 2008, edited by Die Präsidentin des Landtages Nordrhein-Westfalen. Vol. 19. Schriften des Landtags Nordrhein-Westfalen 19. Düsseldorf, Germany.

(belief that the crisis was cyclical not structural) led to persistent attempts to modernize the old structures of the Ruhr area, instead of turning to new economic possibilities.

The Ruhr area case study shows that economic reorientation worked best when new projects were related to the existing industries in the cities of the Ruhr area. Large projects from distant sectors did not prove as successful. An important step was therefore the inclusion of local stakeholders and increased regionalization in structural policy decision making. This enabled the region to benefit from its endogenous potentials and at the same time, in comparison to top-down decision making, reduced local resistance. In order to achieve that, especially in a federal state like Germany, it was necessary to create an organizational structure that represents the different cities of the Ruhr area as one entity with respect to the different political levels.

The following structural policy programs focused more on ecological and cultural aspects and increased the entrepreneurial activity in the Ruhr area (Hospers 2004, 154f). These efforts changed the perception of the Ruhr area from the outside as well as from the inside and helped it move beyond the image of a dirty industrial area. The share of people working in the secondary sector decreased from 58% in 1976 to 26% in 2014, whereas the share in the tertiary sector increased from 42 to 74%, respectively.[38]

3.5   Lessons Learned from the Past Transformation Process

Germany's example illustrates that even within a single country coal reduction pathways vary strongly and require different measures. Table 3.1 lists some of the main differences between the reduction in hard coal and lignite production in Germany.

Even though the circumstances are different, the German historical experience, however, also shows that regardless of the specificity of each reduction, certain identical dimensions need to be addressed to enable a "just transformation" [based on the "just transition" concept by the International Labour Organization, in order to create social justice (ILO 2015, 6)]. Figure 3.12 illustrates important aspects following the concept of the just transformation, which need to be addressed when a region transforms itself from a fossil fuel-based economy to a low-carbon society in a just way. They can be divided into aspects that account mainly for the mining regions and others that have to be dealt with on a national or even supranational level. The figure lists important areas that should be addressed by policy makers in future transformations, while actually implemented measures will vary for each case study

---

[38]Own calculations based on: Regionalverband Ruhr. 2017b. 'Zeitreihe Zur Erwerbstätigkeit'. http://www.metropoleruhr.de/fileadmin/user_upload/metropoleruhr.de/Bilder/Daten___Fakten/ Regionalstatistik_PDF/Erwerbstaetigkeit/05_Zeit_Ewt_SVB14.pdf

Table 3.1 Differences between the Ruhr and Lusatian mining regions

	Ruhr area	Lusatian region
Main energy carrier	Hard coal	Lignite
Type of mining	Deep mines	Open cast mines
Follow-up costs	High costs over a long time period with an uncertain end date ("eternity costs"); mainly for water management in the former mines	Costs for renaturation and reuse of the vast areas of destroyed land (cost and time period easier to predict compared to hard coal)
Environment/ Population	Most densely populated area in Germany, >5 million people	Rural/peripheral area, ~1.1 million people
Phase-out consequences for energy system	Limited; coal demand covered with imports from overseas	No imports, mining and power plants are coupled; potential "domino-effects"
Time period phase-out	1957–2018; Long, continuous process	Since 1989; Rapid reduction in only a few years with follow-up consequences
Employment in mining	1957: ~500,000 1967: ~230,000 1977: ~150,000 2016: ~5,800	1989: ~80,000 1999: ~8000 2016: ~5000
Civil society	Protests against coal reduction in the mining regions; Strong connection and identification with jobs in hard coal production	Very little ecological concerns before reunification; Reduction in coal dominated by reunification effects
Reasons for mining reduction	Comparably cheap imported oil and hard coal	Reunification, inefficient and costly production compared to West Germany
Labor productivity	Increased more than fourfold since 1950	Almost constant from 1950s to 1990; then steep increase catching up with Western German standards
Replacement of jobs	Focus on education, the service industry and becoming a "knowledge society"; strong social security net, however, also a strong increase in unemployment	Replacement of coal jobs difficult due to the economic and political breakdown also outside the lignite sector

Source: Based on Agora Energiewende (2017), RAG-Stiftung (2015, 1), Sachverständigenrat für Umweltfragen (2017), Verein der Kohleimporteure e.V (2017), Statistik der Kohlenwirtschaft (2017c, 2017d), Goch (2009, 12) and Matthes (2000, 238) (Agora Energiewende. 2017. 'Eine Zukunft für die Lausitz—Elemente eines Strukturwandelkonzepts für das Lausitzer Braunkohlerevier'. Berlin, Germany: Agora Energiewende. https://www.agora-energiewende.de/fileadmin/Projekte/2017/Strukturwandel_Lausitz/Agora_Impulse_Strukturwandel-Lausitz_WEB.pdf; RAG-Stiftung. 2015. 'RAG-Stiftung: Geschäftsbericht 2015'. Essen. http://www.rag-stiftung.de/ueber-uns/jahresabschluesse/; Verein der Kohleimporteure e.V. 2017. 'Jahresbericht 2017—Fakten und Trends 16/17'. Hamburg, Germany. http://www.kohlenimporteure.de/publikationen/jahresbericht-2017.html; Statistik der Kohlenwirtschaft e.V. 2017c. 'Steinkohle—Belegschaft im Steinkohlebergbau'. Steinkohle. 2017. https://kohlenstatistik.de/18-0-Steinkohle.html and Statistik der Kohlenwirtschaft e.V. 2017d. 'Braunkohle'. Statistik der Kohlenwirtschaft. 9 September 2017. https://kohlenstatistik.de/19-0-Braunkohle.html)

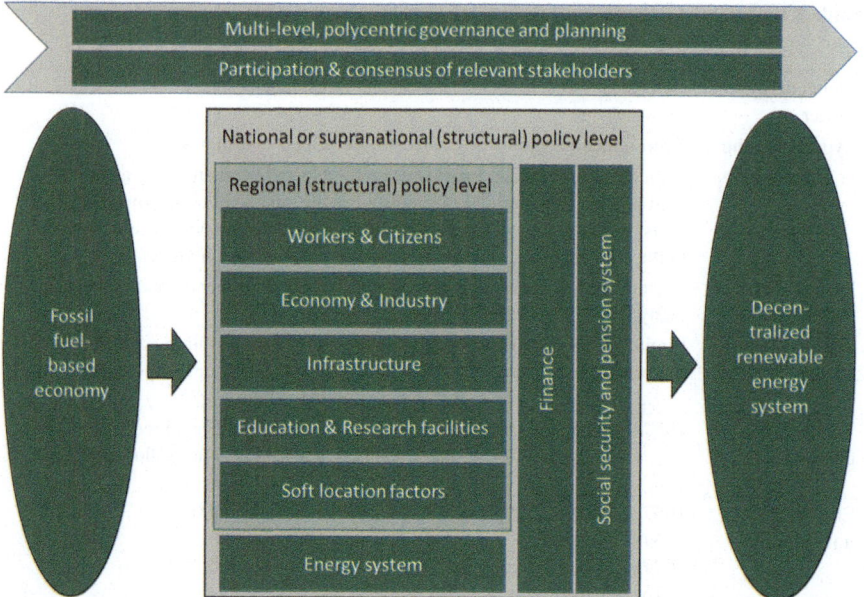

Fig. 3.12 Dimensions for a just transformation in coal regions. Note: The size of each area does not implicate any valuation in terms of financial volume or importance of the dimensions

depending on regional specifics. In both the Ruhr area and Lusatia some of the aspects were addressed successfully while others were neglected. While important lessons can be identified through the two transformation pathways, room for improvement is left and neither region has yet completed the process. In the following, important German experiences with structural policies on both the regional and national (supranational) level are highlighted. The whole process of the just transformation away from fossil fuels should be guided by multi-level, polycentric governance and planning, encouraging the different political levels to interact with each other in order to plan and implement effective strategies. Further-more, planning and decision making should include a high degree of participation of all relevant stakeholders and deliver consent-based solutions to increase acceptance and to tap endogenous potentials.

3.5.1 Regional Level

Workers and Citizens

A just transformation needs to guarantee social security of mining workers and give them and the regions a perspective beyond coal mining: That means that lost jobs in the mining industry need to be replaced with new comparable jobs in other industries and sectors. In the Ruhr area, all employees entered either new employment or the pension fund (in combination with early retirement measures). Additional anticipative measures included retraining of workers.

Regardless of the success of policies to ensure a more moderately-paced decline, the Ruhr area struggled for a long time to create new jobs, especially due to the coal and steel crises coinciding with the end of a period of strong economic growth. The inability to create new jobs was mainly caused by the resistance of the mining companies, problems for the domestic industries caused by increasing competition due to globalization and the misjudgment of the true nature of the coal and steel crises, which prevented action towards a transformation.

The Lusatian region faced and still faces the challenges of demographic changes and migration (to West Germany), which is a common problem for rural areas. The situation for Lusatia was aggravated as not only the employment in the mining sector broke down but the whole economic and political system. Also, the reduction was not accompanied by direct subsidies for the mining sector to enable a moderate decline, as it had been the case for hard coal in the Ruhr area. Instead, measures like early retirement were implemented to ease tension on the labor market in entire East Germany. Consequently, Lusatia faces even stronger problems than the Ruhr area to attract new businesses providing local jobs.

Anticipative elements like retraining and an early communication of phase-out plans can ease the disruptiveness of upcoming changes, by helping former coal miners to stay in the labor market and to prevent future erroneous education and employment choices. Furthermore, the job decrease in an up-coming phase-out could, be organized along the age structure. A move to renewable energies poses not only a challenge for the mining regions but might also be an opportunity: In 2016, the German renewable energy industry employed around 334,000 people, compared to only around 160,000 in 2004 (IRENA 2017).[39] Although not all former coal workers will simply be able to move to the renewables sector, it can enable regions to continue to play an important role in the energy sector.

Economy and Industry
In the Ruhr area, the economy shifted from the primary sector to the tertiary sector ("knowledge society"). The economy in the Ruhr area is now more diversified thanks to a reorientation towards a more participative approach in structural policy making, the polycentric coordination of national, state and regional policy making, and the majority of subsidies going to industries and sectors other than the coal and steel industries. In particular, the Ruhr area experimented with different structural policy instruments and governance structures, as previously only a limited amount of experience with structural change existed.

---

[39]Burger, Andreas, Benjamin Lünenbürger, David Pfeiffer, and Benno Hain. 2015. 'Klimabeitrag für Kohlekraftwerke—Wie wirkt er auf Stromerzeugung, Arbeitsplätze und Umwelt'. Positionspapier. Dessau-Roßlau: Umweltbundesamt. https://www.umweltbundesamt.de/publikationen/klimabeitrag-fuer-kohlekraftwerke

A struggle during the developments of structural policy in the Ruhr area, was to identify the right system of governance to lead the transformation. The first large projects initiated on a federal level were ineffective and faced regional and local resistance. Over time, decision making and planning shifted to a more regional level, to include the endogenous local potentials and to enhance approval of the transformation by increasing participation of the stakeholders. The implementation of an institution representing the Ruhr area as an entity has helped to coordinate national funding but is still not fully capable of creating a coherent strategy for all cities in the Ruhr area, possibly leading to further exploitation of its (economic) potential.

The Lusatian economy still suffers from the structural break which occurred after the reunification, when many industries collapsed. The region additionally faces the challenges many rural areas have (demographic changes, a lack in infrastructure, emigration, etc.). Due to a failure to develop alternative industries, the local economy in some communities still heavily depends on lignite production. Investment support for new businesses often comes from outside the region. As a result, over the decades, projects have heavily relied on this financial and intellectual support, resulting in repeated closures of projects that cannot sustain themselves when the support ends.

Infrastructure

In the Ruhr area, infrastructure investments were a crucial aspect of the first structural policy program "Development Program Ruhr", since the "new economy" beyond the mining industry relied on an enhanced mobility of the people. The area now plays a major role in the logistic sector due to its links to economic centers within Europe. However, on a regional level, there is still room for improvement, especially in the public transportation systems. A major part in the programs of "Reconstruction East" after reunification consisted of infrastructure programs. Due to the condition of the existing infrastructure and the urgency to renew it, many projects were realized without a sufficient planning phase (demographic changes and economic development were not taken into account properly). This has resulted in a situation where many infrastructure projects are now not being used to their full potential. Besides the traffic infrastructure, the regions (especially Lusatia) need high-speed data connections in order to create an attractive environment for companies.

Education and Research Institutions

Education and research institutions can play an important role in order to enable a shift from a mining area towards a more knowledge based society. In 1965 the Ruhr area was devoid of a single university; the opening of several new universities enhanced the attractiveness of the region for companies as well as for citizens, constituting an important location factor. In 2014, 22 universities existed with more than 250,000 students.[40] The deployment of the universities enabled a shift from the mining economy towards an economy which is based on high-value adding

---

[40]Kriegesmann, Bernd, Matthias Böttcher, and Torben Lippmann. 2015. Wissenschaftsregion Ruhr: (Langfassung): wirtschaftliche Bedeutung, Fachkräfteeffekte und Innovationsimpulse der Hochschulen und außeruniversitären Forschungseinrichtungen in der Metropole Ruhr. Essen: Regionalverband Ruhr.

sectors (such as the lead markets in the Ruhr area) with increased demand for highly skilled workers and research-based innovation. The universities and research facilities need to be integrated into networks of companies and other institutions in order to create competitive and resilient structures which keep companies in the region and attract new ones. In Lusatia, only two universities exist, concentrating skills in these cities. However, due to a lack of related skilled jobs, migration after completing a degree remains a problem.

Soft Location Factors
Soft location factors like cultural and leisure time possibilities, but also environmental issues (air pollution levels, clean rivers, etc.) play an important role in the public perception of a region. They increase the quality of life in the region and can convince people to stay in or to move to a region. Migration is not only caused by better job options but also because of higher cultural potential of regions. In the Ruhr area, the aspect of soft location factors was neglected for a long time but with the "Action Program Ruhr" and the "IBA Emscher Park" these issues were addressed. Former industrial sites were transformed into landmarks and cultural sites in order to conserve the identification with the region but also to enable a shift towards a new, more future oriented perception. The entire migration effect is not likely to be due to soft location factors, but must be seen as a combination with job and study opportunities, trends coinciding with the new focus on living quality: Net migration turned after the "Action Program Ruhr" and the "IBA Emscher Park". Within 8 years (1987–1995) 247,000 people migrated (net) to the Ruhr area, whilst net migration stabilized after a new downward trend after IBA Emscher Park. As a comparison, net migration from 1977 until 1986 was minus 158,000.[41] For Lusatia, the pending renaturation, hence, not only poses a challenge but also an opportunity to increase the attractiveness of the region.

3.5.2 National and Supranational Level

Energy System
In Germany, the decline in coal production affects electricity and heat generation. The reduction in hard coal production starting in the 1950s was replaced (and also caused) by comparatively cheap hard coal and oil imports. The decrease in the domestic production therefore had little immediate consequences for hard coal-fired power plants. However, the reduction of lignite mining in East Germany caused a decline in lignite based electricity generation of almost 40 TWh between 1989 and 1995 (which corresponds to a decline of approximately 40% of the East German gross electricity generation). In 2017, Germany generated more than 35% of its

---

[41]Regionalverband Ruhr. 2017a. 'Bevölkerungsbewegung in der Metropole Ruhr—Zeitreihe seit 1962'. 2017. www.metropoleruhr.de/fileadmin/user_upload/metropoleruhr.de/05_MR_Sonstige/Excel/Statistik/Bevoelkerung/Zeitreihe_Bevoelkerungsbewegung.xlsx

electricity with renewable energies and exported more than 50 TWh of electricity. To prevent lock-ins and resistance to a coal phase-out, timely investments in alternative electricity and heat generation are important, guaranteeing energy security, grid stability and affordable energy prices. The deep integration of local electricity markets into national and EU markets facilitates the transformation where not every region needs to be energy self-sufficient. It can be attractive for former coal regions to use their expertise in the energy sector and to move towards renewable energies, energy storages or other innovative energy solutions.

Finance

A just transformation requires financial resources and a fair distribution of the responsibilities for the costs. Germany therefore financed most of the subsidies for the Ruhr area with the national budget. As future coal-phase outs are mostly a political decision due to global climate change concerns, costs should not be born only by the regions but by the whole country or even the supranational level.

The consideration of the finances includes, besides the structural policy and social policies, sufficient measures to guarantee the polluter-pays principle, in line with the German mining law. In both the hard coal and lignite phase-out, the state is at risk to bear shares of the (eternity) costs. For hard coal mining, a foundation to secure the provisions was implemented, however whether the funds will be sufficient remains to be seen. In East Germany, the state bore the full costs for the recultivation. After the reunification of German, the responsibilities for environmental damages were socialized whereas the lignite companies were privatized. In a future lignite phase-out, Germany (and other countries) need to implement measures which secure the polluters-pay principle. Possibilities include the introduction of a public fund (analog to the nuclear sector in Germany), a foundation (analog to hard coal sector in Germany) or laws to protect at least the provisions which mining companies have built up so far from insolvency. Securing sufficient funds needs to be ensured as fast as possible before the regular mining business ends (see also Oei et al. 2017).

Transfer to Other Countries and Future Phase-outs

As the hard coal and lignite reduction have shown, the situations differ from case to case and therefore policies guiding the transformation need to be adjusted to the respective circumstances. In East Germany the circumstances were unprecedented since the drastic reduction in lignite mining was accompanied (and caused) by the German reunification, resulting in a breakdown of the whole economic and political system. This should be taken into account when assessing the consequences of the lignite reduction and the effect of structural policies.

The case of the Ruhr area is special, as future reductions in coal mining are not likely to be granted a similarly long time period for a phase-out. Germany conserved (a shrinking share) of its hard coal production for more than 60 years with significant subsidies. The implementation of subsidies on a similar scale in other countries and future phase-outs is unlikely for several reasons. Firstly, in principle, such a market interference is forbidden after 2018 at least for all EU countries by European law. Secondly, the total amount of subsidies for domestic hard coal is difficult to quantify as they consist of direct subsidies, infrastructure investments, labor market interventions, etc. Nevertheless, Germany spent more than 330 billion € on direct and indirect hard coal subsidies, to an extent that will be hard for other countries to replicate. Thirdly, the subsidies in the Ruhr area supported a German company that was interconnected on the regional level, since politicians were holding positions within the firm and cities were shareholders of the company. The acceptance of change among the citizens of the Ruhr area (and Germany) might therefore have been somewhat more difficult than for other countries where foreign mining companies often exploit the resources and export them abroad. Fourthly, ecological consciousness and especially concerns about potentially devastating consequences of climate change are widespread, making subsidies to extend the lifetime of fossil fuel exploration more difficult.

In the past, a strong identification and pride existed among workers (and entire regions) with the manly, tough and often dangerous mining job, thought to be essential for economic development. This, along with the influence of powerful unions, helped to prevent a faster transformation away from coal. However, the perception of coal mining as an attractive and necessary job is fading, which might facilitate the move away from coal in other countries, especially when other well paid jobs are available.

The structural policy of the Ruhr area showed that single large projects were not able to replace the mining (and steel) industry and instead faced resistance within the region. Former mining cities had individual needs that needed to be addressed independently. Therefore, the level of decision making shifted more and more from a centralized national level to a regional one. Today, there exists an institution which conceptualizes development strategies for the entire region, coordinating bottom-up strategies from within the various cities themselves. Such an institution might help to guide future phase-outs as well to limit the bureaucratic friction and improve the participation of relevant stakeholders. Especially in Lusatia, people feel left behind and not taken seriously of (inter-)national policy. Therefore, a stronger participation of various stakeholders including civil society is necessary to achieve better policy outcomes and public acceptance (Morton and Müller 2016).

From an energy system's point of view, the transformation has become easier and cheaper for other countries than it was for Germany in the past. The cost of renewable energy technologies has decreased significantly in the last decade, and is now just a fraction of the price compared to when Germany started deploying photovoltaics and onshore wind on a large scale. The ongoing development and installation of renewables in Germany threatens the economic and technical feasibility of its inflexible coal-fired power plants. Many studies have successfully modelled energy systems that are entirely based on 100% renewables not only for Germany but for the global energy system (Jacobson et al. 2017; Löffler et al. 2017; Henning and Palzer 2012).

Germany's two examples of reducing coal mining provide valuable lessons learned but also illustrate the difficulties of structuring a phase-out without negative consequences for employees, companies and entire regions. An important lesson from Germany's past experience is that it is not only necessary to have policies addressing unemployment, the economy and the energy system, but also measures to improve former coal regions' infrastructure, universities and research facilities as well as soft location factors like culture and environmental health. The German example suggests that implementing a fair and realistic transformation from a fossil fuel-based economy can be managed when city, regional, national and supranational governments work together on designing a phase-out and a multi-level polycentric structural policy mix. The upcoming remaining transformation can succeed when past experiences with structural policies and social security systems are considered, along with the incorporation of affordable alternative forms of energy generation and other promising innovative sectors providing new job opportunities for people in the affected regions.

3.6   Conclusions

One necessary (though not sufficient) element to succeed the energiewende is to reduce the use of fossil fuels, and in particular to stop coal electrification and heat generation within the next two decades or so. Even though the German government, various stakeholders, and civil society are still debating about the appropriate instruments to accompany this process, there is a broad consensus that coal will not remain an essential energy source during the low-carbon transformation.

The move away from coal is, however, not only due to climate policies. The large scale deployment of renewable energy threatens the economic viability of coal-fired power plants, as renewables increase competition and lower whole-sale electricity prices. The existing economic situation in Germany will prevent the construction of new coal-fired power plants. Therefore, eventually, coal would be phased-out at the latest at the end of the power plants' lifetimes. However, to achieve a coal phase-out in line with climate protection commitments, climate policy measures need to be introduced to accelerate the decline of coal. A commission called 'growth, structural change and employment' was installed in 2018, with the aim to decide on an end date for coal in Germany and to develop a strategy on how to guide and manage the phase-out.

The good news for Germany is that by far the largest share of coal's decline has already been managed. This chapter therefore serves as an interesting case study to analyze the history of the phase-out of hard coal mining in the Ruhr area and the reduction of lignite mining in East Germany as a result of Germany's reunification. Germany's two examples of reducing coal mining provide valuable lessons learned but also illustrate the difficulties of structuring a phase-out without negative consequences for employees, companies and entire regions. The Ruhr area serves as an interesting example since it shows how the shift away from coal was delayed by the powerful influence of coal mining and steel industries. However, it shows that a phase-out is possible and that potential negative effects can be managed effectively: The perception of the Ruhr area changed from the old industrial area towards a region with a more diversified and strong economy with an increasing quality of life.

In East Germany, and especially in Lusatia, the structural change was stronger, as not only the coal sector experienced a rapid decline but the entire (political) economy was in a difficult state due to the German reunification.

An important lesson is that it is not only necessary to implement policies addressing unemployment, the economy and the energy system, but also measures to improve former coal regions' infrastructure, universities and research facilities as well as soft location factors like culture and ecology. The German case study suggests that implementing a fair and realistic transformation from a fossil fuel-based economy can be managed when city, regional, national and supranational governments work together on designing a multi-level polycentric structural policy mix to guide the phase-out.

Despite having specific regional characteristics, Germany's experience provides valuable lessons learned for the last step of the coal phase-out in Germany and for other regions or sectors in various countries with a phase-out ahead: Learning from the past could help to prevent the repetition of mistakes, and ensure that previously successful policies might be implemented in a similar fashion. The upcoming transformation can succeed by considering past experiences with structural policies and social security systems, along with the incorporation of affordable alternative forms of energy generation and other promising innovative sectors providing new job opportunities and a just transition.

References

Campbell, Stephanie, and Lars Coenen. 2017. *Transitioning Beyond Coal: Lessons from the Structural Renewal of Europe's Old Industrial Regions*. Melbourne: Australian National University.

Goch, Stefan. 2009. Politik für Ruhrkohle und Ruhrrevier - von der Ruhrkohle AG zum neuen Ruhrgebiet. In *Kumpel und Kohle - Der Landtag NRW und die Ruhrkohle 1946 bis 2008*, ed. Die Präsidentin des Landtages Nordrhein-Westfalen, vol. 19, 125–165. Düsseldorf: Schriften des Landtags Nordrhein-Westfalen.

Henning, Hans-Martin, and Andreas Palzer. 2012. *100% Erneuerbare Energien für Strom und Wärme in Deutschland*. Freiburg: Fraunhofer-Institut für Solare Energiesysteme ISE.

Hospers, Gert-Jan. 2004. Restructuring Europe's Rustbelt. *Intereconomics* 39 (3): 147–156.

ILO. 2015. *Guidelines for a Just Transition Towards Environmentally Sustainable Economies and Societies for All*. Geneva: International Labour Organization.

IRENA (2017) *Renewable Energy And Jobs - Annual Review 2017*. Masdar City, Abu Dhabi

Jacobson, Mark Z., Mark A. Delucchi, Zack A.F. Bauer, Savannah C. Goodman, William E. Chapman, Mary A. Cameron, Cedric Bozonnat, et al. 2017. 100% clean and renewable wind, water, and sunlight all-sector energy roadmaps for 139 countries of the world. *Joule* 1 (1): 108–121.

Kahlert, Joachim. 1988. *Die Energiepolitik der DDR - Mängelverwaltung zwischen Kernkraft und Braunkohle*. Bonn: Verlag Neue Gesellschaft GmbH.

Löffler, Konstantin, Karlo Hainsch, Thorsten Burandt, Pao-Yu Oei, Claudia Kemfert, and Christian von Hirschhausen. 2017. Designing a model for the global energy system—GENeSYS-MOD: an application of the open-source energy modeling system (OSeMOSYS). *Energies* 10 (10): 1468.

Matthes, F.C. 2000. *Stromwirtschaft und deutsche Einheit: Eine Fallstudie zur Transformation der Elektrizitätswirtschaft in Ost-Deutschland*. Berlin: Germany.

Morton, Tom, and Katja Müller. 2016. Lusatia and the coal conundrum: the lived experience of the German energiewende. *Energy Policy* 99: 277–287.

Oei, Pao-Yu, Hanna Brauers, Claudia Kemfert, Christian von Hirschhausen, Dorothea Schäfer, and Sophie Schmalz. 2017. Climate protection and a new operator: the eastern German lignite industry is changing. *DIW Econ Bull* 7 (6/7): 63–73.

Rehfeld, Dieter. 2013. *Clusterpolitik, intelligente Spezialisierung, soziale Innovationen - neue Impulse in der Innovationspolitik*. Research report no. 04/2013, Institut Arbeit und Technik (IAT), Forschung Aktuell, Gelsenkirchen, Germany.

Sachverständigenrat für Umweltfragen. 2017. *Kohleausstieg jetzt einleiten*. Stellungnahme, Berlin.

Part II
The Energiewende at Work in the Electricity Sector

Chapter 4
Greenhouse Gas Emission Reductions and the Phasing-out of Coal in Germany

Pao-Yu Oei

> *"Governments, business and regions all around the world are moving beyond coal. Electricity generation from coal is declining. This is an irreversible trend towards clean power, also here in Europe. (...) All Europeans should benefit from this transition, and no region should be left behind when moving away fossil fuels."*
> *Miguel Arias Cañete, Commissioner for Climate Action and Energy, European Commission at the Launch of the Platform for Coal Regions in Transition in Strasbourg, 11 December 2017 (http://europa.eu/rapid/press-release_IP-17-5165_en. htm)*

This chapter summarizes work by our research program on "The Future of the Lignite Industry" and the Junior Research Group "CoalExit", see for other publications https://www.diw.de/sixcms/detail.php?id=diw_01.c.429667.en; as well as the Junior Research Group "CoalExit" https://www.wip.tu-berlin.de/menue/nachwuchsforschungsgruppe_coalexit/parameter/en/; it has been updated from Chapter 3 of my dissertation (Oei 2015). I thank Hanna Brauers, Clemens Gerbaulet, Leonard Göke, Franziska Holz, Christian von Hirschhausen, Claudia Kemfert, Roman Mendelevitch, and Felix Reitz for cooperation and comments, the usual disclaimer applies.

P.-Y. Oei (✉)
Junior Research Group "CoalExit", Berlin, Germany

TU Berlin, Berlin, Germany

DIW Berlin, Berlin, Germany
e-mail: pyo@wip.tu-berlin.de

81

4.1   Introduction

The reduction of greenhouse gas (GHG) emissions, in particular CO_2, is a major objective of the German energiewende. There has been broad consensus on this goal for many years now—in contrast to the continuing discussion over the proposed shutdown of Germany's nuclear power plants. The German government's Energy Concept 2010 already aimed at a 80–95% reduction of GHG by 2050 (compared to the base year 1990) (BMWi and BMUB 2010). This is in line with the objectives of other countries such as the UK and France (80% reduction by 2050) and the EU (80–95% reduction by 2050).

In contrast to other sectors such as transport, agriculture, and heating, the electricity sector is capable of reducing CO_2 emissions at relatively moderate cost. This is due to available low-cost alternatives, in particular renewable energy sources. Numerous studies have shown a pathway to almost complete decarbonization of Germany's electricity generation by 2050, some of the most prominent being the German government's periodic "lead studies" (*Leitstudien*) (Nitsch 2013), and studies by the the Federal Environmental Agency (Umweltbundesamt: Klaus et al. 2010) and the German Advisory Council on the Environment (SRU 2011, 2015). Also, several studies have successfully modelled energy systems that are entirely based on 100% renewables not only for Germany but for the global energy system (Henning and Palzer 2012; Jacobson et al. 2017; Löffler et al. 2017). Likewise, modeling results published in the European Commission's Energy Roadmap (EC 2011) suggest that the electricity sector could be 97% decarbonized by 2050. However, this assumes a major shift in the electricity mix away from fossil fuels towards low-carbon generation technologies. In fact, when excluding the option of carbon capture, transport and storage (CCTS) technologies, achieving ambitious climate objectives in Germany (and elsewhere) implies phasing out both hard coal and lignite.

This chapter provides an overview of Germany's GHG emission reduction targets in the electricity sector and the progress achieved so far. The electricity sector has the potential to lead the way in decarbonization, provided that the appropriate regulatory framework is in place. Due to insufficient price signals that can be expected to persist for the next decade, the European Emissions Trading System (EU-ETS) will not be able to achieve this objective on its own but will require support from appropriate national instruments. A variety of such measures are currently being discussed, and some have already been implemented in Germany (Oei et al. 2014a, b; Oei 2015; Strunz et al. 2015).

The chapter is structured as follows: The next section gives an overview of Germany's GHG emission reduction targets and their relation to European targets. Germany has played a leading role in European efforts to reduce GHGs so far, and is continuing to work in this direction (−40% by 2030 and a similar −80 to 95% target for 2050). Section 4.3 focuses on coal-fired electricity generation and its problematic role in the German energy sector in light of the fact that the continued use of coal would render the GHG reduction targets unachievable when excluding the option of

CCTS. We differentiate between hard coal, which is being phased out gradually for economic reasons (lack of competitiveness), and lignite, which is particularly CO_2-intensive and has high external costs but is still competitive. Section 4.4 discusses the influence of the EU-ETS as well as various additional national instruments, including a CO_2 emissions performance standard (EPS), a CO_2 floor price, and a phase-out law. The analysis is informed by extensive field work, and policy consulting.[1] In Section 4.5, we show that a medium-term coal phase-out is compatible with resource adequacy in Germany. The resulting structural change in the affected local basins can be handled through additional schemes, thus posing no major obstacle to the phase-out of coal. Section 4.6 concludes.

4.2 GHG Emissions Targets and Recent Trends in Germany

4.2.1 German GHG Emissions Targets for 2050

Combating climate change through GHG emissions reduction has a long tradition in Germany. Chancellor Helmut Kohl announced the first CO_2 reduction target of 25% by 2005 (base year 1990) at the first international climate conference in Berlin in 1995. Two years later, Germany signed the Kyoto Protocol, pledging a 21% reduction in GHG emissions from base year 1990 to 2012. This reduction target shows Germany's contribution to the burden-sharing agreement within the European Union, as it lies significantly above the overall European reduction of 8%. In 2007, Germany announced the target of 40% lower GHG emissions in 2020 compared to 1990. The government also strongly supported the target of a 20% reduction by 2020 set by the European Union in its 2008 energy and climate package, and it tried (unsuccessfully) to increase the overall European target to 30% in the subsequent years (Hake et al. 2015). The German Energy Concept 2010 then set the long-term GHG reduction targets that became a pillar of the energiewende (base year: 1990): -40% by 2020, -55% by 2030, -70% by 2040, and -80 to 95% by 2050 (BMWi and BMUB 2010).

Generally speaking, GHG emissions are decreasing in Germany, but significant efforts are required to maintain this downward trend. Figure 4.1 shows annual GHG emissions in Germany since 1990, divided into sectors governed under the EU-ETS

---

[1]Hirschhausen, C. von, & Oei, P.-Y. (2013). Gutachten zur energiepolitischen Notwendigkeit der Inanspruchnahme der im Teilfeld II des Tagebau Welzow-Süd lagernden Kohlevorräte unter besonderer Berücksichtigung der Zielfunktionen der Energiestrategie 2030 des Landes Brandenburg (Politikberatung kompakt No. 71). Berlin, Germany: Deutsches Institut für Wirtschaftsforschung (DIW); and Hirschhausen, C. von, & Oei, P.-Y. (2013). Gutachten zur energiewirtschaftlichen Notwendigkeit der Fortschreibung des Braunkohlenplans "Tagebau Nochten" (Politikberatung kompakt No. 72). Berlin, Germany: Deutsches Institut für Wirtschaftsforschung (DIW).

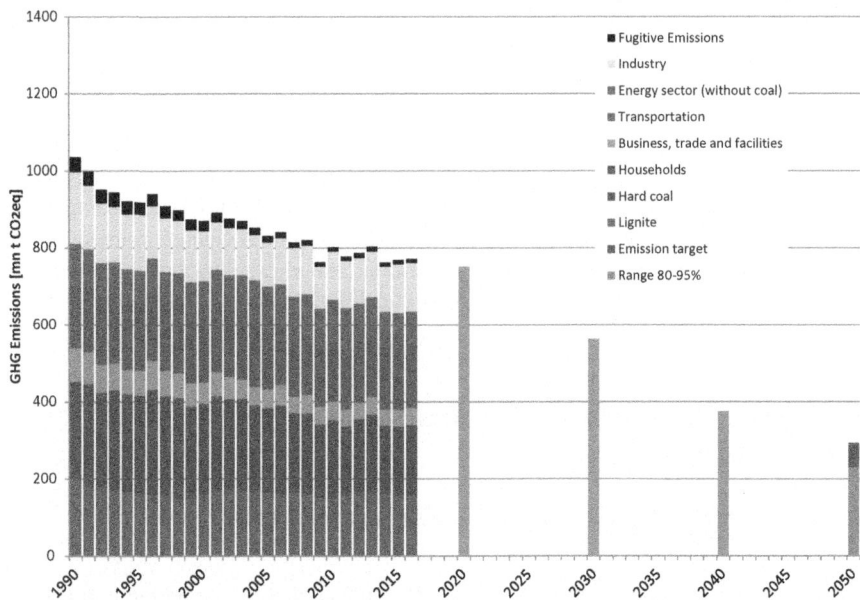

Fig. 4.1 GHG emissions and emission targets in Germany from 1990 to 2050. Source: Umweltbundesamt (2018). Nationale Trendtabellen für die deutsche Berichterstattung atmosphärischer Emissionen 1990–2016, Stand 01/2018

(i.e., electricity, steel, energy-intensive industries) and so-called "non-ETS" sectors. It further distinguishes between hard coal and lignite, and indicates the reduction path to 2050 (−80 to 95%). The overall decline in GHG emissions is particularly evident in two major reduction periods: (1) the economic recession in East Germany after reunification (1990–1994), and (2) the global economic and financial crisis (2008–2010). However, since 2013 emissions are relatively unchanged, and the 2020 target can no longer be reached.

Achieving a long-term GHG emissions reduction of up to 95% by 2050 in Germany requires drastic measures across all emitting sectors. Figure 4.2 shows the distribution of GHG emissions in Germany across different sectors in 1990 and 2012 compared to two different reduction scenarios for 2050, assuming 80% and 90% GHG emissions reductions, respectively. All sectors will need to reduce their emissions up to 2050, but their reduction potentials vary depending on existing mitigation options.

Within the energy sector, electricity generation is responsible for the lion's share of GHG emissions (around 75%). Low-carbon alternatives are already in place in the electricity sector, the most significant being renewable wind and solar technologies. These continue to benefit from declining costs. Other energy sector emissions come from refineries that have much higher specific abatement costs compared to coal power plants. Overall, the energy sector is expected to contribute the largest absolute as well as relative reduction share of −86%/−99% compared to the base year 1990.

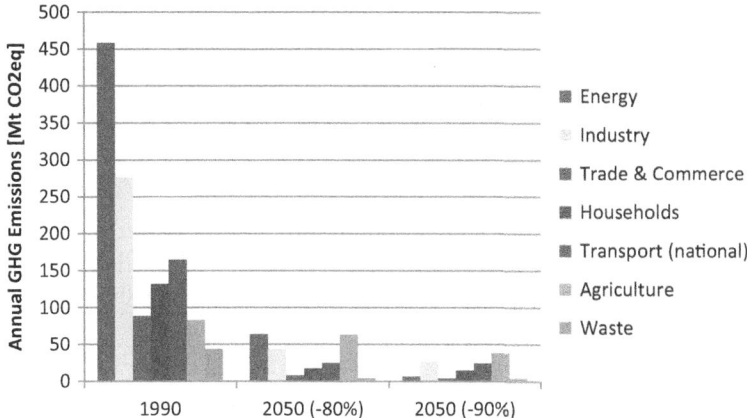

Fig. 4.2 Distribution of German greenhouse gas (GHG) emissions by sector. Source: Öko-Institut, and Fraunhofer ISI. (2014). Klimaschutzszenario 2050 – 1. Modellierungsrunde (Studie im Auftrag des Bundesministeriums für Umwelt, Naturschutz, Bau und Reaktorsicherheit). Berlin, Germany

Equivalent reductions are needed in the industrial sectors ($-84\%/-90\%$), trade and commerce ($-91\%/-95\%$), household consumption ($-87\%/-89\%$),[2] transport ($-85\%/-85\%$), and waste ($-90\%/-91\%$). In the latter sectors, emissions reductions are possible but require more specific action and entail higher costs.[3] GHG emissions from agriculture, in particular nitrogen oxide (NO_x) from fertilizers and methane (CH_4) in livestock farming, are the most difficult to reduce and will therefore be the biggest emitters in 2050. Their reduction levels in the 80% reduction scenario remain close to 2012 levels at around -25%. Projections in the 90% reduction scenario account for a 54% reduction of agriculture at best (Öko-Institut and Fraunhofer ISI 2014).

4.2.2 Ambitious Targets at the Federal State Level in Germany

The low-carbon transformation requires a multitude of instruments at different levels of government, from global to local. Germany provides a good example of this

---

[2]See also Michelsen, Neuhoff and Schopp (2015): Using Equity Capital to Unlock Investment in Building Energy Efficiency? DIW Economic Bulletin 19/2015. p. 259–265. DIW Berlin, Germany.

[3]See Projektionsbericht der Bundesregierung (2015), pursuing to regulation NO. 525/2013/EU; BMVI (Hg.) (2014): Verkehrsverflechtungsprognose 2030. Los 3: Erstellung der Prognose der deutschlandweiten Verkehrsverflechtungen unter Berücksichtigung des Luftverkehrs. Intraplan Consult, BVU Beratergruppe Verkehr+Umwelt, Ingenieurgruppe IVV, Planco Consulting; Oeko-Institut & Prognos (2009).

polycentric approach: the federal states (*Länder*) are playing a key role in the energiewende as both drivers and implementers of global and national climate policies. In fact, all 16 federal states have defined their own climate targets, and some of them are now legally binding. Baden-Württemberg, Bremen, North Rhine-Westphalia (NRW), and Rhineland-Palatinate have all signed laws to reduce their GHG emissions in line with concrete targets for 2020 and 2050. Similar agreements or draft laws exist in other federal states (see Table 4.1).

Brandenburg aims at a 72% CO_2 emissions reduction by 2030 (base year 1990) while Saxony intends to reduce CO_2 emissions by 25% by 2020 (base year 2009). These goals are of particular relevance, as electricity production in these two states is based mainly on lignite. Federal states in northern Germany rely primarily on increasing wind power capacities to reduce GHG emissions in the power sector. Bavaria and Baden-Württemberg in the South, on the other hand, are planning to replace their nuclear and coal capacities with a mix of photovoltaic (PV) and gas power plants. All of Germany's federal states, however, have at least some kind of climate agreement targeting emissions reductions, the expansion of renewable energy sources, and the improvement of energy efficiency.

4.2.2.1 Low-carbon Transformation and the Phasing-out of Coal

The low-carbon transformation and the move towards renewables is now a broadly accepted goal in most countries of the Western world. The main challenge for national and international climate policies lies in the need to continuously phase out the remaining global coal-fired power generation.[4] Part of the difficulty in making this transition from fossil-fuel-based electricity generation to renewables lies in the widespread failure to consider the negative externalities of fossil fuels in the costs of power generation. The list of externalities ranges from global effects such as global warming to local contamination from pollutants such as NO_x, SO_2, mercury, small particles, and noise emissions. It also includes groundwater contamination and water pollution (e.g., iron oxides) as well as the relocation of towns and villages to make way for mines, resulting in some cases in thousands of people losing their homes. The New Climate Economy report (2014) has highlighted the negative externalities of coal, and several studies have shown that the monetized negative externalities from coal-fired electricity generation often exceed electricity prices.[5]

---

[4]This section is based on a comprehensive study by Oei et al. (2014a, 2014b) on phasing out coal, in particular lignite.

[5]These costs are paid by society and are therefore not taken into account by the polluting entity. See Ecofys (2014): Subsidies and costs of EU energy. Study for the European Commission; Climate Advisors (2011): The Social Cost of Coal: Implications for the World Bank. Washington, USA; and EC (2003): External Costs. Research results on socio-environmental damages due to electricity and transport. Brussels, Belgium.

Table 4.1 Overview of climate protection laws (top) and other agreements or drafts (bottom) by German federal states

Federal state	GHG Target 2020 (base: 1990)	GHG Target 2050 (base: 1990)
Baden-Württemberg	−25%	−90%
Bremen	−40%	−80 to 95%
North Rhine-Westphalia	−25%	−80%
Rhineland-Palatinate	−40%	−90%
Other climate agreements or drafts for planned climate protection laws		
Bayern	Below 2t CO_2 annually per person until 2050	
Berlin	−40% until 2020, −60% until 2030, −85% until 2050 (base: 1990)	
Brandenburg	−55/62% until 2030 (base: 1990)	
Hamburg	−50% until 2030 (base: 1990)	
Hessen	−30% until 2020, −40% until 2025, −90% until 2050 (base: 1990)	
Lower Saxony	−40% until 2020, −80 to 95% until 2050 (base: 1990)	
Mecklenburg-Western Pomerania	−40% until 2020 (base: 1990)	
Saarland	−80% until 2050 (base: 2005)	
Saxony	−25% until 2020 (base: 2009)	
Saxony-Anhalt	−40% until 2020, −80%/−95%	
Schleswig-Holstein	−40% 2020, −55% 2030, −70% 2040, −80 to 95% 2050 (base: 1990)	
Thuringia	−60/70% until 2030, −70/80% until 2040, −80/90% until 2050 (base: 1990)	

Source: Information based on climate policies of the individual federal states. Baden-Württemberg: http://bit.ly/1KLWkYO; Bremen: http://bit.ly/1PdkBwX; NRW: http://bit.ly/1KLWcZl; Rhineland-Palatinate: http://bit.ly/1dNlWJP; Bayern: https://bit.ly/2s9g3Qk; Berlin: https://bit.ly/2x7fsnP; Brandenburg: https://bit.ly/2GLI99x; Hamburg: https://bit.ly/2s03PtZ; Hessen: http://bit.ly/1c5R9H0; Lower Saxony: http://bit.ly/1yJ0QBk; Mecklenburg-Western Pomerania: http://bit.ly/1EQfLhd; Saarland: https://bit.ly/2saivpu; Saxony: http://bit.ly/1Cc4CJ6; Saxony-Anhalt: https://bit.ly/2GKDIff; Schleswig-Holstein: http://bit.ly/1JQmcFe; Thuringia: https://bit.ly/2s55P4h;. Last accessed May 24, 2018

In the absence of abatement technologies such as CCTS (discussed in more detail in Chaps. 10 and 11), decarbonization of the electricity sector implies phasing out coal altogether (von Hirschhausen et al. 2012). The consensus on the need to phase out coal goes beyond the expert energy community and now extends across the political and social mainstream, as shown by statements from the Group of Seven (Leader of the G7 2015),[6] Pope Francis (2015), and the Islamic community (IICCS 2015). Similarly, the Intergovernmental Panel on Climate Change (IPCC) stated in

---

[6]Leader of the G7. (2015). Leaders' Declaration G7 Summit, June 7–8, 2015. Schloss Elmau, Germany.

its Fifth Assessment Report that it sees no long-term prospects for coal-based power generation (IPCC 2014).

4.3   Significant CO_2 Emissions from Hard Coal and Lignite in Germany

As a traditional coal producing and consuming country, Germany had a pre-energiewende energy mix that was very CO_2-intensive and dominated by hard coal and lignite (see Fig. 4.3). In 2016, coal-fired electricity generation emitted 240 Mt of CO_2, which is equivalent to 80% of total CO_2 emissions from power generation in Germany. Additional pressure for CO_2 mitigation will come with the planned closure of the remaining nuclear power plants by 2022 (9.52 GW in 2018 still operational), creating the need to find substitutes for this power in the electricity mix. Against this backdrop, Germany is running the risk of falling short on its CO_2 emissions reduction targets. The Federal Ministry for the Environment, Nature

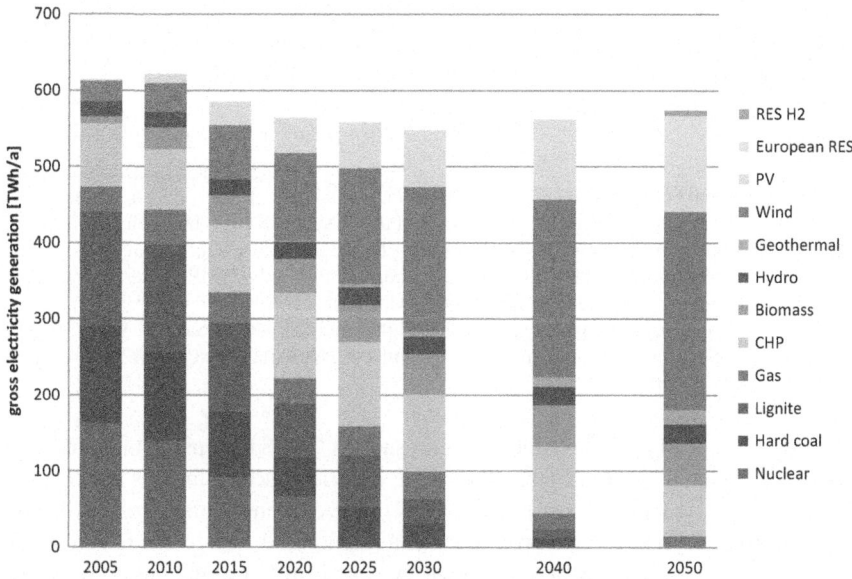

Fig. 4.3 Generation mix in the German electricity sector from 2005 to 2050. Source: BMU. 2012. "Langfristszenarien und Strategien für den Ausbau der erneuerbaren Energien in Deutschland bei Berücksichtigung der Entwicklung in Europa und global." Schlussbericht BMU-FKZ 03MAP146. Stuttgart, Germany: Deutsches Zentrum für Luft- und Raumfahrt (DLR), Stuttgart Institut für Technische Thermodynamik, Fraunhofer Institut (IWES), Kassel Ingenieurbüro für neue Energien (IFNE)

Conservation, and Nuclear Safety[7] and the German Advisory Council on the Environment (SRU 2015) have both stated that for Germany to meet these targets, a coal phase-out will have to take place in the 2040s (see Fig. 4.3).

4.3.1 Electricity Generation from Hard Coal

In 2017, a total of 93 TWh of electricity was generated by the 25 GW of Germany's hard-coal-fired power plants (compared to 112 TWh in 2016). Most of these plants are located near rivers in North Rhine-Westphalia (around 13 GW) and Baden-Württemberg (around 5 GW) or near the coast of the North Sea and the Baltic Sea. The majority of hard coal power plants that were still active in 2018 were constructed in the 1980s. Only 2.3 GW of new capacities came online between 1990 and 2010. But in the 2010s, Germany's major energy utility companies RWE, E.ON, Vattenfall, and Steag again began investing in new hard-coal-fired units, based on an underestimation of the speed of the energiewende and an overestimation of future demand (see Fig. 4.4) (Kungl 2015). Increasing shares of renewable energy sources (from 9% in 2004 to 26% electricity production in 2014) reduced the residual electricity demand. The resulting overcapacities of conventional power plants together with decreasing EU-ETS certificate prices and low global coal prices caused lower wholesale prices and reduced the load factor of the entire fleet. The average load factor of hard-coal-fired power plants dropped to 42% in 2017 (from 50% in 2005) compared to an unchanged high load factor of 80% for lignite power plants. As a result, operators faced some sizeable impairment losses on hard-coal-fired power plants. In addition, stricter environmental regulations, construction problems, and opposition from affected residents delayed the construction of some new coal power plants. Rising costs led to some of these projects being shelved. Low wholesale electricity prices also resulted in the closure of several older units that had become unprofitable due to low efficiency rates. This pattern is very likely to continue in the near future: Older, less efficient hard-coal-fired units will be the first ones to be overtaken in the merit order by gas-powered units if the price of CO_2 allowances increases. The overall setting makes retrofitting hard coal power plants uneconomical and therefore leads to a continuous market-driven phase-out of hard coal electricity in Germany (Oei et al. 2014b; Göke et al. 2018).

---

[7]BMU. 2012. "Langfristszenarien und Strategien für den Ausbau der erneuerbaren Energien in Deutschland bei Berücksichtigung der Entwicklung in Europa und global." Schlussbericht BMU-FKZ 03MAP146. Stuttgart, Germany: Deutsches Zentrum für Luft- und Raumfahrt (DLR), Stuttgart Institut für Technische Thermodynamik, Fraunhofer Institut (IWES), Kassel Ingenieurbüro für neue Energien (IFNE).

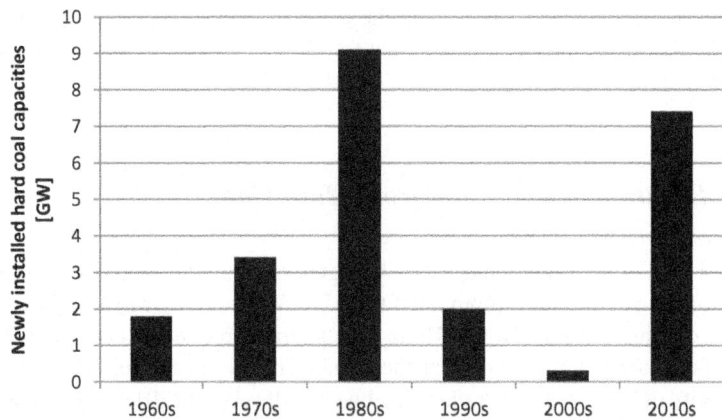

Fig. 4.4 Installation of hard coal power plants in Germany. Source: Own graph based on BNetzA (2018) power plant database. BNetzA. 2018. *Kraftwerksliste,* February 2018. Bonn, Germany: Bundesnetzagentur

4.3.1.1 Electricity Generation from Lignite

In 2015, more than 60 lignite-fired units with an overall capacity of around 20 GW are located mainly in the Rhineland (around 10 GW), in central Germany (around 3 GW), and in Lusatia (around 7 GW) (see Fig. 4.5). In 2017, lignite-based power generation declined slightly compared to the previous years, totaling around 147.5 TWh in 2014 (22.5% of electricity generation).

A rapid reduction of lignite power generation appears inevitable in Germany in light of the long-term climate targets agreed upon at the national, European, and global levels. As a resource that emits 1161 g CO_2/kWh per unit of electricity produced, lignite is by far the largest producer of greenhouse gas emissions in the German energy mix (hard coal: 902 g CO_2/kWh; natural gas: 411 g CO_2/kWh).[8] With annual emissions of 157 $MtCO_2$, lignite makes up around 47% of the emissions of the German power sector and is therefore incompatible with GHG reduction targets of 80–95% by 2050. Analyses of power plant and grid capacity for the mid-2020s also show that lignite will become less important in Germany's energy mix in the future (see Gerbaulet et al. 2012[9] and Mieth et al. 2015).

---

[8]The average CO_2 emission factors refer to power consumption for the year 2010, see UBA (2013): Entwicklung der spezifischen Kohlendioxid-Emissionen des deutschen Strommix in den Jahren 1990 bis 2012. Petra Icha, Climate Change 07/2013. More modern plants, in contrast, emit around 940 g/kWh for lignite, 735 g/kWh for hard coal, and 347 g/kWh for natural gas-based power plants, see UBA (2009): Klimaschutz und Versorgungssicherheit. Entwicklung einer nachhaltigen Stromversorgung, Climate Change 13.

[9]Gerbaulet, Clemens, Jonas Egerer, Pao-Yu Oei, and Christian von Hirschhausen. 2012. "Abnehmende Bedeutung der Braunkohleverstromung: weder neue Kraftwerke noch Tagebaue benötigt." DIW Wochenbericht 79 (48): 25–33.

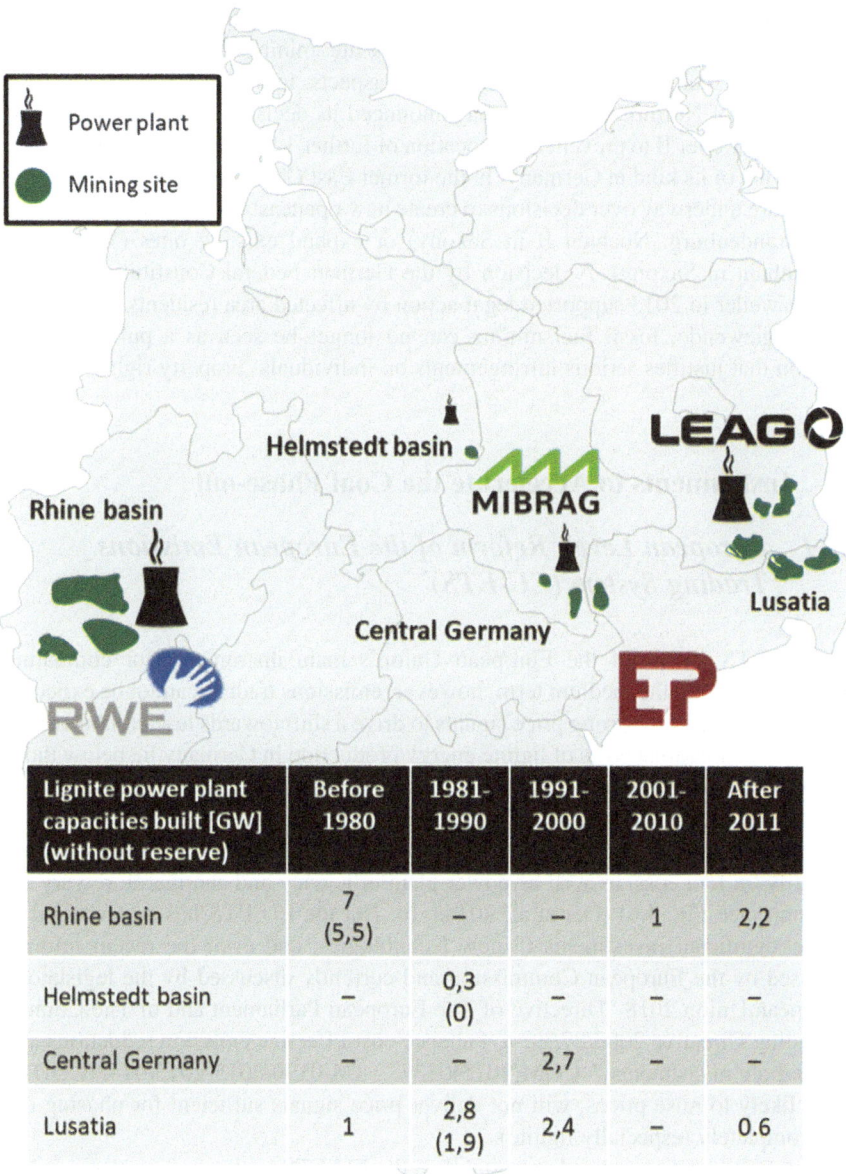

Lignite power plant capacities built [GW] (without reserve)	Before 1980	1981– 1990	1991– 2000	2001– 2010	After 2011
Rhine basin	7 (5,5)	–	–	1	2,2
Helmstedt basin	–	0,3 (0)	–	–	–
Central Germany	–	–	2,7	–	–
Lusatia	1	2,8 (1,9)	2,4	–	0.6

Fig. 4.5 Remaining lignite basins, power plants and operators in Germany. Source: Own illustration based on BNetzA (2018) power plant database. BNetzA. (2018). *Kraftwerksliste, 2018.* Bonn, Germany: Bundesnetzagentur

Given the uncertain future of lignite-based power generation, it is hardly surprising that there is controversy surrounding lignite mining districts on issues of employment, reallocation, and environmental aspects. In March 2014, the coalition government of Northrhine-Westphalia announced its decision to reduce the mining area at Garzweiler II to prevent the relocation of further 1400 residents. This decision was the first of its kind in Germany. In the former East German federal states, similar debates are underway over decisions to create new opencast mines (Welzow-Süd TF II in Brandenburg, Nochten II in Saxony) or expand existing ones (Vereinigtes Schleenhain in Saxony). A decision by the German Federal Constitutional Court on Garzweiler in 2013 supported legal action by affected area residents. In times of the energiewende, fossil fuel mining can no longer be seen as a public interest decision that justifies serious infringements on individuals' property rights.[10]

4.4 Instruments to Accelerate the Coal Phase-out

4.4.1 European Level: Reform of the European Emissions Trading System (EU-ETS)

The EU-ETS is one of the European Union's main instruments for combating climate change. In the medium term, however, emissions trading cannot be expected to provide sufficiently strong price signals to drive a shift towards low-carbon energy sources. The marginal costs of lignite energy production in Germany lie below those of combined-cycle gas turbine (CCGT) power plants as long as CO_2 prices do not exceed €40–50/tCO_2. The prices of switching from older hard coal power plants to new gas power units are in the range of €20–40/tCO_2. These switch prices depend primarily on fuel costs as well as power plant efficiency and can therefore vary for each unit (see Fig. 4.6) (Oei et al. 2014a, b). But the EU-ETS has so far failed to induce significant investments in new technologies; and even the recent reforms proposed by the European Commission and currently discussed by the legislators, (European Union 2018 "Directive of The European Parliament and of The Council Amending Directive 2003/87/EC to enhance cost-effective emission reductions and low-carbon investments." COM(2015)0337 – C8-0190/2015-2015/0148(COD).), while likely to raise prices, will not deliver price signals sufficient for phasing out coal completely (especially lignite).

Thus, while action is needed to stabilize the EU-ETS in the medium term, it has also become clear that it cannot be the only instrument to promote decarbonization at the European level. In 2013, the structural surplus of certificates exceeded the

---

[10]See Ziehm (2014): "Neue Braunkohlentagebaue und Verfassungsrecht – Konsequenzen aus dem Garzweiler-Urteil des Bundesverfassungsgerichts." Expert report commissioned by Alliance '90/ The Greens.

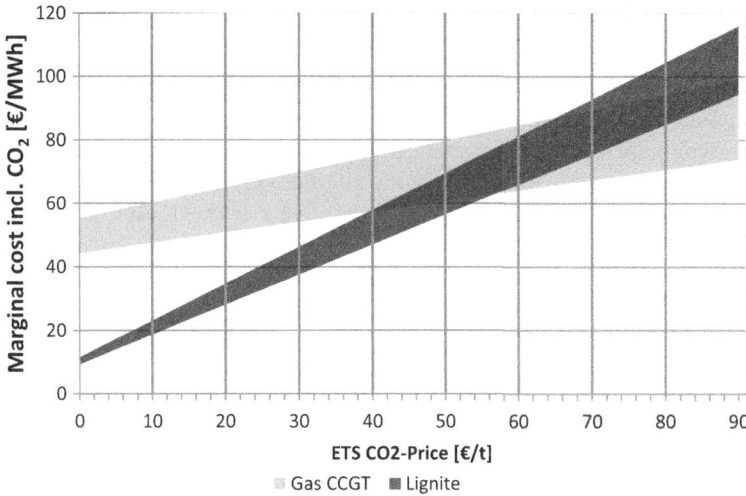

Fig. 4.6 Marginal cost of lignite and gas-fired combined-cycle gas turbine (CCGT) power generation depending on the CO_2 price. Source: Oei et al. (2014a, b)

allowances for more than 2 bn tCO_2. The EC (2014)[11] expects the surplus to remain at the same magnitude at least until the end of the third trading period in 2020. Canceling this surplus would have been an important signal for the EU to maintain its credibility and capacity to steer the EU-ETS. This proposal, however, did not receive sufficient political support on the EU level due to opposition from countries including Poland. The newest reform of the EU-ETS instead includes a regular stocktaking with regard to the Paris agreement and a strengthening of the linear reduction factor of the emissions cap from 1.74% to 2.2%. It also comprises a reform of the market stability reserve, leading to a quicker reduction of the banking surplus by a doubling of the intake rate and a cancellation of those allowances in the market stability reserve that exceed the previous year's emissions (from 2023). Finally, national governments may reduce the amount of allowance they auction if they implement additional national measures that lead to the closure of electricity generation capacity within their territory. Especially the last point would allow additional national measures taken with regard to coal generation capacities to be fully effective as an emission reducing policy also at the European level. Nevertheless, the expected certificate surplus leads us to believe that the European system will have a limited impact on compliance with short- and medium-term national emissions targets. For this reason, additional national instruments are under discussion that could be introduced in parallel to emissions trading.

---

[11]EC. (2014). *Questions and answers on the proposed market stability reserve for the EU emissions trading system.* Brussels, Belgium: European Commission.

4.4.2   Towards More Specific Climate Instruments

It is clear that if the GHG targets set out by the German government are to be met, additional action will be required in all sectors, including electricity. The government's thinking on this topic has evolved, and as the discussion on GHG targets has developed, its approach has become more specific. Previously the German government either focused on the overall EU-ETS targets at the European level or on national non-ETS targets, but now the discussion also includes specific national targets for the electricity sector. The "grand coalition" of Christian Democrats and Social Democrats governing the country agreed on a Climate Action Plan (*Aktionsprogramm Klimaschutz 2020*) in 2014 to counteract the rise of emissions in 2012–2014 and restrict coal usage. Moreover, according to an analysis by Agora Energiewende, a reduction of lignite and hard coal-based power generation by 62% and 80% by 2030 is needed to achieve the climate targets.[12] Reducing power sector emissions also plays a major role in the national Climate Protection Plan 2050 (*Klimaschutzplan 2050*) published in 2016.

The German government is therefore using different instruments to combat climate change at the national and EU levels, and has developed a variety of different mechanisms to complement these instruments. The aim of this polycentric approach is not to establish mutually exclusive instruments and mechanisms, but rather to take action in several areas simultaneously. The German government cites three possible courses of action: greater commitment outside the framework of the EU-ETS, a focus on an ambitious structural reform of the EU-ETS, and accompanying measures within the context of the energiewende.[13]

Support for a reduction of coal power plant's utilization has been expressed by different players such as Energie Baden-Württemberg (EnBW) and 70 municipal utilities.[14] These companies would profit from higher load factors for their gas utilities and the rise in wholesale electricity prices. The energy-intensive industries, on the other hand, benefit from low wholesale prices and are therefore opposed to any measures that might lead to a price increase. The major argument from these branches of industry is a fear of deindustrialization, as Germany would no longer be able to compete with lower production costs in foreign countries. Various studies,

---

[12]See Graichen and Redl (2014): Das deutsche Energiewende-Paradox: Ursachen und Herausforderungen; Eine Analyse des Stromsystems von 2010 bis 2030 in Bezug auf Erneuerbare Energien, Kohle, Gas, Kernkraft und CO_2-Emissionen. Agora Energiewende. Berlin.

[13]BMUB. 2014. "Aktionsprogramm Klimaschutz 2020: Eckpunkte Des BMUB." Berlin, Germany: Bundesministerium für Umwelt, Naturschutz, Bau und Reaktorsicherheit.

[14]Handelsblatt (2015): Stadtwerke gegen RWE http://www.handelsblatt.com/politik/deutschland/klimaabgabe-plaene-stadtwerke-gegen-rwe/11677972.html; Süddeutsche Zeitung (2015): Dicke Luft in der Strombranche http://www.sueddeutsche.de/wirtschaft/klimaschutz-dicke-luft-in-der-strombranche-1.2502249, last accessed September 20, 2016.

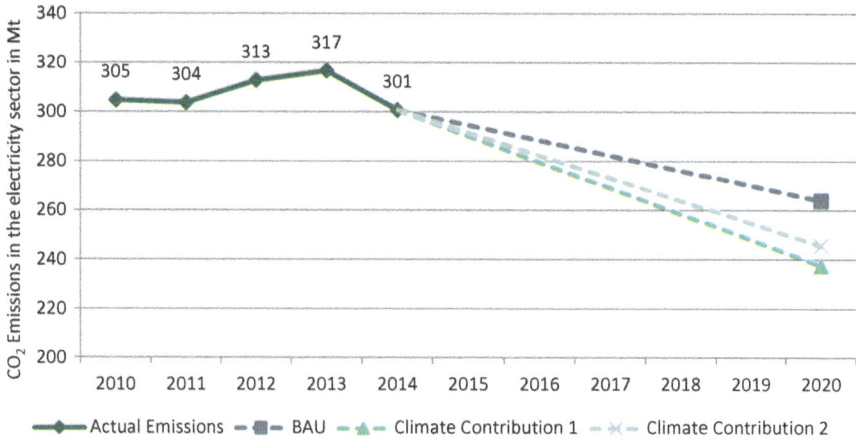

Fig. 4.7 CO_2 emissions in Germany for two different climate contribution settings compared to a BAU scenario without any levy. Source: Oei et al. (2015a, b)

however, have shown that a moderate increase in electricity prices would have only limited effects on the competitiveness of German industry.[15]

4.4.2.1   Plans for a "Climate Contribution Levy" and the Introduction of a "Coal Reserve"

Along the lines of a minimum CO_2 price, the German Ministry for Economy and Energy (BMWi) in 2015 proposed the introduction of a "climate contribution" to achieve a reduction of 22 $MtCO_2$, in addition to the reduction foreseen in the business-as-usual or BAU scenario (from the *Projektionsbericht* submitted to the EU). The "climate contribution" was an additional financial levy paid by power plant operators to the German state addressing primarily old and CO_2-intensive coal power plants. A level of 18 €/tCO_2, in combination with a free allocation of 3–7 $MtCO_2$/GW of plant capacity (depending on the age of the plant) would have been appropriate to assure a 22 $MtCO_2$-reduction by 2020. Figure 4.7 shows the effects of different parameterizations of the climate contribution and the corresponding effect on the reduction of CO_2-emissions compared to the BAU scenario without the fee. A reduction of the climate contribution, e.g. in the range of 12–16€/tCO_2, and/or an increase of the free allocation to older power plants, would weaken the effects. The climate contribution includes the option for power operators to emit beyond their free allocation levels when decommissioning additional EU-ETS CO_2-certificates (Oei et al. 2015b). The introduction of the climate contribution, similarly to most of

---

[15]See Agora Energiewende (2014): Comparing Electricity Prices for Industry. Analysis. An Elusive Task – Illustrated by the German Case. Berlin; and Neuhoff et al. (2014): Energie- und Klimapolitik: Europa ist nicht allein. (DIW Wochenbericht Nr. 6/2014) DIW Berlin.

the other described additional measures would have mainly affected older and CO_2-intensive lignite power plants in NRW and Lusatia (see Fig. 4.7).

However, in 2016 the BMWi shelved the idea of the climate contribution and instead introduced a "coal reserve" aimed at reducing CO_2 emissions by around 10 mn t until 2020. The reserve consists of eight relatively old lignite units with an overall capacity of 2.7 GW (see Fig. 4.5). The operators agreed that all units would be shut down entirely after the reserve period of four years. The technical requirements for this "coal reserve" such as an early notification period of 11 days prior to plant activation as well as the existing overcapacities in the German and European electricity market, however, make it unlikely that this reserve will ever be activated. In addition, the majority of units would have been closed anyhow in the upcoming years and therefore the reserve payments of in total 1.6 bn € can be seen as a scrappage bonus (see Table 4.2).

Table 4.2 Newly introduced "coal reserve" in Germany

Owner	Unit	Power [MW]	Age in 2020	Reserve start (shut down after 4 years)	Particularities
Mibrag/ EPH	Buschhaus	352	35	10/2016	Was moved into reserve in 09/2016 as the mining site was fully exploited. Next site is 150 km away, resulting in higher variable costs.
RWE	Frimmersdorf P	284	54	10/2017	Last two (out of eight) units; facing economic problems for several years.
	Frimmersdorf Q	278	50	10/2017	
	Niederaußem E	295	50	10/2018	Were already listed in the official list of expected closures "Kraftwerksliste Bundesnetzagentur zum erwarteten Zu- und Rückbau 2015 bis 2019" with the closing date 2019
	Niederaußem F	299	49	10/2018	
	Neurath C	292	47	10/2019	Similar efficiency factors as other 300 MW units and near its technical lifetime.
LEAG/ EPH (Vattenfall)	Jänschwalde E	465	33	10/2018	Most recent units at the site Jänschwalde (start of operation of the six units 1981–1989); it is sometimes easier to start shutting down the last units first.
	Jänschwalde F	465	31	10/2019	

Source: Oei, Pao-Yu, Clemens Gerbaulet, Claudia Kemfert, Friedrich Kunz, and Christian Hirschhausen. 2016. "'Kohlereserve' vs. CO_2-Grenzwerte in der Stromwirtschaft – Ein modellbasierter Vergleich." Energiewirtschaftliche Tagesfragen 66 (1/2): 57–60; The plant in Jänschwalde was bought by LEAG in 09/2016 from Vattenfall

4.4.2.2 Continuous Need for a Coal Phase-out

To negotiate the details of a German coal phase-out, a non-partisan, structured dialog process with key stakeholders has been suggested.[16] Concrete aspects to achieve a coal phase-out at the latest by 2040 include:

- The establishment of a 'Round Table on a National Consensus on Coal' with key stakeholders, similar to the approach taken with nuclear power.
- A set end date for coal as well as a phase-out trajectory enshrined in law.
- No new construction of power plants, no additional lignite mines and no more related relocations.
- A cost-efficient decommissioning plan with flexibility options between lignite mining regions and operators to avoid domino effects (between mines and power plants).
- The creation of a foundation for the follow-up costs of lignite mining, payed for by the operators.
- The implementation of a 'Structural Change Fund' over €250 million, payed for by the federal budget, to support affected regions.
- Safeguarding security of supply, as well as the economic competitiveness of the German economy and in particular the energy intensive industry.
- CO_2 certificates which are set free are retired immediately to strengthen the EU ETS.

The focus on phasing-out coal would need to be accompanied by an acceleration of renewable energy capacity expansion, as well as support for lignite regions, that need to cope with the coal exit. A fund for structural changes would need to provide both financial as well as capacity building support, to strengthen the economy, science and research, improve infrastructure and help civil society adapt to the changes (see Agora Energiewende 2017[17] and Herpich et al. 2018).

In the 2017 general election campaign, Angela Merkel promised that Germany would meet its 2020 GHG emission reduction target of −40% compared to 1990 levels. To achieve these emission reductions, older and more inefficient coal-fired power plants would need to be closed by 2020. In the climate protection plan 2050, sectoral targets for 2030 have been set: The energy sector will need to reduce emissions to 175–183 million t CO_2-eq from 358 mt CO_2-eq in 2014.[18] Cutting the sector's emissions by half will require further coal-fired power plant closures (80% of electricity generation CO_2 emissions can be attributed to coal[19]).

---

[16]Agora Energiewende. 2016. Elf Eckpunkte für einen Kohlekonsens. Konzept zur schrittweisen Dekarbonisierung des deutschen Stromsektors (Langfassung). Impulse, Berlin.

[17]Agora Energiewende. 2017. Eine Zukunft für die Lausitz: Elemente eines Strukturwandelkonzepts für das Lausitzer Braunkohlerevier. Impulse, Berlin.

[18]BMUB. 2014. "Aktionsprogramm Klimaschutz 2020: Eckpunkte Des BMUB." Berlin, Germany: Bundesministerium für Umwelt, Naturschutz, Bau und Reaktorsicherheit.

[19]See Umweltbundesamt. 2017. "Entwicklung der Kohlendioxid-Emissionen der fossilen Stromerzeugung nach eingesetzten Energieträgern." Available online.

Germany's coal sector is not only being challenged by domestic regulations but also a growing global movement tackling climate change and coal. E.g., in 2017, during COP 23 in Bonn, the "Powering Past Coal" alliance has been announced. The United Kingdom and Canada, as well as more than 20 other states and regions pledged to end coal consumption. On a European level, the "Coal Regions in Transition Platform" addresses upcoming changes in former coal mining regions. Another aspect that could reduce economic viability for coal in Germany, is French President Macron's initiative for an EU-wide minimum CO_2-price.

4.4.3 National Level: A Variety of Instruments

An important aspect of the German coal phase-out will be the choice of an appropriate policy instruments to structure the phase-out process. Some countries in the EU and North America have taken the initiative by adopting complementary measures: the UK (CO_2 emissions performance standards, EPS, and a CO_2 price floor), the USA (EPS and an additional retirement plan for older plants), and Canada (EPS). In the following, we analyze policies designed to reduce German power sector GHG emissions in general and to phase out coal in particular. Possible accompanying measures to reduce coal-based power generation in Germany include minimum fuel efficiency or greater flexibility requirements, national minimum prices for CO_2 emissions allowances, capacity mechanisms, a residual emissions cap for coal-fired power plants, emissions performance standards, and policies regulating transmission grids (see Table 4.3). In Germany, these could be implemented in parallel to the desired EU-ETS reform and are described in more detail below.

4.4.3.1 Emissions Performance Standards

In addition to the EU-ETS, another means of tackling the emissions problem is the introduction of CO_2 limits in the form of an EPS. Following Canadian and Californian initiatives, the UK has already incorporated this measure into an amendment of its Energy Act adopted in December 2013.[20] The UK EPS prevents the construction of new unabated coal-fired power plants, that is, units that do not make use of CCTS. The Canadian EPS also applies to existing power plants when they reach the age of 45–50, depending on the year of their commissioning. The introduction of an EPS in EU Member States (and thus also in Germany) is in conformity with European Law as set out in Article 193 of the Treaty on the Functioning of the European Union (TFEU).[21]

In a study on the potential effects of an EPS in Germany, we quantified the effects of a CO_2 emissions limit of 450 g CO_2/kWh for newly constructed as well as

---

[20]The Parliament of Great Britain. Energy Bill, HL Bill 30. The Stationary Office, London, UK (2013).

[21]See Ziehm and Wegener (2013): Zur Zulässigkeit nationaler CO_2-Grenzwerte für dem Emissionshandel unterfallende neue Energieerzeugungsanlagen. Deutsche Umwelthilfe. Berlin.

Table 4.3 Possible instruments for reducing coal-based power generation

Proposed measure	Expected effect	Possible advantages	Possible shortcomings
EU-ETS reform	Price signal through: introduction of market stability reserve (MSR) in 2019 instead of 2021; 900 mn EUA from backloading directly in MSR; increase of intake rate to 24% until 2023; invalidation of certificates in the reserve; possibility for voluntary reduction of auctioned certificates in case of national policy-induced power plant closures	EU-wide instrument; thus, no cross-border leakage effects targets several sectors besides electricity	Structural reforms uncertain from today's perspective; the timing of the impact is unpredictable due to high surplus of certificates
CO_2 floor price	CO_2 certificates would become more expensive	Investment security for operators	Feasible prices probably too low to result in a switch from lignite to natural gas in the short term
Minimum efficiency	Closure of inefficient power plants	More efficient utilization of raw materials	Open-cycle gas turbines (OCGT) could also be affected; complex and time-consuming test and measurement processes
Flexibility requirements	Closure or singling out of inflexible power plants	Better integration of fluctuating renewable energy sources	Combined-cycle gas turbines (CCGT) could also be affected; complex and time-consuming test and measurement processes
Coal phase-out law	Maximum production [TWh] or emissions allowances [tCO_2] for plants	Fixed coal phase-out plan & schedule investment security	Outcome of auctioning of allowances would be difficult to predict
Emissions performance standard (per unit; for new plants and retrofits)	Restrictions for new plants and retrofits (without CO_2 capture) [<x g/MWh]	Prevention of CO_2-intensive (future stranded) investments	Minor short-term reduction in emissions
Emissions performance standard (emissions cap for existing plants)	Reduce load factor for depreciated coal-fired power plants (e.g., >30y) [<x g/MW]	Preservation of generation capacities	Negative impact on economic efficiency of power plants might lead to closure of older blocks

(continued)

Table 4.3 (continued)

Proposed measure	Expected effect	Possible advantages	Possible shortcomings
Capacity mechanisms or reserve for coal plants	Incentive for construction of less CO_2-intensive power plants when including environmental criteria	Support for gas power plants or moving of coal power plants into a reserve to reduce their emissions and prevent supply bottlenecks	Difficult to establish criteria that are in line with EU state aid laws if payments should only be given to selected units
Reduced transmission grid expansion	Increased congestion might prohibit lignite electricity generation in times of high renewable energy production	Redispatch of less CO_2-intensive capacities; lower investment costs for transmission lines	Transmission grids might be needed for renewables in the long run
Climate contribution fee	Additional levy for old CO_2-intensive power plants	Limiting output of most CO_2-intensive generation facilities; preserving capacities; compatible with EU-ETS	Older units might become uneconomical if the fee is too high

Source: Updated from Oei et al. (2014a, b) (with original references)

retrofitted plants (Ziehm et al. 2014).[22] This provision would put a halt to the construction of new coal-fired power plants. In addition, existing plants that have been in operation for 30 years or more could be subject to an annual emissions cap.[23] Such regulations are aimed particularly at the oldest and least efficient power plants. In this case, the performance standard involves limiting the maximum net annual emissions to ~3000 t CO_2/MW.[24] Depending on the given emissions factor and efficiency of individual plants, this is equivalent to a load factor of around 90–100% for CCGT power plants, 40–50% for hard-coal-fired power plants, and around 30–40% for lignite power plants. Separate regulations would be applicable to combined heat and power (CHP) plants. In this scenario, hard-coal-fired power plants with a total output of around 10.5 GW and lignite plants with around 9.5 GW would be affected by a regulation for existing plants starting in 2015. The annual power generation of these plants would thus fall by 45 TWh. The net emissions reduction effect depends on whether these generation volumes are replaced by additional renewable capacities, gas generation with lower CO_2

---

[22]Ziehm, C., Kemfert, C., Oei, P.-Y., Reitz, F., & v. Hirschhausen, C. von. (2014). Entwurf und Erläuterung für ein Gesetz zur Festsetzung nationaler CO_2-Emissionsstandards für fossile Kraftwerke in Deutschland (Politikberatung kompakt No. 82). Berlin, Germany: DIW Berlin.

[23]According to plans for the phase-out of nuclear energy in Germany, the 30-year limit is calculated based on amortization plus a given profit realization period.

[24]Calculation basis: gas power plant emissions data (450 g CO_2/kWh), the total annual operating hours at 80% capacity: 450 g CO_2/kWh × 8760 h × 0.8 = 3154 t CO_2/MW.

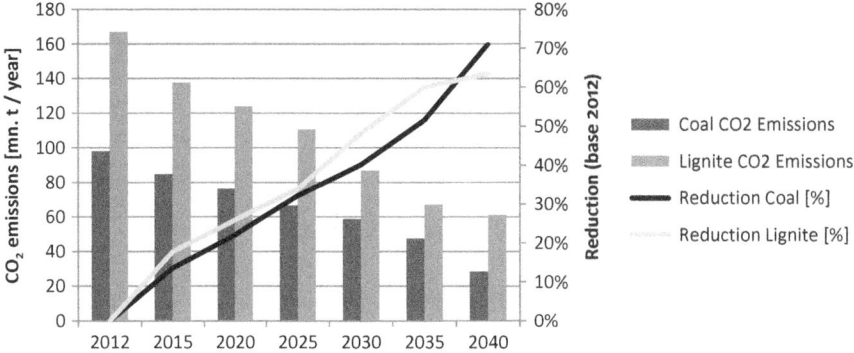

Fig. 4.8 Effect of an Emissions Performance Standard (EPS) on coal electrification in Germany. Source: Ziehm, C., Kemfert, C., Oei, P.-Y., Reitz, F., & v. Hirschhausen, C. von. (2014). Entwurf und Erläuterung für ein Gesetz zur Festsetzung nationaler CO_2-Emissionsstandards für fossile Kraftwerke in Deutschland (Politikberatung kompakt No. 82). Berlin, Germany: DIW Berlin

volumes, or an increase of newer unrestricted hard coal units.[25] The number of coal-fired power plants falling under this regulation would increase over time since neither retrofit measures nor the construction of new plants would be allowed. The implementation of an EPS therefore leads to a continuous reduction of coal generation as well as CO_2 emissions (see Fig. 4.8).

4.4.3.2 Carbon Floor Price

To enhance the efficiency of the EU-ETS, a minimum price for CO_2 emissions could be set at the EU level. However, national governments could also set their own individual minimum prices to help meet climate targets. In 2013, for example, the UK introduced an additional tax on CO_2 emissions in the power sector known as the Carbon Price Floor (CPF). Together, the tax and CO_2 price create a "minimum price" for CO_2 emissions. For the 2013/2014 financial period, the minimum price was £16 (around €20) for each ton of CO_2 emitted.[26]

In Germany, the introduction of a minimum CO_2 price in the form of an additional tax on the purchase of CO_2 emissions allowances, as proposed in a bill

---

[25]A reduction of German production also reduces net exports and consequently increases generation and emissions in neighboring countries. A more recent study shows that the net CO_2 reduction effect in the European electricity sector is around 50% of the German reduction when introducing a national EPS (Oei et al. 2015a).

[26]See HM Revenue & Customs (2014): Carbon price floor: reform and other technical amendments. Originally, the CPF was to increase linearly to 30 £/t by 2020/2021, but this figure was frozen at 18 £/t for the rest of the decade. The reason for this decision was the large gap between the CPF and the CO_2 price in the EU-ETS scheme, which might have had a negative impact on the competitiveness of the UK's domestic industry.

by Alliance '90/The Greens, would be possible.[27] Under energy tax laws in Germany, power plant operators are exempt from the existing energy tax, and plans are in place to remove this tax altogether. In all likelihood, however, a government-fixed minimum price on CO_2 emissions would have very little impact on coal-based power generation unless switch prices to gas are being met (see Fig. 4.6).

4.4.3.3 Minimum Efficiency and Greater Flexibility Requirements

Innovations in the energy sector have focused on increasing efficiency levels. This was intended mainly, however, to promote competition and not to create regulatory measures. However, further advances due to coal pre-drying or retrofit measures would only lead to insignificant increases in efficiency of a few percent. In Germany, for instance, a bill to introduce a minimum efficiency level put forward in the German Bundestag by the parliamentary group Alliance '90/The Greens in 2009 failed.[28] The bill proposed an amendment to the Federal Immission Control Act (*Bundesimmissionsschutzgetz*, BImSchG) that would have required all newly built power plants to have a minimum efficiency of 58%. Existing hard coal and lignite power plants would have had to have a minimum efficiency factor of 38 and 36%, respectively. In 2020, these figures were to be increased to 40 and 38%. The existing legal hurdle for efficiency requirements was also to be removed. At 40% efficiency and above, the introduction of minimum efficiency levels for power plants, including existing plants, would affect more than 10 GW of lignite and 10 GW of hard coal capacity in Germany. However, if a general, non-technology-specific minimum efficiency requirement were to be introduced, this would affect not only coal-fired power plants but also open-cycle gas turbines (OCGT) that have similar efficiencies to coal-fired power plants. Owing to their flexibility, however, open-cycle gas turbines are an essential part of an energy mix based on a high percentage of fluctuating renewable energy sources.

Given the steady increase in the share of RES in the German energy mix, the flexibility of conventional power plants is becoming increasingly important. The key benchmarks for flexibility are the short-term ability to change production levels, minimum must-run generation, the start-up as well as ramping times, and the minimum run-time of a power plant. Irrespective of what fuel is used, steam power plants in particular face certain technical restrictions. CCGT plants use the waste heat generated by the gas turbine to fuel a secondary steam process and

---

[27] A Climate Change Act bill recently proposed by the parliamentary group Alliance '90/The Greens calls for the introduction of a minimum price for CO_2 similar to that in the UK. According to the bill, the CO_2 price was to start at €15/t in 2015 and increase by €1/t per annum up to 2020, See Deutscher Bundestag (2014): Entwurf eines Gesetzes zur Festlegung nationaler Klimaschutzziele und zur Förderung des Klimaschutzes (Klimaschutzgesetz), Bundestag printed paper 18/1612.

[28] See Deutscher Bundestag (2009): Neue Kohlekraftwerke verhindern – Genehmigungsrecht verschärfen: Beschlussempfehlung und Bericht des Ausschusses für Umwelt, Naturschutz und Reaktorsicherheit.

Table 4.4 Technical properties of natural gas and coal power plants

	Ramp-up [h]	Min load [%]	Efficiency at full nominal power P_n [%]	Efficiency at 50% nominal power P_n [%]
OCGT	<0.1	20–50	30–35	27–32
CCGT normal	0.75–1.0	30–50	58–59	54–57
CCGT flexible	0.5	15–25	>60	52–55
Coal normal	2–3	40	42–45	40–42
Coal flexible	1–2	20	45–47	42–44

Source: VDE (2012)

therefore reach higher efficiency values. They are not, however, as flexible as OCGT, which run without steam. Both the minimum generation (must-run) and the maximum start-up times of CCGT plants are therefore similar to those of coal-fired power plants (see Table 4.4) (VDE 2012).[29]

Minimum efficiency and flexibility requirements would affect either OCGT or CCGT power plants in addition to coal-fired power plants. These instruments are therefore not ideally suited to reducing coal-based power generation unless they are introduced as fuel-specific.

4.4.3.4 Coal Phase-out Law

A coal phase-out law sets a fixed phase-out schedule based on (1) a limit for full load hours or (2) CO_2 emissions. A specific scenario on the basis of full load hours for coal power plants was described in a study conducted by Ecofys.[30] The alternative option is CO_2 allowances that are allocated to the individual power plants on the basis of "historical" emissions (free allocation) or by means of individual auctions. A coal phase-out law can include the option of transferring remaining full load hours or CO_2 emissions from one power plant to another. Transferring run-times between lignite plants also affects the extraction in the respective open-cast mines, which could result in additional relocations of people living in the area. A conceivable solution would be to impose requirements that a transfer of emissions permits is only allowed if the new configuration does not lead to a higher number of needed relocations.

---

[29]VDE. (2012). Erneuerbare Energie braucht flexible Kraftwerke – Szenarien bis 2020. Frankfurt am Main, Germany: VDE Verband der Elektrotechnik Elektronik Informationstechnik e.V. – Energietechnische Gesellschaft im VDE (ETG).

[30]See Klaus et al. (2012): Allokationsmethoden der Reststrommengen nach dem Entwurf des Kohleausstiegsgesetzes – Verteilung der Reststrommengen und Folgenabschätzung für den Kohlekraftwerkspark; Studie von Ecofys im Auftrag von Greenpeace.

4.4.3.5 Introducing Capacity Mechanisms

Elements of climate policies can be taken into account in the design of capacity mechanisms. Capacity mechanisms such as a capacity reserve include payment for selected capacities to secure resource adequacy of electricity generation. One example is the German Climate Action Plan 2050 announced in 2015, which includes an explicit reference to coal policy and provides a platform for negotiations with the operators to reduce CO_2 emissions.[31] The configurations of capacity mechanisms strongly affect the energy mix and, consequently, the CO_2-intensity of future power generation. Discussions surrounding capacity mechanisms therefore have to take climate policy into account. Put simply, the more the existing power plant fleet is being supported, the more CO_2-intensive the future fleet will be. Having an instrument to promote less CO_2-intensive gas power plants (for example, by establishing minimum flexibility requirements or EPS as criteria), however, would help make these plants more profitable.[32]

It would also be possible to transfer coal-fired power plants into a capacity reserve of some kind. Such a reserve would help cut emissions while retaining capacity. In turn, investment incentives for gas power plants would increase, and power plant operators would be given compensation for complying with the given capacity requirements. We use a detailed model of the German electricity market to simulate a range of different scenarios of closing down coal power plants (Reitz et al. 2014).[33] The main scenario consists of the additional closure of 3 GW of hard coal and 6 GW of lignite plants, leading to about 23 Mt of avoided CO_2 emissions. Lignite power would lose substantially (-40 TWh), whereas natural gas would benefit ($+26$ TWh). Hard coal, too, would slightly increase generation ($+13$ TWh). A second scenario assumes a shutdown of 3 GW of hard coal and 10 GW of lignite capacities resulting in an emission reduction of 35 $MtCO_2$ (see Fig. 4.9). With increasing wholesale prices, the EEG surcharge declines, so that consumer prices would be less affected than the wholesale price.[34] We conclude that a structured shutdown of old and inefficient coal plants facilitates the accomplishment of GHG reduction goals,

---

[31]In the Netherlands, for example, agreements were made with individual operators, who, after a Dutch tax on coal electrification was abolished, agreed to the closure of several older coal-fired power plants with a total capacity of 3 GW until 2017.

[32]See Matthes et al. (2012): Fokussierte Kapazitätsmärkte. Ein neues Marktdesign für den Übergang zu einem neuen Energiesystem. Öko-Institut e.V. – LBD-Beratungsgesellschaft mbH – RAUE LLP. Berlin.

[33]Reitz, F., Gerbaulet, C., Kemfert, C., Lorenz, C., Oei, P.-Y., & v. Hirschhausen, C. (2014). Szenarien einer nachhaltigen Kraftwerksentwicklung in Deutschland (Politikberatung kompakt No. 90). Berlin, Germany: Deutsches Institut für Wirtschaftsforschung.

[34]The effects of this modeling approach, however, focus on Germany only. Including the neighboring countries would lead to a small shift of production and emissions from Germany to its neighbors.

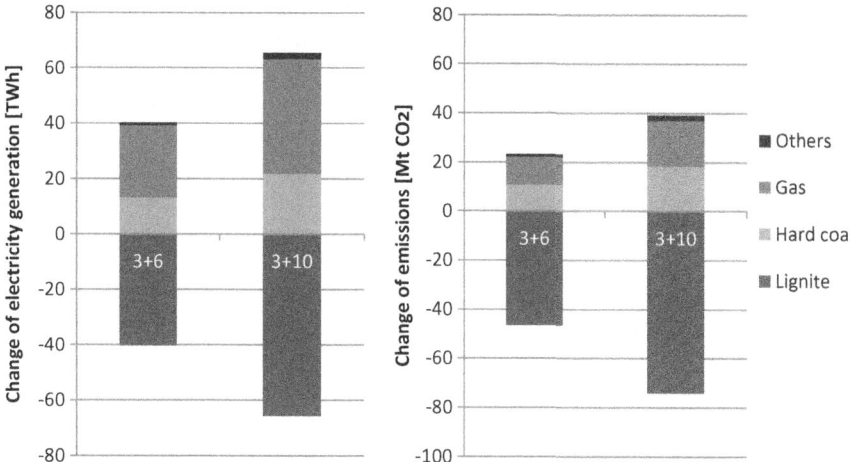

Fig. 4.9 Change of electricity generation (left) and CO_2 emissions (right) in the different scenarios (shutdown of 3 GW hard coal and 6/10 GW of lignite) in the year 2015. Source: Reitz et al. (2014)

while at the same time improving the market situation and preventing the need for CO_2-intensive and expensive capacity mechanisms.[35] Emissions of other pollutants such as NO_x, SO_2, small particles and mercury are also reduced. In addition, less coal electrification reduces the need for new mines, resulting in a double dividend for affected residents and the environment.[36]

4.4.3.6 Transmission Corridors and Lignite Basins

Limiting available transmission capacity and thus constraining the access of lignite basins to faraway electricity consumers is yet another instrument to reduce lignite power generation. Brancucci (2013) and Abrell and Rausch (2016) use both bottom-up and top-down perspectives to show that an increase of high-voltage electricity lines favors coal electrification if external costs are not sufficiently internalized. The argument applies to the German situation as well. In fact, discussions in Germany

---

[35]The German Ministry for Economy and Energy (BMWi) decided in November 2015 to move 2.7 GW of old lignite capacities into a reserve for climate reasons. An analysis shows that this reserve, however, is too small to reach Germany's 2020 climate targets (Oei et al. 2015a). Oei, P.-Y., Gerbaulet, C., Kemfert, C., Kunz, F., & Hirschhausen, C. (2016). "Kohlereserve" vs. CO_2-Grenzwerte in der Stromwirtschaft – Ein modellbasierter Vergleich. Energiewirtschaftliche Ta-gesfragen, 66(1/2), 57–60.

[36]This study only analyzes the situation in Germany. It neglects that a reduction of German production also reduces net exports and consequently increases generation and emissions in neighboring countries. More recent studies show that the net CO_2 reduction effect in the European electricity sector is around 50% of the German reduction when introducing national measures (Oei et al. 2015a).

center around the planning of three high-voltage direct current (DC) lines that were supposed to transport wind energy generated in the North to demand centers in the South (see Chap. 8): Two of the three originally planned corridors had their starting points in the lignite regions of NRW and eastern Germany. They would therefore enable consistently high lignite power generation, even at times of high wind generation in Northern Germany. The excess electricity could then be exported to neighboring countries, replacing foreign gas power plants. The higher CO_2 output would, however, increase German as well as the European GHG emissions. In a study on the low-carbon energy strategy of the State of Bavaria, Mieth et al. (2015)[37] provided bottom-up calculations of the effects of an additional HVDC line from the lignite basins of East Germany to Southern Germany. They confirmed an effect discussed previously in the literature: The new line would lead to about 10 TWh more lignite electrification.

In this context, Germany also emerged as the first country in which the CO_2 intensity of electricity was explicitly capped by the network regulator. In fact, the ten-year network development plan (TYNDP) for Germany, which had 2015 as its base year, was the first to include specific CO_2 targets for network planning. Today, electricity transmission planners have concrete CO_2 targets that they have to consider in their calculations, and they must also align the planning of new lines with the objectives of the Energiewende (Mieth et al. 2015). Caps have been fixed at 187 Mt of CO_2 for 2025 and 134 Mt for 2035, and correspond to the reduction target of -55% in 2030 (compared to 1990). This target reflects a proportional reduction of the electricity sector and should be increased as emission reductions in other sectors become possible but still require more specific action and higher costs.

4.5 Effects on Resource Adequacy and Structural Change

There is no doubt that if the German government is serious about its climate targets, coal will have to be gradually phased out of the electricity mix as CCTS is no viable option for Germany. This section looks at two potential effects of the coal phase-out on resource adequacy and structural change in the major coal regions.

4.5.1 Coal Plant Closures and Resource Adequacy

A German coal phase-out would have various effects on electricity generation, wholesale and consumer prices, as well as revenue streams. These effects depend

---

[37]Mieth, R., Gerbaulet, C., von Hirschhausen, C., Kemfert, C., Kunz, F., & Weinhold, R. (2015). *Perspektiven für eine sichere, preiswerte und umweltverträgliche Energieversorgung in Bayern* (No. 97). Berlin, Germany: DIW Berlin: Politikberatung kompakt.

Table 4.5 Generation capacities in Germany up to 2035

in GW	2013	2020	2025	2035
Nuclear	12.1	8.1	–	–
Lignite	21.2	20.0	12.6	9.1
Hard coal	25.9	26.0	21.8	11.1
Gas	26.7	19.2	25.4	32.7
Hydro	3.9	4.0	4.0	4.2
Wind onshore	33.8	52.2	63.8	88.8
Wind offshore	0.5	6.5	10.5	18.5
Biomass	6.2	7.2	7.4	8.4
Solar	36.3	48.2	54.9	59.9
Pumped Hydro	6.4	7.8	8.3	12.5
Others	4.7	2.2	2.8	2.4
Total	165.6	201.4	211.5	247.6

Source: BNetzA. (2014). *Genehmigung des Szenariorahmens 2025 für die Netzentwicklungsplanung und Offshore-Netzentwicklungsplanung.* Bundesnetzagentur

on the chosen instruments and their specifications. Some general findings, however, are very similar across all of the options (Oei et al. 2014a, b; Oei et al. 2015b). The following section therefore presents some representative modeling results up to 2035. They assume a gradual phase-out of coal generation capacities with no retrofits according to the scenario framework of the BNetzA (Table 4.5). In Oei et al. (2015a, b), we developed two scenarios to analyze different policy instruments, both of which assume the same power plant capacities:

- the *green* scenario includes a fee on electricity from coal, in the spirit of the "climate contribution" that restricts the load factors of older coal power plants;
- the *black* scenario, a business-as-usual (BAU) scenario.

Germany steadily increased its electricity exports in recent years, reaching an all-time high of 35 TWh in 2014. This has led to decreased gas electricity production in neighboring countries. Modeling results show that this rise in export quantities will continue in the black BAU scenario to figures above 50 TWh. Such a rise also implies increasing congestion at cross-border interconnectors. A gradual coal phase-out would halt rising exports in 2020 slightly above the level of 2014 and reduce line congestion. Germany would still remain an exporter of electricity with a volume of around 10 TWh in 2035 (see Fig. 4.10).

The effect of the gradual coal phase-out on wholesale electricity prices is relatively low since Germany is integrated into the central European electricity grid. The price increase remains in the range of €2–3/MWh (0.2–0.3 cents/kWh). The price effect on households and small industry consumers will be dampened by a simultaneous reduction of the renewables levy ("EEG-Umlage"); the overall rise is likely to be in the range of €1–2/MWh (0.1–0.2 cents/kWh). At under €40/MWh up to 2035, the wholesale electricity price lies still below prices from the years 2010 to 2012. The coal phase-out therefore has little effect, contrary to some media coverage, on

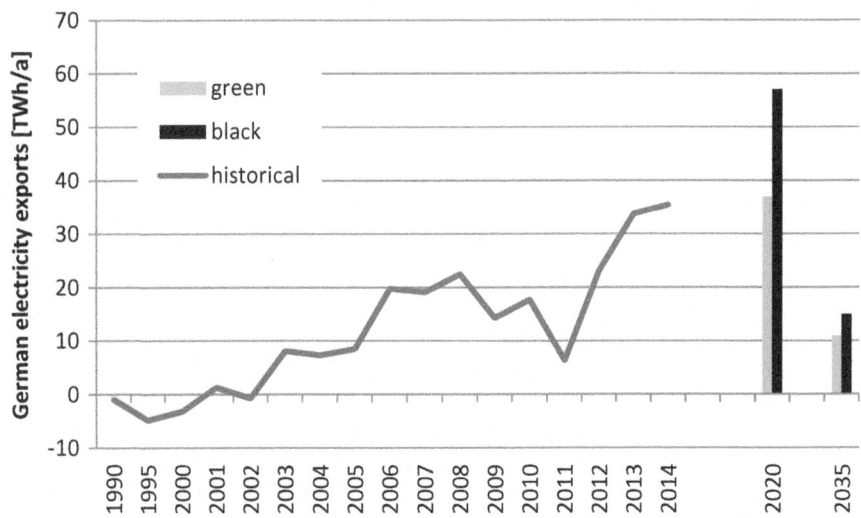

Fig. 4.10 Electricity exports from Germany to its neighboring countries. Source: Oei et al. (2015a, b)

the competitiveness of German energy-intensive firms. Neuhoff et al. (2014)[38] show that in any case, electricity prices contribute less than 5% of overall production costs for most sectors. Additional factors that have a stronger effect are resource prices for hard coal, gas, and oil. Prices in 2015 for all these resources are still below 2008—that is, pre-crisis—values and therefore favor these firms. The increased wholesale price for the post-2020 period in the modelling runs also represents a benefit to the majority of utilities, providing additional revenues for all remaining generation capacities: The overall annual benefit adds up to around €500 million. Mostly newer hard coal plants as well as some natural gas plants benefit from this effect (in addition to nuclear power plants in 2020). For older and more CO_2-intensive coal plants, the reduction of full load hours might overcompensate for the price effect (see Fig. 4.11).

The low level of wholesale electricity prices up to 2035 is an indicator of the existing overcapacities in the European electricity sector.[39] This effect is still visible in 2035 despite the shutdown of all remaining German nuclear power plants in 2023 and the assumed gradual coal phase-out (20 GW in 2035 compared to 46 GW in 2013). Modeling the implementation of an additional climate levy (green scenario) secures the set CO_2 targets for 2020 and 2035 without endangering the security of supply at any point. Germany even remains an electricity exporter in the range of

---

[38]Neuhoff, K., Acworth, W., Dechezlepretre, A., Sartor, O., Sato, M., & Schopp, A. (2014). Energie- und Klimapolitik: Europa ist nicht allein (DIW Wochenbericht No. 6/2014). Berlin, Germany: DIW Berlin.

[39]Additional effects are the low EU-ETS CO_2 certificate and global coal prices.

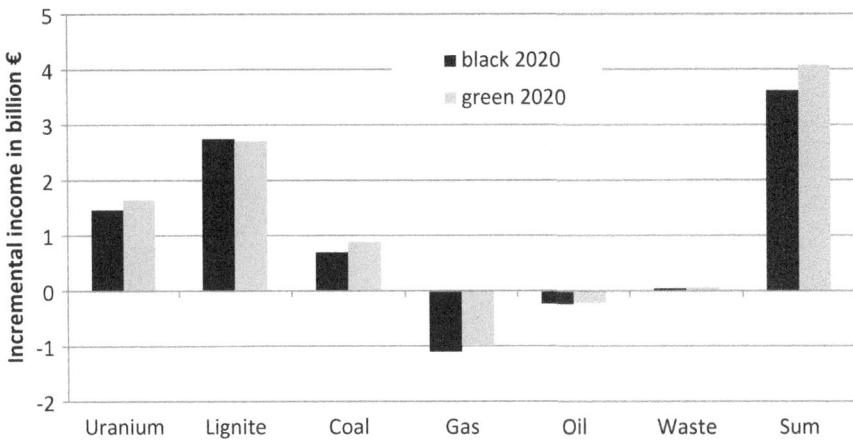

Fig. 4.11 Revenue from electricity sales in 2020. Source: Oei et al. (2015a, b)

~10 TWh in 2035. The majority of utilities in Germany but also abroad even profit from the limitation of coal-fired electricity generation in the green scenario.

All modeling results depend on future assumptions and were therefore tested by running sensitivity analyses more than 600 times with respect to input parameters such as full load hours of renewables, EU-ETS CO_2 prices, and different variations of the climate levy. One major influence, however, that is often not sufficiently included in national discussions are developments in neighboring countries and the counter-effects in Germany. The ENTSO-E (2014) published four visions that resemble possible European development pathways, and these were represented in various modelling runs. The visions vary on the integration of the European electricity market as well as to their contribution to the climate targets for 2050.[40] The results show that the long-term decline of German CO_2 emissions (301 Mt in 2014) are influenced to a greater extent by developments in neighboring states (difference between visions: 20–26 Mt) than by the presence or absence of an additional national instrument (difference between black and green scenario: 3–9 Mt). It is therefore in the interest of Germany that other neighboring countries also take action and complement the EU-ETS with national instruments to enable a generation portfolio in line with the European climate targets (Visions 3 and 4) (Fig. 4.12).[41]

---

[40]Vision 1 "Slow Progress" assumes little European integration and delayed climate action. The second vision, "Money Rules," also does not assume the achievement of climate targets but is based on increased European integration. The climate targets of the Roadmap 2050 are reached in the third vision, "Green Transition," as well as in the fourth vision, "Green Revolution." "Green Transition," in contrast to "Green Revolution," assumes little European integration.

[41]This is also due to the fact that the visions assume different generation capacities in the other countries. Generation capacities for Germany, however, were left constant throughout all runs.

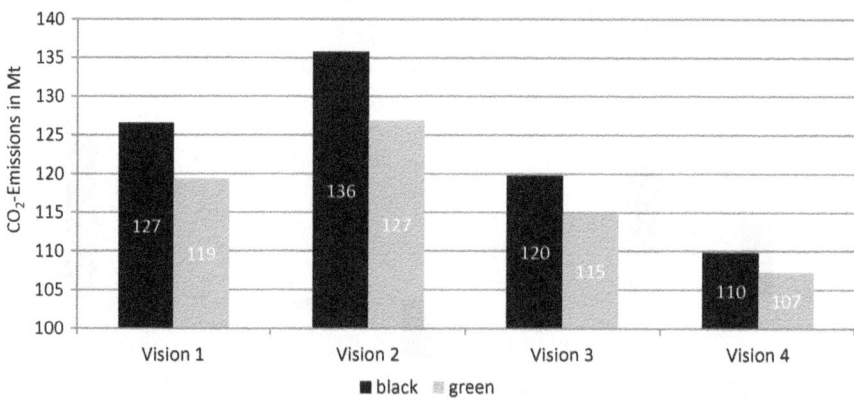

Fig. 4.12 German CO_2 emissions in 2035 depending on developments in neighboring countries.
Source: Oei et al. (2015a, b)

4.5.2 Regional Structural Changes Almost Complete

4.5.2.1 Aggregate Employment Effects

When considering the structural change at the level of the lignite mining basins, one
has to recall the last three decades (discussed in more detail in Chap. 3). This was a
period of constant structural change in West Germany and particularly sweeping
changes in East Germany following reunification. In the 1980s, the lignite industry
still accounted for more than 350,000 direct and indirect jobs. The transformation
process after German reunification and continuous industrialization, however, led to
radical reorganization of this sector. The resulting steep decline in employment to
50,000 jobs in the lignite industry in 2002 marked the beginning of a lignite mining
phase-out especially in the former East Germany. This occurred when the
energiewende was just beginning (Statistik der Kohlewirtschaft 2018).[42]

The decline of employment in hard coal mining was even greater, falling from as
many as 600,000 direct employees in the 1950s to 30,000 in 2005. In 2013 there
were only 10,000 employed people in this sector, including older employees in semi-
retirement. The closure of the last deep-cast mines of the RAG Deutsche Steinkohle
AG in NRW in 2018, when production subsidies expired in line with EU state aid
law, marked the next step in the German coal phase-out (Statistik der
Kohlewirtschaft 2018) (Fig. 4.13).[43]

---

[42]Statistik der Kohlenwirtschaft (2018). Belegschaft im Steinkohlenbergbau der Bundesrepublik
Deutschland. Essen, Bergheim. https://kohlenstatistik.de/files/arbeiter_u_angestellte_3.xlsx.
Retrieved July 17, 2018.
[43]Ibid.

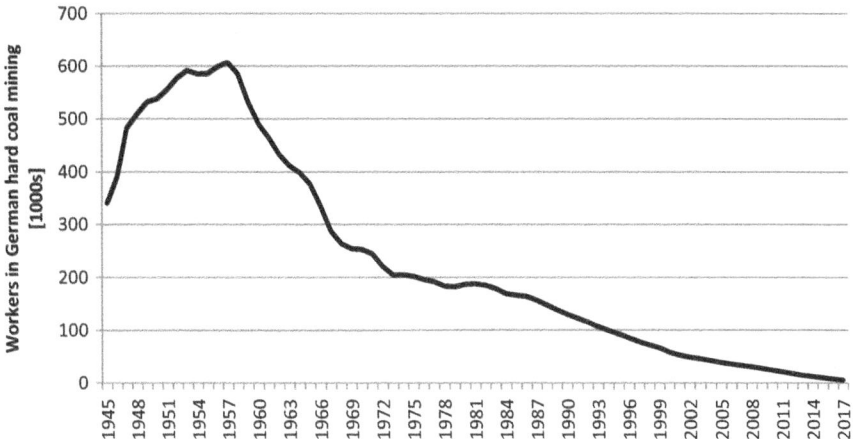

Fig. 4.13 Workers in German hard coal mining from 1945 to 2017. Source: Statistik der Kohlenwirtschaft (2018). Belegschaft im Steinkohlenbergbau der Bundesrepublik Deutschland. Essen, Bergheim. https://kohlenstatistik.de/files/arbeiter_u_angestellte_3.xlsx. Retrieved July 17, 2018

Overall, while West Germany witnessed a gradual decline in employment in the 1990s, East Germany saw a radical drop in employment in the early part of that decade but has undergone a continuous but less steep decline since then. Thus, although the remaining coal phase-out will be challenging, one can conclude that structural change in the affected regions has already largely taken place (Herpich et al. 2018).

The coal phase-out has two major effects on employment in the electricity sector: First, a decrease in jobs in mining and coal-fired electricity generation, and second, as a counter-effect, an increase in jobs in the renewables sector. Jobs in the renewables sector exist in different stages of the value chain (e.g., invention, construction, and maintenance) as well as throughout the country (the North specializing more in wind power, the South in PV). Due to the success of the energiewende in Germany and abroad, employment figures rose to more than 338,600 in 2013 (Lehr et al. 2015). The renewables sector has consequently become the most important electricity sector in terms of employment, overtaking the coal sector in the last two decades (see Fig. 4.14).

Employment effects of the energiewende differ, however, in specific regions. The positive effects of newly created jobs in the renewables sector are spread relatively evenly across the county. Jobs in the coal and in particular in the lignite sector, however, are mostly concentrated in the mining regions and are also better paid on average. As a result, most regions of Germany profit significantly from the energiewende, while the situation in NRW and Lusatia is more complex.

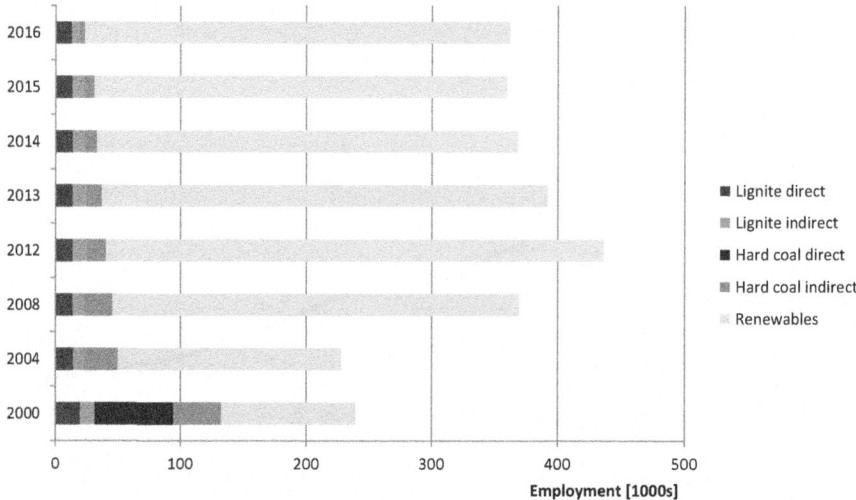

Fig. 4.14 Employment in the coal and renewables sector from 1998 to 2013. Source: Own calculations based on Lehr et al. (2015) and Statistik der Kohlenwirtschaft e.V. (2018). An additional 5000 employees were working in German hard coal power plants in 2014. Their number, however, is not shown here due to a lack of data for the previous years.

4.5.2.2   Regional Effects

In all German lignite and hard coal mining regions, mining activities and power plant operations have declined dramatically in recent decades. Shutting down all remaining mines and plants by 2040 should be organized in a way that minimizes the social impacts as much as possible so as not to undermine public support for the energiewende. This is possible as more than two thirds of the employees in the coal sector are aged 45 and older (Statistik der Kohlewirtschaft 2018).[44] Shutting down the plants when most of the staff have reached retirement age causes few layoffs. Also, a large number of workers are and still will be working in the sector of renaturation in the mining regions for decades and will therefore even profit from the closing of plants and mining sites. New jobs are needed especially in the affected regions to secure job opportunities for future generations. The energiewende will enable this transition towards more sustainable jobs in the service sector, tourism, and in particular in the renewable energy sector. In 2015, in fact, even in those federal states with lignite mining (NRW, Brandenburg, Saxony, and Saxony-Anhalt), more people are already employed in the renewable energy sector than in the coal industry (Lehr et al. 2015).

---

[44]Statistik der Kohlenwirtschaft e.V. (2018). Datenübersichten zu Steinkohle und Braunkohle in Deutschland 2018. Retrieved May 15, 2018, from http://www.kohlenstatistik.de/.

4.6   Conclusion

Coal-fired power plants are responsible for around a third of total CO_2 emissions in Germany. Failure to reduce the persistently high level of coal-based power generation puts Germany's climate targets at risk and undermines the potential for a sustainable and successful energiewende. The government has consequently published the national Climate Protection Plan 2050 (*Klimaschutzplan 2050*) in 2016, in which power generation plays a major role. Furthermore, the scenario framework proposed by the German regulator (BNetzA) suggests a reduction of CO_2-emissions to 187 Mt (2025) and 134 Mt (2035). This can be achieved through a reduction of most of the lignite power plant production and a continuing increase in the share of renewables. All of Germany's federal states have made commitments to similar climate targets. The government of North Rhine-Westphalia was the first to limit the use of the existing mine Garzweiler, preventing the relocation of a further 1400 residents. In eastern Germany, too, there is no need to open up new lignite mines.

Current prices for CO_2 emissions allowances in the European Emissions Trading System (EU-ETS) make a market-driven transformation from coal to less CO_2-intensive energy sources such as natural gas unlikely in the near future. Missing the 2020 climate targets of −40% CO_2 reduction compared to 1990, however, also puts the long-term targets and therefore the entire energiewende in jeopardy. This is where additional national instruments to accompany the EU-ETS come into play, some of which are already being implemented in various countries. An analysis of the options discussed indicates that:

- A national CO_2 floor price would presumably not be sufficient to effect a switch from lignite to natural gas in the near future.
- The introduction of a national CO_2 emissions performance standard (EPS) for new and existing fossil-fired power plants could be contemplated as a specific means of reducing coal-based power generation, e.g., taking into account the plant age structure.
- Minimum efficiency and flexibility requirements for power plants do not directly aim at a reduction of CO_2 emissions and, depending on specifics, would also affect gas power plants.
- A coal phase-out law with fixed production or emissions allowances for coal-fired power plants could prescribe a schedule for phasing out coal-based power generation in Germany and therefore provide investment security for all affected parties.
- Older plants could be integrated into a capacity reserve to compensate the operators and at the same time prevent scarcity of generation capacity.
- The discussed "climate contribution" fee for old coal power plants, as proposed by the German Ministry for Economy and Energy in 2015, would have been a cost cost-efficient instrument; it would also have been compatible with the EU-ETS, as certificates are taken from the market and no leakage effect occurs.

- Future electricity transmission planners now have concrete CO_2 targets that need to be respected in their calculations and will influence the planning of new lines in a way that is more closely aligned with the goals of the energiewende.

From a European perspective, interaction between the German and European power sectors will intensify in the future. Modelling analysis on the basis of the European Scenario Outlook and Adequacy Forecast (SOAF) confirms that aggregate CO_2 emissions in the European power sector will only meet the climate targets if at least some of Germany's neighboring countries take action as well, complementing the EU-ETS with national instruments to reduce their CO_2 emissions.

The EU-ETS, however, is and remains a central component of EU policy on combating climate change, despite its currently limited steering capacity. The introduction of the market stability reserve as well as the adjustment of the reduction factor are therefore important—but insufficient—changes to strengthen the EU-ETS. Cancelling the existing surplus of more than 2 bn allowances would be an important additional signal to retain the credibility of the EU-ETS and bolster European climate policy. A strengthened EU-ETS supplemented by national instruments would provide a suitable framework to ensure the continuous reduction of greenhouse gases in line with national and European climate targets.

Limiting German GHGs and meeting the climate target automatically implies a coal phase-out in Germany in the 2030s. The coal phase-out in Germany is a process that has already started with the country's continuous industrialization since the 1950s—long before the energiewende began. A further step was the German reunification, which led to a radical contraction of the lignite industry in East Germany. This chapter shows that an overall phase-out is possible without jeopardizing resource adequacy at any point. The majority of power sector actors, including but not limited to renewables and gas operators, even profit from such a trend. The resulting net employment effects differ across regions and sectors but are expected to be positive for all regions.

References

Abrell, Jan, and Sebastian Rausch. 2016. Cross-country electricity trade, renewable energy and European transmission iinfrastructure policy. *Journal of Environmental Economics and Management* 79 (September): 87–113.

Agora Energiewende. 2017. *Eine Zukunft für die Lausitz: Elemente eines Strukturwandelkonzepts für das Lausitzer Braunkohlerevier*. Berlin: Impulse.

BMWi, and BMUB. 2010. Energiekonzept für eine umweltschonende, zuverlässige und bezahlbare Energieversorgung. Berlin, Germany.

Brancucci Martínez-Anido, Carlo. 2013. Electricity Without Borders – The Need for Cross-border Transmission Investment in Europe. Proefschrift/Dissertation, The Netherlands: Technische Universiteit Delft.

EC. 2011. Energy roadmap 2050. Communication from the Commission to the European Parliament, the Council, the European Economic and Social Committee and the Committee of the Regions. Brussels, Belgium: European Commission.

EC. 2014. *Questions and answers on the proposed market stability reserve for the EU emissions trading system*. Brussels: European Commission.

ENTSO-E. 2014. *Scenario Outlook & Adequacy Forecast 2014–2030*. Brussels, Belgium: European Network of Transmission System Operators for Electricity.

Gerbaulet, Clemens, Jonas Egerer, Pao-Yu Oei, Judith Paeper, and Christian von Hirschhausen. 2012. Die Zukunft der Braunkohle in Deutschland im Rahmen der Energiewende. DIW Berlin, Politikberatung kompakt 69. Berlin, Germany: Deutsches Institut für Wirtschaftsforschung (DIW).

Göke, Leonard, Martin Kittel, Claudia Kemfert, Pao-Yu Oei, and Christian von Hirschhausen. 2018. Scenarios for the coal phase-out in Germany – A model-based analysis and implications for supply security. DIW Weekly Report 28/2018. Berlin, Germany: DIW Berlin, German Institute for Economic Research.

Hake, Jürgen-Friedrich, Wolfgang Fischer, Sandra Venghaus, and Christoph Weckenbrock. 2015. The German Energiewende – history and status quo. *Energy* 92 (Part 3: Sustainable Development of Energy, Water and Environment Systems): 532–546.

Henning, Hans-Martin, and Andreas Palzer. 2012. *100% Erneuerbare Energien für Strom und Wärme in Deutschland*. Freiburg, Germany: Fraunhofer-Institut für Solare Energiesysteme ISE.

Herpich, Philipp, Hanna Brauers, and Pao-Yu Oei. 2018. *An Historical Case Study on Previous Coal Transitions in Germany*. IDDRI and Climate Strategies.

IICCS. 2015. *Islamic Declaration on Global Climate Change*. Istanbul, Turkey: International Islamic Climate Change Symposium.

IPCC. 2014. *Climate Change 2014: Impacts, Adaptation, and Vulnerability. Part A: Global and Sectoral Aspects. Contribution of Working Group II to the Fifth Assessment Report of the Intergovernmental Panel on Climate Change*. Cambridge, UK: Cambridge University Press.

Jacobson, Mark Z., Mark A. Delucchi, Zack A.F. Bauer, Savannah C. Goodman, William E. Chapman, Mary A. Cameron, Cedric Bozonnat, et al. 2017. 100% Clean and renewable wind, water, and sunlight all-sector energy roadmaps for 139 countries of the world. *Joule* 1 (1): 108–121.

Klaus, T., C. Vollmer, K. Werner, H. Lehmann, and K. Müschen. 2010. *Energieziel 2050: 100% Strom aus erneuerbaren Quellen*. Dessau-Roßlau, Germany: Umweltbundesamt (UBA).

Kungl, Gregor. 2015. Stewards or sticklers for change? Incumbent energy providers and the politics of the German energy transition. *Energy Research & Social Science* 8 (July): 13–23.

Leader of the G7. 2015. *Leaders' Declaration G7 Summit*, June 7–8, 2015. Schloss Elmau, Germany.

Lehr, Ulrike, Dietmar Edler, Marlene O'Sullivan, Frank Peter, and Peter Bickel. 2015. Beschäftigung durch erneuerbare Energien in Deutschland: Ausbau und Betrieb, heute und morgen. Studie im Auftrag des Bundesministeriums für Wirtschaft und Energie. Berlin, Germany: GWS, DLR, Prognos, ZSW, DIW Berlin.

Löffler, Konstantin, Karlo Hainsch, Thorsten Burandt, Pao-Yu Oei, Claudia Kemfert, and Christian von Hirschhausen. 2017. Designing a model for the global energy system—GENeSYS-MOD: an application of the Open-Source Energy Modeling System (OSeMOSYS). *Energies* 10 (10): 1468.

Mieth, Robert, Richard Weinhold, Clemens Gerbaulet, Christian von Hirschhausen, and Claudia Kemfert. 2015. Electricity grids and climate targets: new approaches to grid planning. *DIW Economic Bulletin* 5 (6): 75–80.

Neuhoff, K., Acworth, W., Dechezlepretre, A., Sartor, O., Sato, M., and Schopp, A. 2014. *Energie- und Klimapolitik: Europa ist nicht allein* (DIW Wochenbericht No. 6/2014). Berlin, Germany: DIW Berlin.

New Climate Economy. 2014. *Better Growth, Better Climate – Executive Summary*. The Synthesis Report. Washington, DC, USA: The Global Commission on the Economy and Climate.

Nitsch, Joachim. 2013. "Szenario 2013" – eine Weiterentwicklung des Leitszenarios 2011. Stuttgart, Germany: Deutsches Zentrum für Luft- und Raumfahrt (DLR).

Oei, Pao-Yu. 2015. *Decarbonizing the European Electricity Sector – Modeling and Policy Analysis for Electricity and CO_2 Infrastructure Networks*. Berlin, Germany: Technische Universität Berlin.

Oei, Pao-Yu, Claudia Kemfert, Felix Reitz, and Christian von Hirschhausen. 2014a. *Braunkohleausstieg – Gestaltungsoptionen im Rahmen der Energiewende*, Politikberatung kompakt. Vol. 84. Berlin, Germany: DIW.

Oei, Pao-Yu, Felix Reitz, and Christian von Hirschhausen. 2014b. *Risks of Vattenfall's German Lignite Mining and Power Operations – Technical, Economic and Legal Considerations*, Politikberatung kompakt. Vol. 87. Berlin, Germany: DIW Berlin — Deutsches Institut für Wirtschaftsforschung e. V.

Oei, Pao-Yu, Clemens Gerbaulet, Claudia Kemfert, Friedrich Kunz, and Christian von Hirschhausen. 2015a. *Auswirkungen von CO_2-Grenzwerten für fossile Kraftwerke auf den Strommarkt und Klimaschutz*, Politikberatung kompakt. Vol. 104. Berlin, Germany: DIW.

Oei, Pao-Yu, Clemens Gerbaulet, Claudia Kemfert, Friedrich Kunz, Felix Reitz, and Christian von Hirschhausen. 2015b. *Effektive CO_2-Minderung im Stromsektor: Klima-, Preis- und Beschäftigungseffekte des Klimabeitrags und alternativer Instrumente*, Politikberatung kompakt. Vol. 98. Berlin, Germany: DIW.

Öko-Institut, and Fraunhofer ISI. 2014. Klimaschutzszenario 2050-1. Modellierungsrunde (Studie im Auftrag des Bundesministeriums für Umwelt, Naturschutz, Bau und Reaktorsicherheit). Berlin, Germany.

Pope Francis. 2015. *Laudato Si: On Care for Our Common Home – Encyclical Letter of Pope Francis*. Rome, Italy: CreateSpace Independent Publishing Platform.

Reitz, Felix, Clemens Gerbaulet, Claudia Kemfert, Casimir Lorenz, Pao-Yu Oei, and Christian von Hirschhausen. 2014. *Szenarien einer nachhaltigen Kraftwerksentwicklung in Deutschland*, Politikberatung kompakt. Vol. 90. Berlin, Germany: DIW.

SRU. 2011. Pathways towards a 100% Renewable Electricity System. Special Report. Berlin, Germany: Sachverständigenrat für Umweltfragen.

———. 2015. *The Future of Coal through 2040. Comment on Environmental Policy*. Berlin, Germany: Sachverständigenrat für Umweltfragen.

Statistik der Kohlenwirtschaft. 2018. Belegschaft im Steinkohlenbergbau der Bundesrepublik Deutschland. Essen, Bergheim. https://kohlenstatistik.de/files/arbeiter_u_angestellte_3.xlsx. Retrieved July 17, 2018.

Strunz, Sebastian, Erik Gawel, and Paul Lehmann. 2015. Towards a general 'Europeanization' of EU member states' energy policies? *Economics of Energy & Environmental Policy* 4 (2): 143–159.

VDE. 2012. Erneuerbare Energie Braucht Flexible Kraftwerke – Szenarien Bis 2020. VDE-Studie. Frankfurt am Main, Germany: VDE Verband der Elektrotechnik Elektronik Informationstechnik e.V. - Energietechnische Gesellschaft im VDE (ETG).

von Hirschhausen, Christian, Johannes Herold, and Pao-Yu Oei. 2012. How a 'low carbon' innovation can fail – tales from a 'lost decade' for Carbon Capture, Transport, and Sequestration (CCTS). *Economics of Energy & Environmental Policy* 1 (2): 115–123.

Ziehm, C., Kemfert, C., Oei, P.-Y., Reitz, F., & Hirschhausen, C. von. 2014. Entwurf und Erläuterung für ein Gesetz zur Festsetzung nationaler CO2-Emissionsstandards für fossile Kraftwerke in Deutschland (Politikberatung kompakt No. 82). Berlin, Germany: DIW Berlin.

Chapter 5
Nuclear Power: Effects of Plant Closures on Electricity Markets and Remaining Challenges

Friedrich Kunz, Felix Reitz, Christian von Hirschhausen, and Ben Wealer

"This technology is a lousy choice."
Henry Cordes, CEO of the Public Nuclear Utility EWN
(Energiewerke Nord)
(Source: ARD: http://www.infosperber.ch/Umwelt/die-
wahren-Kosten-der-Atomenergie-Atomdeal-Endlagerung,
accessed August 29, 2016. Translated from the original in
German: "Atomtechnologie ist ein gigantischer Griff ins
Klo.")

This chapter is based on two branches of research: one on the longer-term system adequacy of the German and European electricity markets in the context of the nuclear plant closures, see Kunz et al. (2011), Kunz and Weigt (2014), and Goeke, et al. (2018), and another branch dealing with the institutional aspects of decommissioning and waste storage (see Wealer et al. 2015; and the chapter on decommissioning in the recent World Nuclear Industry Status Report, see Schneider et al. 2018). We thank Pao-Yu Oei, Jan-Paul Seidel, Alexander Weber, and Hannes Weigt for contributions and discussions; the usual disclaimer applies.

F. Kunz
Tennet, Bayreuth, Germany

F. Reitz
Europe beyond Coal, Berlin, Germany

C. von Hirschhausen · B. Wealer (✉)
TU Berlin, Berlin, Germany

DIW Berlin, Berlin, Germany
e-mail: bw@wip.tu-berlin.de

5.1   Introduction

Nuclear power has been a major topic of energy policy debate in Germany since the 1950s, and it was a key issue in all energiewende discussions. For many anti-nuclear activists, the closure of nuclear power stations was the most important aspect of the energiewende; in fact, the anti-nuclear movement goes back to the 1960s, even before the emergence of the environmental movement. As shown in Chap. 2, Chancellor Merkel's March 2011 "Declaration of the nuclear moratorium" was a very important turning point of the energiewende.

The March 2011 closure of seven nuclear power plants (the oldest in Germany) and the Krümmel reactor sparked an intense debate over the economic effects this might have, particularly in terms of prices and supply security. Proponents of the closures argued that supply security was not at risk, whereas opponents insisted that an "electricity gap" would result. This discussion has since been resolved: The plants were closed at a time of high overcapacities on German and European markets, and the economic effects of their closure were almost imperceptible. The electricity industry had also had time to prepare since the first closure decision in 2002: Because further closures were anticipated, more fossil-fuel powered plants came online in Germany between 2011 and 2013 than nuclear power plants were closed during the same period.

After 2013, the discussion turned to the other issues of dismantling the old nuclear facilities, storing the radioactive waste, and defining new corporate strategies. These challenges had been largely neglected since the 1960s in Germany (and worldwide) due to the low political priority of these issues and a lack of incentives for nuclear plant operators. However, it soon became clear that not only did the costs of these processes exceed the provisions made by companies, but also the survival of the four utilities (E.ON, RWE, Vattenfall, EnBW) was at stake: had the "polluter-pays" principle been applied rigorously (which it should have been according to §9 of the Atomic Energy Act), all four would have had to file for bankruptcy. Thus, the nuclear sector posed severe challenges to both the political decision makers and the corporate strategists.

This chapter therefore focuses on the two central issues arising with the closure of nuclear power plants in Germany: (1) the effects on German and European electricity markets; and (2) the complex process of decommissioning old plants and finding suitable solutions for storing radioactive waste and adapting corporate strategies to the new challenges. The next section describes the main steps in closing all of Germany's nuclear power plants between 2011 and 2022. Section 5.3 discusses the effects of plant closures on German and European electricity markets based on a survey of the literature and our own modeling results on the moratorium. Our quantitative modeling confirms our hypothesis—also spelled out in a paper from June 2011 (Kunz et al. 2011)—that thanks to the significant overcapacities, the effects of both the moratorium (March 2011) and the final plant closures (by 2022 at the latest) have been modest and can be absorbed relatively easily through generation of power from other sources including renewable electricity, both in Germany and some neighboring European countries. Section 5.4 addresses what we consider to be the most important challenges of the final phase of nuclear power—decommissioning and storage—which have not received sufficient attention anywhere in the world. We identify the technical, financial, and institutional challenges of this process that are likely to continue well into the next century and also look at the implications for corporate strategies and diversification. Section 5.5 concludes.

5.2   The Timetable for Closing Nuclear Power Plants in Germany (2011–2022)

As described in detail in Chap. 2, there have been heated debates about nuclear power in Germany since the 1950s, and these debates gained new urgency with the September 2010 lifetime extension decision on nuclear power plants by Germany's ruling conservative government. On March 14, 2011, shortly after the nuclear catastrophe in Fukushima, Chancellor Merkel swiftly brought an end to this policy by announcing a moratorium on lifetime extensions. This decision was followed by legislation confirming the closure of all nuclear power plants in Germany: the 13th Amendment of the Atomic Energy Act was passed on June 17, 2011, with overwhelming support from all parties.

Figure 5.1 identifies the concrete steps in the closure process, sometimes also called "phase-out," between 2011 and 2022. Two distinct periods can be identified:

• The March 14, 2011, "moratorium" imposed by Chancellor Angela Merkel and later approved by the federal government and set down in binding legislation, foresaw the immediate shutdown of the seven oldest NPPs, with about 8.8 GW in total: Brunsbüttel, Unterweser, Biblis A, Biblis B, Philippsburg 1, Neckarwestheim 1, and Isar/Ohu 1. In addition, the NPP in Krümmel (near Hamburg), which had already been out of service since 2007, is generally considered part of the "moratorium plants."
• In addition, the law defines fixed closure dates for the remaining reactors (12.7 GW in total) between 2015 and 2022. Concretely, the following timetable was set: Grafenrheinfeld by December 2015 (taken off the grid in June 2015); Gundremmingen B (taken off the grid in December 2017); Philippsburg 2 (2019); Gundremmingen C, Grohnde, and Brokdorf (2021) followed by Isar 2, Emsland, and Neckarwestheim, all by the end of 2022.

The Atomic Energy Act of 2011 thus reversed the September 2010 lifetime extension of nuclear power plants and constitutes an essential, if not the single most important element of the energiewende. The act confirmed the ban on the construction of new nuclear power plants, passed in 2000, and ended all prospects of reviving the German nuclear power industry even in the distant future. Matthes (2012) provides an analysis of the decisions and their expected economic effects.

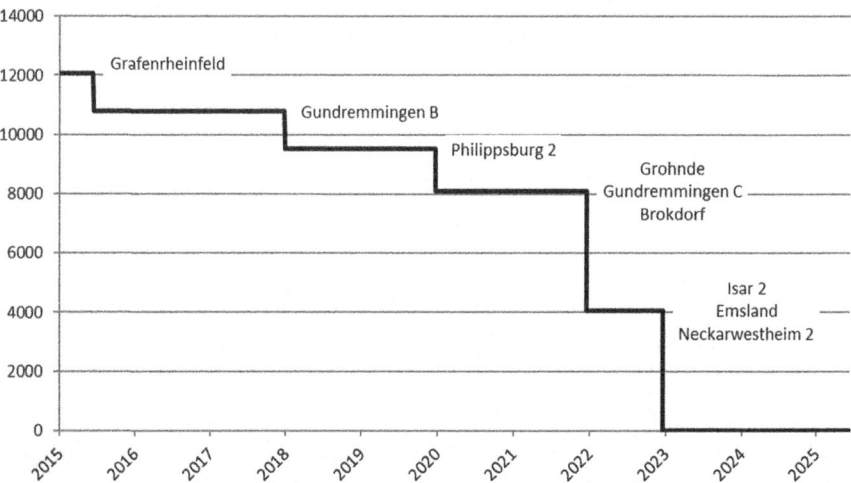

Fig. 5.1 Closure of nuclear power plants in Germany, 2011–2022. *Source:* 13. Revision of the Law on Nuclear Power (Atomgesetz, 2011)

The nuclear phase-out may come even earlier than 2022. In fact, the political will to achieve this goal has convinced the energy industry that phase-out is unavoidable and that no amount of lobbying will alter this decision.[1] Faced with the high costs of dismantling power plants and developing waste storage sites, the utilities that own nuclear capacities have modified their strategies and are now trying to accelerate the closure process. In a proposal published in spring of 2014, the utilities proposed to hand over the nuclear power plants and the financial reserves to a public fund that would take responsibility for the decommissioning process. The early closure of the Grafenrheinfeld NPP in June 2015 instead of December 2015 also indicates that operators may no longer consider keeping nuclear plants open to be advantageous. The very low price levels predicted for 2020 also suggest that operators anticipate the closure of the plants and plan not to be paying the operating expenses by that time.

---

[1]In its coalition agreement, after the 2013 federal elections the German government, made up of the CDU/CSU and SPD (Christian Democratic Union/Christian Social Union and Social Democratic Party), explicitly confirmed to phase out nuclear energy; see Coalition Agreement between CDU, CSU and SPD, *Deutschlands Zukunft gestalten* (2013, p. 59): "Wir halten am Ausstieg aus der Kernenergie fest. Spätestens 2022 wird das letzte Atomkraftwerk in Deutschland abgeschaltet." "We remain committed to a nuclear phase-out. The last nuclear power station in Germany will be shut down by 2022 at the latest." [own translation].

5.3 Energy Economic Effects of Nuclear Power Plant Closures

5.3.1 Review of Existing Studies

The comprehensive survey by Kunz and Weigt (2014) of studies on the energy economic effects of nuclear plant closures in Germany provides extensive information from ad-hoc analyses of the early energiewende period, including Traber and Kemfert (2012), Matthes (2012), Bruninx et al. (2013), and Nestle (2012). While the studies that immediately followed the March 2011 decision were split between optimistic and pessimistic assessments of supply security, later studies were much more positive, noting overcapacities throughout all of Europe and therefore little cause for concern. Matthes et al. (2011), for instance, analyzed the short-term impacts of the moratorium and concluded that the seven oldest plants could be shut down immediately and another five in the near term, and that capacity construction and demand management would compensate for these closures. Clearly, the most visible effect was the replacement of nuclear power by fossil fuels and renewables. The total additional abatement needed by options other than nuclear after the phase-out was predicted to be in the range of 40–100 Mt CO_2. As this represented only a small share of total ETS emissions, models also predicted rather modest emission price increases of a few euros per ton. With respect to cross-border flows, studies showed that reduced nuclear generation would lead to a reduction of Germany's large export surplus. Using a stylized model of the German electricity system, Bruninx et al. (2013) showed that nuclear generation was being replaced by other existing technologies.

In a subsequent study, Matthes (2012) calculated "the relatively low cost of German's nuclear phase-out," which he explained by the high level of non-nuclear generation activities; the potential price effect of the plant closures was expected to be "around €5/MWh or less for a few years around 2020" (Felix Christian Matthes 2012, 42). This is in line with other assessments of a modest and short-term price effect. In fact, German utilities had anticipated the closure of nuclear power plants, i.e., the "first" phase-out decision of 2000, and had invested heavily in fossil plant capacity during the first decade. Thus, in addition to the overcapacity already prevailing in 2011, net additions of 9 GW were expected to come online by 2013.[2] There was also no capacity shortage in other Central or West European countries.

---

[2] See BNetzA (2011, p. 7): Genehmigung des Szenariorahmens zum NEP 2012. Bonn, Germany. Last accessed August 27, 2012 at http://www.netzausbau.de/SharedDocs/Downloads/DE/2023/SR/Szenariorahmen2023_Genehmigung.pdf?__blob=publicationFile.

5.3.2 The First Model-Based Network Analysis...

In May 2011, only two months after the moratorium, Kunz et al. (2011) published a model-based analysis of the potential effects of both the moratorium and what was then called "phase-out" on electricity flows and prices in Germany and its neighboring countries; for details, see Kunz et al. (2011). In a comprehensive European model, we provided a quantitative estimation of the network effects using our network model of the European electricity market ELMOD (Leuthold et al. 2012). Calibrating the model to a December weekday, we estimated that Germany would compensate for most of the capacity reduction through increased generation from coal and gas-fired units; see Fig. 5.2. Nevertheless, net exports would decrease as a result of more imports from neighboring countries. In case of the full closure of plants, the changed dispatch would lead to some shifts in the European power flows; for instance, Italy would have to rely on more domestic production to compensate for the reduced imports from Northern Europe.

In this study, we also looked at potential price effects resulting from the closure of nuclear power plants in Germany and in Central and Western Europe. Figure 5.3 shows the estimated effect of the nuclear moratorium on wholesale electricity prices on a representative day of the year. In off-peak hours, the estimated price effect was almost negligible since the European market had ample capacities available. The average estimated increase in peak hours in the moratorium case was €3–5/MWh.

5.3.3 ...Is Confirmed By Real-world Developments ...

Looking back at first model-based assessment from 2011, the prediction of only modest effects from the closure of nuclear power in Germany can be confirmed. Even though the closure of the moratorium plants and the "regular" closures (starting in 2015) proceeded as planned, the German and the Central and West European electricity markets remained in balance, and prices fell significantly instead of increasing, reflecting substantial overcapacity throughout the 2010s.

Kunz and Weigt (2014, 19–21) provide qualitative and quantitative evidence on developments between 2010 and 2013, highlighting the modest impact of the nuclear plant closures on electricity prices and trade as well as on the net export situation. As early as September 2011, Germany went from being a net importer to a large net exporter. The shutdown of the seven oldest nuclear power plants induced a short-term shift from nuclear to fossil fuel generation along with a decrease of German electricity exports. Furthermore, the change in generation implied increasing electricity market prices as nuclear generation was replaced with more expensive generation technologies, which essentially shifted the merit order to the left. Historical generation in Germany is depicted in Fig. 5.4, showing that reduced nuclear generation in spring 2011 was replaced by fossil and renewable generation as well as

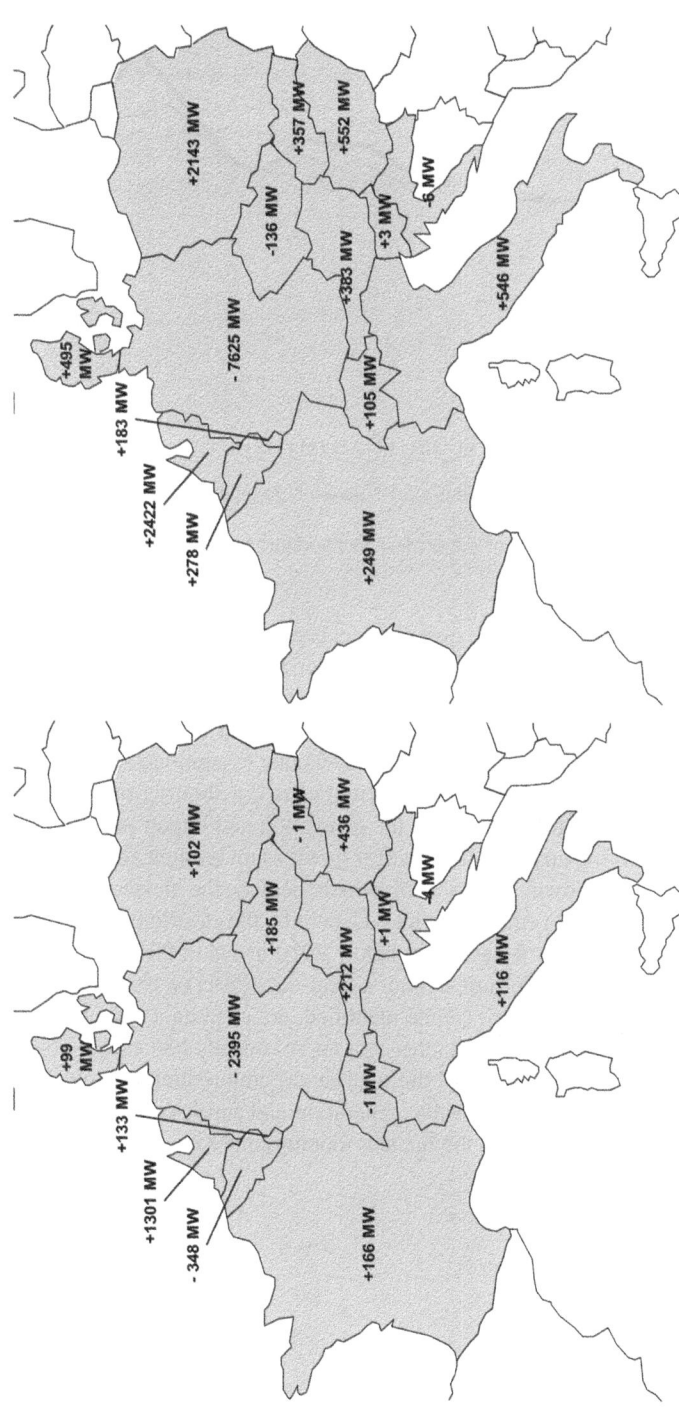

Fig. 5.2 Change of dispatch in the case of the nuclear moratorium (left) and full closures (right). *Source:* Kunz et al. (2011, 8)

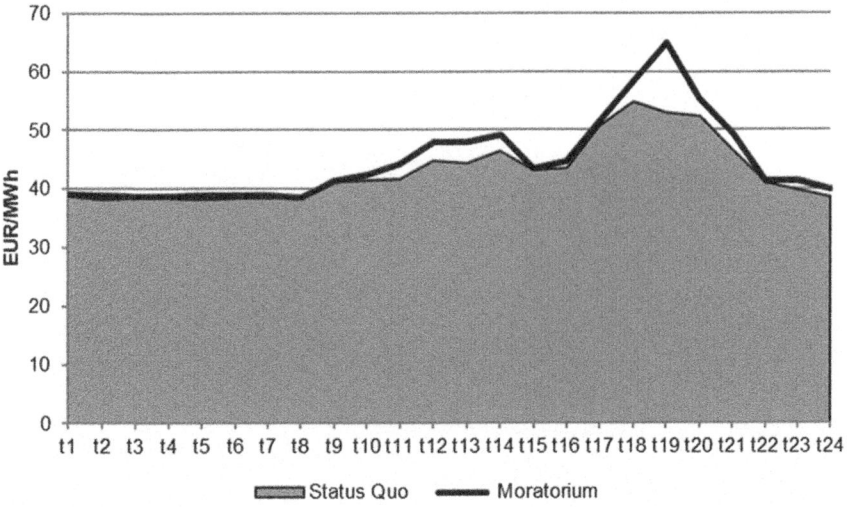

Fig. 5.3 Estimated price effects of the nuclear moratorium, compared to the status quo (in €/MWh). *Source:* Kunz et al. (2011, 9)

increased imports.[3] However, the import effect persisted for only a few months in the summer of 2011, a time of year usually characterized by lower exports due to both lower load and power plant maintenance. In winter 2011, the import situation changed again and domestic generation again exceeded load. In 2012 and 2013, German net exports increased above 2010 levels due to significantly higher renewable generation. Thus, the nuclear moratorium caused a short-term generation shift but no underlying structural change in the generation and import patterns.

Rather than driving up prices, the nuclear moratorium brought about a substantial price decrease, as shown in Fig. 5.5, which depicts the development of both electricity prices and CO_2 emission prices. Generally, the electricity price was higher in 2011 than in 2010, but it decreased below 2010 levels in 2012 and 2013. With respect to the nuclear moratorium, neither a clear-cut impact on electricity prices nor an underlying structural change can be identified, not even on a daily level. For the emission price, a temporary price increase of approximately €2/t can be observed in spring 2011, which may be due to the nuclear decommissioning. However, other impacts like the increase of renewable generation and surplus emission allowances affected prices much more than the nuclear moratorium (Kunz and Weigt 2014).

---

[3]In comparison to 2010, fossil and renewable generation increased by 18 TWh and net exports were lowered by 11 TWh to 6 TWh in 2011.

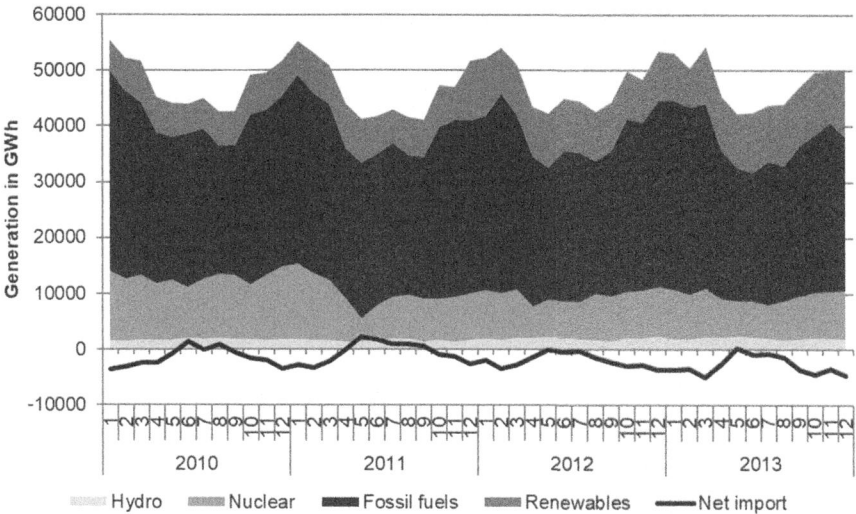

Fig. 5.4 Monthly electricity generation and import in Germany (2010–2013). *Source:* ENTSO-E (2014)

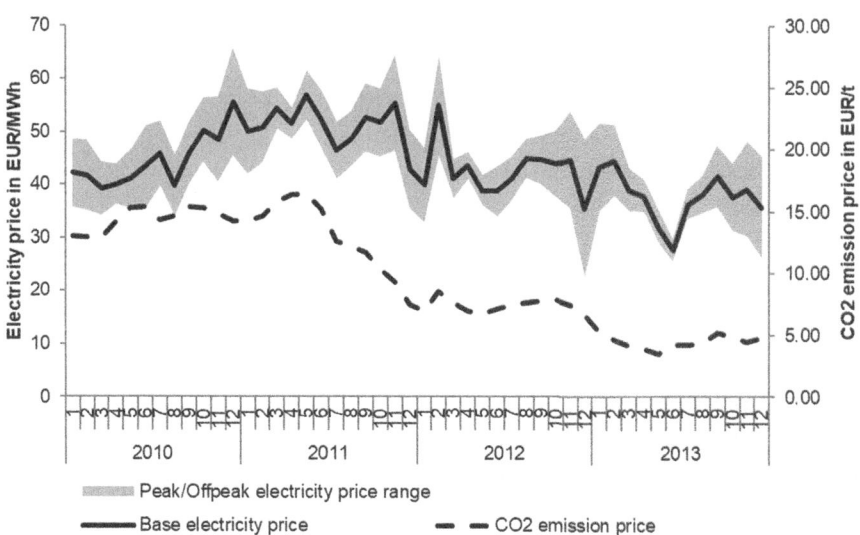

Fig. 5.5 Monthly average electricity and CO_2 emission price in Germany (2010–2013). *Source:* EEX (2014)

5.3.4   ... and No Shortage to Be Expected After 2022

It is too early to make reliable forecasts for the mid-2020s, but our model-based analysis suggests that Germany—and Central and Western Europe as well—will be able to absorb the closure of the remaining nuclear power plants. A first indication is that the closure of the Grafenrheinfeld NPP, in June 2015, had no noticeable effect on the electricity system, either on prices or on security of supply. Both in Germany and in other European countries, there were sufficient reserves available on the electricity market to compensate for the loss of the 1.3 GW net output (von Hirschhausen et al. 2015b).

In a study published in 2013, we reported that the German electricity system could cope with the loss of nuclear capacity based on calculations that included detailed network modeling.[4] Our results were confirmed in a subsequent study commissioned by the Central and Western European network operators for the period of 2020/21 that also concluded that there were no significant capacity bottlenecks expected in Germany.[5]

Our own updated model calculations indicate that even after 2022, when all of the remaining nuclear power plants will have been shut down, the security of supply in Germany and its neighboring countries will still be guaranteed under current plans. The analysis used both the scenario framework for Germany and the System Outlook and Adequacy Forecast (SOAF), the capacity planning scenario of the European Transmission System Operators (ENTSO-E) for 2025. According to ENTSO-E's 2015 capacity planning, capacities of 367 GW of conventional capacity were expected for Germany and its neighboring countries. In the context of the Central European electricity market, the German electricity supply was expected to be secure even during peak load hours when Germany would become a net importer.

The electricity price forecast for the wholesale market depends heavily on CO_2 and fuel prices. A comparison of the ordered price duration curve of the SOAF standard scenario and a scenario with reduced commodity prices for 2025 indicates a slight increase in prices overall. The average prices shown vary from €34 to 47/MWh (Fig. 5.6). CO_2 emissions would increase moderately depending on the scenario.

The analysis is based on a bottom-up approach and calculates a capacity balance for the year 2023. This is based on the capacity stock in 2013 and takes into account capacities under construction as well as expected plant decommissioning by 2023. Furthermore, planned and unplanned plant outages, intermittency of renewables (wind and solar), a contribution of peak load capacity from Austria and Luxembourg

---

[4]See Kunz et al. (2013): Mittelfristige Strombedarfsdeckung durch Kraftwerke und Netze nicht gefährdet. DIW Wochenbericht 80(48), pp. 25–37.

[5]See Pentalateral Energy Forum (2015): Pentalateral Generation Adequacy Probabilistic Assessment. Support Group Generation Adequacy Assessment, Final Report, March 5, 2015 online: http://www.bmwi.de/BMWi/Redaktion/PDF/G/gemeinsamer-versorgungssicherheitsbericht, property=pdf,bereich=bmwi2012,sprache=de,rwb=true.pdf, retrieved May 13, 2015.

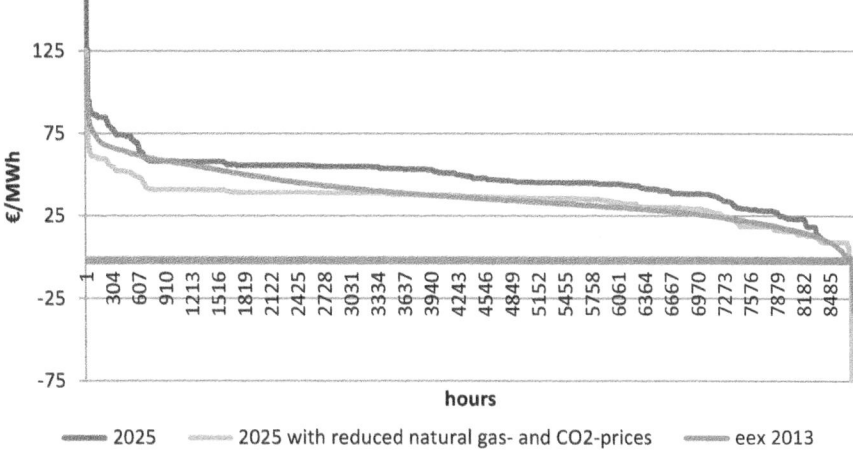

Fig. 5.6 Hourly wholesale prices in Germany (2013 and scenario for 2025). *Source:* Own depiction

as well as estimated load management between 4 and 10 GW are considered. It indicates that forecasted available capacities of conventional plants as well as an increasing share of renewables will suffice for resource adequacy.

5.4 Additional Challenges: Dismantling NPPs and Storing Nuclear Waste

In addition to the energy economic impacts of the German nuclear power policy, the accelerated closure of plants has also revealed a very interesting impact of the energiewende on corporate strategy and public policy: Dismantling nuclear power plants and developing a coherent strategy for long-term storage of nuclear waste are both extremely expensive undertakings. Both of these challenges had been largely ignored by the industry and policy makers in the early phase of nuclear power generation from the 1960s through the 1980s. But the financial, technical, and institutional challenges of the post-closure phase have grown substantially since then in Germany, and are beginning to appear in countries including France and the UK as well (Brunnengräber et al. 2015). The German example confirms what some countries experienced earlier and shows what others may experience in the near future: The technical and financial challenges have grown so large as to challenge the very survival of established utilities. Indeed, the high current and expected costs of dismantling and long-term storage would imply high financial risks for utilities if the "polluter-pays-principle," embodied in §9 of the German Atomic Energy Law, were actually applied. This underscores the need for solutions that can resolve the conflict between public policy goals ("the polluter pays, not the taxpayer") and corporate strategies in the long term by transferring a maximum of costs to the public sector and diversifying towards more sustainable business models.

5.4.1 Dismantling Nuclear Power Plants Delayed ...

Germany's nuclear legacy entails a number of challenges. The shutdown of nuclear power plants is followed by their decommissioning and dismantling. The Atomic Energy Act allows for two basic approaches to dismantling:

- The first is "immediate dismantling," in which the dismantling of the power plant begins immediately after the five-year post-operational period. Although this process generally takes about two decades, it can take longer and become more expensive if sufficient repository capacity for the radioactive waste is not secured in time. This already appears likely.
- In the second dismantling option known as "long-term enclosure," the power plant is closed off for several decades. The control area and in particular the nuclear reactor are not dismantled until after this time.

One advantage of the immediate dismantling option is that the operational nuclear power plants' staff and expertise can continue to be deployed. An argument in favor of the enclosure option is that the radioactivity decreases most in the first decades after the shutdown, which simplifies the subsequent dismantling and potentially reduces the volumes of waste.

The German record on dismantling nuclear power plants is weak, and further delays and cost overruns are expected. To monitor the long-term process of dismantling, we have begun a periodical survey of the progress on dismantling and of remaining challenges (Wealer et al. 2015, 2017; Schneider et al. 2018). Table 5.1 summarizes the current state of the dismantling process for German nuclear power plants, and Fig. 5.7 provides an overview of the nuclear power plants and waste storage sites in Germany. Only the prototype thorium high-temperature reactor THTR-300 and the plant in Lingen, which closed in 1977, have opted for the long-term enclosure strategy. Only three reactors or 140 MW have been successfully released from regulatory control.[6] Of these early prototype reactors only Kahl operated for a longer period of time (24 years, closed in 1985, completed in 2010, 25 years later). Of the commercial reactors only Würgassen and Gundremmingen-A have de facto completed decommissioning; although, both sites cannot be released from regulatory control as parts of the buildings are used for interim storage of wastes awaiting the commissioning of the disposal facility Konrad. Decommissioning of Stade (closed in 2003) was thought to be achieved by 2014, but ongoing difficulties due to unexpected contamination keep delaying the project. The Soviet-style plant in Lubmin, however, which closed immediately after

---

[6]BWR VAK Kahl, BWR HDR Großwelzheim, and the PHWR Niederaichbach.

Table 5.1 Overview of nuclear power plant dismantling in Germany (as of late 2017)

Reactor	Shareholder	Federal State	Reactor type	Net Capacity [Mega-watt]	Commercial start date	Permanent shutdown date	Decommissioning strategy	Begin
Biblis A	RWE AG	Hessen	PWR 2. Gen.	1167	26.02.1975	06.08.2011	Immediate dismantling	End 2017
Biblis B	RWE AG	Hessen	PWR 2. Gen.	1240	31.01.1977	06.08.2011	Immediate dismantling	End 2017
Brokdorf	80% E.ON; 20% Vattenfall	Schleswig-Holstein	PWR	1410	22.12.1986	Expected. 31.12.2021		
Brunsbüttel	66.6% Vattenfall; 33,3% E.ON	Schleswig-Holstein	BWR (KWU-69)	771	09.02.1977	06.08.2011	Immediate dismantling	2017
Grafenrheinfeld	E.ON	Bayern	PWR	1275	17.06.1982	27.06.2015	Immediate dismantling	2020
Greifswald 1–5	Energiewerke Nord GmbH (EWN)	Mecklen-burg-Vorpom-mern	PWR (VVERV-230/213)	5 × 408	12.07.1974–1.11.1989	24.11.1989–18.02.1990	Immediate dismantling	1995
Grohnde	83,3% E.ON; 16,7% SW Bielefeld	Nieder-sachsen	PWR	1360	01.02.1986	Expected 31.12.2021		
Gundremmingen A	75% RWE; 25% E.ON	Bayern	BWR (AEG-GE)	237	12.04.1967	13.01.1977	Immediate dismantling	1983
Gundremmingen B	75% RWE; 25% E.ON	Bayern	BWR (KWU-72)	1284	19.07.1984	31.12.2017	Immediate dismantling	
Gundremmingen C	75% RWE; 25% E.ON	Bayern	BWR (KWU-72)	1288	18.01.1985	Expected 31.12.2021	Immediate dismantling	
Isar 1/Ohu 1	E.ON	Bayern	BWR (KWU-69)	878	21.03.1979	06.08.2011	Immediate dismantling	
Isar 2/Ohu 2	75% E.ON; 25% SW München	Bayern	PWR (KWU-80)	1410	09.04.1988	Expected 31.12.2022		
Kahl	RWE, Bayernwerk AG		BWR(AEG)	16	14.05.1905	25.11.1985	Immediate dismantling	1988–2010
Krümmel	50% Vattenfall; 50% E.ON	Schleswig-Holstein	BWR (KWU-69)	1346	28.03.1984	06.08.2011	Immediate dismantling	2019/2020 (planned)

(continued)

Table 5.1 (continued)

Reactor	Shareholder	Federal State	Reactor type	Net Capacity [Mega-watt]	Commercial start date	Permanent shutdown date	Decommissioning strategy	Begin
Lingen	RWE	Nieder-sachsen	BWR (AEG-GE)	183	01.10.1968	05.01.1977	Long-term enclosure	
Lingen 2/Emsland	87,5% RWE; 12,5% E.ON	Nieder-sachsen	PWR (KWU-80)	1335	20.06.1988	Expected 31.12.2022		
Mühlheim-Kärlich		Rheinland-Pfalz	PWR	1219	18.08.1987	09.09.1988	Immediate dismantling	2004
Neckar-westheim 1	98,45% EnBW	Ba.-Wü.	PWR2. Gen.	785	01.12.1976	06.08.2011	Immediate dismantling	2017
Neckar-westheim 2	98,45% EnBW	Ba.-Wü.	PWR (KWU-80)	1310	15.04.1989	Expected 31.12.2022		
Obrigheim	KKW Obrigheim GmbH (100% EnBW)	Ba.-Wü.	PWR1. Gen.	340	31.03.1969	11.05.2005	Immediate dismantling	15.09.2008
Philipps-burg 1	EnBW	Ba.-Wü.	BWR (KWU-69)	890	26.03.1980	06.08.2011	Immediate dismantling	2017
Philipps-burg 2	EnBW	Ba.-Wü.	PWR	1402	18.04.1985	Expected 31.12.2019		
Rheinsberg	Energiewerke Nord GmbH (EWN)	Branden-burg	PWR(VVER-70)	62	10.10.1966	01.06.1990	Immediate dismantling	1995
Stade	66,7% E.ON; 33,3% Vattenfall	Nieder-sachsen	PWR(KWU1. Gen)	640	19.05.1972	14.11.2003	Immediate dismantling	2005
Unterweser	E.ON	Nieder-sachsen	PWR	1345	06.09.1978	06.08.2011	Immediate dismantling	
Würgassen	E.ON	Nordrhein-Westfalen	PWR	640	11.11.1975	26.08.1994	Immediate dismantling	1997

Source: von Hirschhausen et al. (2015a, p. 1074); Wealer et al. (2015); Bredberg et al. (2017) and Schneider et al. (2018)

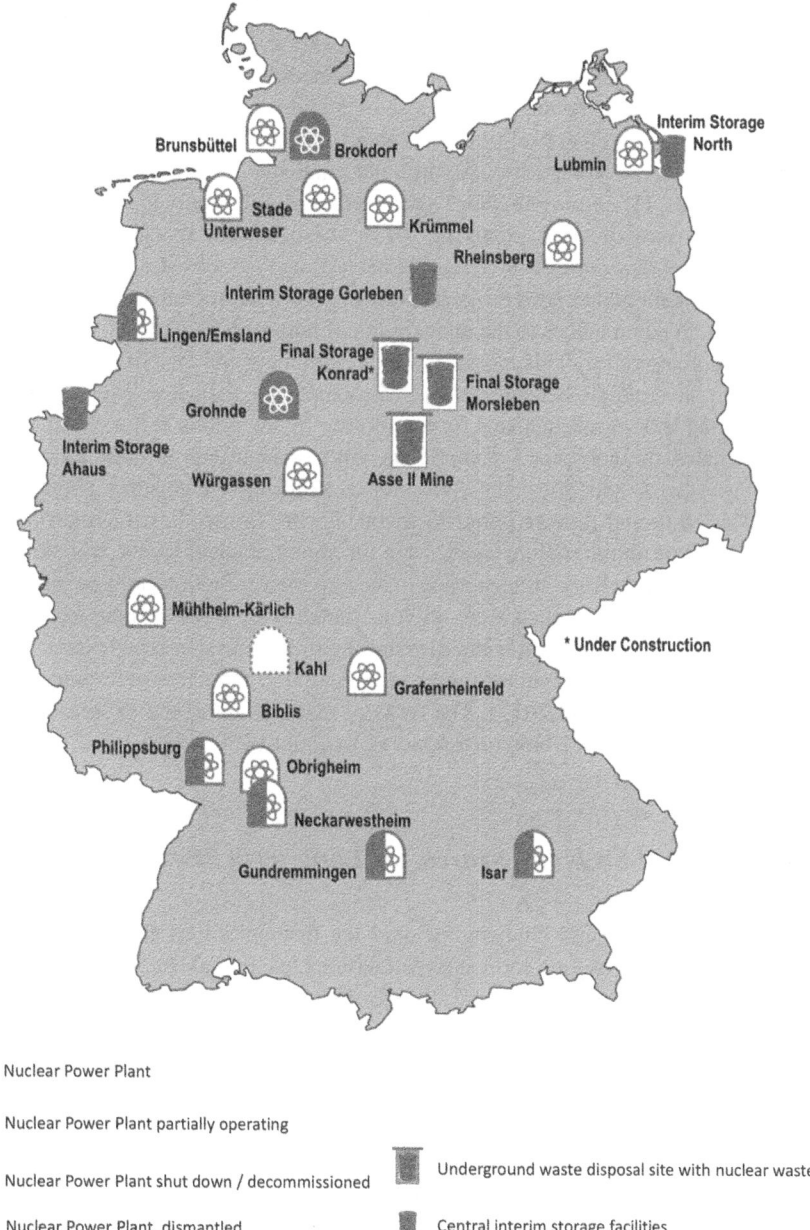

Fig. 5.7 Nuclear power plants and waste storage sites in Germany. *Source:* von Hirschhausen, et al. (2015a, p. 1076)

unification in 1989, is still in the process of dismantling, with completion not expected until the late 2020s (40 years after closure).

With the other plants that have opted for immediate dismantling, serious delays are expected, leading to the risk of cost escalations. In some cases, the lack of transport containers ("castors") has prevented transport between the core and the intermediate on-site storage site. A particular challenge lies in the remaining "special" fuel rods (*Sonderbrennstäbe*) that were not fully burned down after the premature closure of some plants in 2011. These require specific treatment, for which the necessary permits have not yet been issued (see Wealer et al. 2015).

When implementing major projects, planners face the problem of uncertainties in future costs. When it comes to the dismantling of nuclear power plants, the problems are even more complex. This has to do partly with a general lack of experience and partly with the strategic behavior of nuclear companies, their low level of transparency, and the lack of mechanisms for monitoring by the public sector. Standardized cost estimates do not exist.[7] Experience with dismantling is both limited and extremely varied, and does not allow for generalized conclusions about future costs. The costs and time required to dismantle the former East German nuclear power plant at Lubmin, for instance, were far above planned levels. But since this plant was built with Russian technology, the experience there cannot be applied to the remaining reactors, which were all constructed with West German technology. The initial estimate of €1 bn for the dismantling of the Russian-type reactors had to be revised upward several times; they had increased to the range of around €6.5 bn up to 2016 (Wealer et al. 2017). The costs of dismantling of the Würgassen plant (640 MW) were about €1 bn, corresponding to about €1500/kW.[8]

5.4.2 ... and So Is the Search for Long-term Storage

Even more complex and challenging than the dismantling of the nuclear power plants is the search for long-term storage facilities for the radioactive waste. This is largely because the long-term secure storage of highly radioactive waste has not been

---

[7]See: Küchler, S. et al. (2014): Atomrückstellungen für Stilllegung, Rückbau und Entsorgung—Kostenrisiken und Reformvorschläge für eine verursachergerechte Finanzierung. (Nuclear provisions for decommissioning, dismantling, and storage—cost risks and reform proposals for a polluter-pays-principle financing). Study conducted on behalf of the Bund für Umwelt und Naturschutz Deutschland (Friends of the Earth Germany). Online: http://www.bund.net/fileadmin/bundnet/pdfs/atomkraft/140917_bund_atomkraft_atomrueckstellungen_studie.pdf. Retrieved on May 8, 2015. For example, the energy company RWE accounts for only €600 of dismantling costs per kW of nuclear capacity; for Vattenfall, however, this figure stands at €1400/kW.

[8]See: Neue Westfälische online from October 25, 2015: "Rückbau des AKW Würgassen nach 17 Jahren abgeschlossen—Kosten von mehr als einer Milliarde Euro" ("Decommissioning of the Würgassen nuclear power plant completed after 17 years—costs in excess of 1 billion euros"). http://www.nw.de/lokal/kreis_hoexter/beverungen/beverungen/11276380_Rueckbau-des-AKW-Wuergassen-nach-17-Jahren-abgeschlossen.html. Last accessed: October 28, 2015.

seriously addressed since the German nuclear program was launched in 1959.[9] For the final storage, the Federal Ministry for the Environment, Nature Conservation, Building and Nuclear Safety (BMUB) is expecting a volume of around 190,000 cubic meters of low- and mid-level radioactive waste from German nuclear reactors.[10] This waste, along with other radioactive waste, is planned to be stored in the Konrad mine near Salzgitter in Lower Saxony, which is currently being constructed.[11] Since the entire 303,000 cubic meter capacity of the Konrad mine has already been completely allocated, there is no place for the waste that has yet to be retrieved from the Asse II mine—an additional volume of about 175,000–220,000 cubic meters. To accommodate this waste, either another repository would be needed or the Konrad repository would need to be expanded.

For the heat-generating waste—which refers to the high-level radioactive waste that includes spent fuel elements, as well as the radioactive waste that originates from the reprocessing of irradiated nuclear fuel and is solidified in glass canisters—the ministry expects a waste volume of 28,100 cubic meters.[12] For this high-level radioactive waste, there is still no final repository or even a planned repository site. The only attempt to build such a repository for high-level waste took place in Gorleben, where the investigations of a salt dome to house the waste have been ongoing since 1979. The dome has still not been granted a certificate of suitability based on geoscientific studies, and no alternative location has been investigated thoroughly. Until one is found, the high-level radioactive waste is being held in a large number of intermediate on-site storage areas.

The institutional process for identifying long-term storage sites for high-level radioactive waste is sketched out in the Repository Site Selection Act (StandAG) of 2013. It presents a transparent and scientific repository selection process, as well as a process for comparing potential sites that involves extensive public participation. The selection of underground sites for further investigation is scheduled to be completed by 2031, and a final decision on the location of the repository for high-level radioactive waste is to be made by 2031. Only a few years remain to complete these steps, which will be followed by the actual planning of the repository, the

---

[9]Although the commercial use of nuclear power has been permitted in the Federal Republic of Germany since the passage of the Nuclear Energy Act (AtG) in 1959, a safe option for the storage of radioactive waste did not exist at that time. Only with the 1976 adoption of the Fourth Amendment to the Nuclear Energy Act were waste producers obligated to remove the radioactive waste in an "organized" manner.

[10]See: BMUB (2015): Programm für eine verantwortungsvolle und sichere Entsorgung bestrahlter Brennelemente und radioaktiver Abfälle—Nationales Entsorgungsprogramm (Program for responsible and safe management of spent fuel and radioactive waste—National Waste Management Program).

[11]The schedule for the relatively simple Konrad Mine project for low- and mid-level radioactive waste has already experienced significant delays: between the 1982 application for planning approval for the low-level radioactive waste repository and the actual approval of the project in 2007, 25 years elapsed. At the time of writing, Konrad is estimated to open in the late 2020s.

[12]Federal Office of Radiation Protection: Forecasts for future waste volumes, online: http://www.bfs.de/de/endlager/abfaelle/prognose.html, retrieved May 14, 2015.

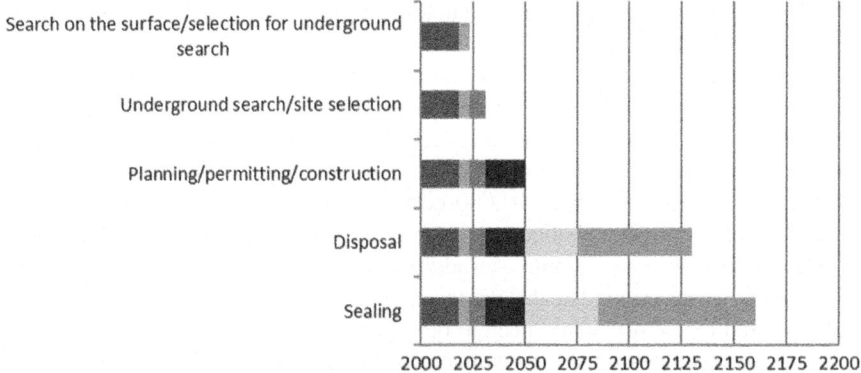

Fig. 5.8 Timetable for developing high-level radioactive waste storage in Germany. *Source:* von Hirschhausen et al. (2015b, 296)

nuclear licensing procedures, possible judicial reviews of the site and commission decisions, and finally, the actual construction of the repository. Thus, it appears that a site for high-level radioactive waste will not be ready before 2050.

Despite these uncertainties, the Commission for the Storage of High-level Radioactive Waste, which was established under the Repository Site Selection Act, has published a timetable for the selection process and the final storage process. Unforeseen developments could delay the final storage process by decades. The schedule of the ongoing process is fraught with uncertainties, as strikingly depicted in Fig. 5.8: the identification of a suitable site (scheduled for 2031), the storage process (starting after 2050), and the completion (towards the end of the century) and final closure of the site are expected to take around a century in total

5.4.3 Financing of the Process Uncertain

Given the high uncertainties related to dismantling plants and storing waste, there are complex issues and open questions about the financing of the project, which is a major liability of the process for decades to come. Over the past decades, the operators of NPPs have created provisions for unforeseeable liabilities arising from the dismantling of NPPs and the storage of radioactive waste. At the end of the 2013 fiscal year, these provisions amounted to nearly €38 bn (Table 5.2).[13]

At just under €22 bn, the estimated costs of dismantling significantly exceeded those of storage (€16 bn). Although the operators of nuclear power plants created these funds, they were regularly redirected to the respective parent companies, where they had then been used in other company divisions. This offered corporations a

---

[13]This section is largely based on von Hirschhausen et al. (2015b).

Table 5.2 Provisions for dismantling nuclear power plants and storage of radioactive waste in Germany (by company)

		Collection by Becker/Büttner/Held based on annual reports			
	Reserves in the nuclear energy sector in 2013 declared to BMWi (€ mn)	Already paid deposits until 2013 (especially Repository Financing Ordinance "EndlagerVLV") (€ mn)	Sum of collum 3 and 4 (€ mn)	of that: for dissmantling (€ mn)	of that: for waste disposal (€ mn)
E.ON	14,607	1134	15,741	10,308	5433
RWE	10,250	790	11,040	4769	6271
EnBW	7664	570	8234	4515	3719
Vattenfall	1652	91	1751	1155	595
Nuclear Power Plant Krümmel	1805	149	1954	900	1054
Sum	35,878	2735	38,720	21,647	17,072

Source: von Hirschhausen et al. (2015b, p. 527)

comparatively convenient financing source for profitable investment opportunities. In 2016, it remained highly questionable whether the current provisions were sufficient to satisfy the storage obligations placed on operators. Also, the question arised whether the value of the provisions was guaranteed up to the settlement date. The provisions were—like equity and debt—bound up in physical assets, and the upheavals that had taken place in the energy market in recent years had shown that the value and profitability of physical assets can change at short notice. For this reason, there was no financing security for the long-term commitments, in particular, in the area of radioactive waste management. In Germany, if an operating company would have become insolvent, the parent company would be liable for the subsidiary (provided that a control and profit transfer agreement exists, or an "unrestricted comfort letter" has been submitted). If a nuclear power plant operator were to become insolvent, there would be an increased risk that the federal government and thus the taxpayers would have to bear the additional dismantling and storage costs.

Clearly, the accumulated provisions (Table 5.2) did not include the risk of cost increases during the whole process of decommissioning, dismantling, and storage up to the closure of the repositories. Because science and technology are constantly evolving, the requirements emanating from the nuclear law and thus the cost of dismantling may ultimately be higher than expected. In addition, the provisions that had been set up did not include risk premiums to cover the potential costs of any necessary recoveries of radioactive waste or redevelopments of the repositories after their closures.

Different organizational models for the financing of dismantling and final storage were discussed in Germany. Maintaining the status quo, in which the NPP operators alone would be responsible for financing the unknown dismantling costs, made little sense due to the many unresolved issues, which included not just the costs but also the liability issues in cases of insolvency. Moreover, the Federal Administrative Court has made it clear that establishing provisions within the responsible companies is not sufficient to ensure the financing of decommissioning and post-closure obligations.[14]

One alternative to the status quo was to create two separate funds: one for decommissioning and dismantling, the other for final storage. However, this separation appeared risky, given the uncertainties in both areas. For example, the creation of a private legal fund by the energy companies for dismantling and decommissioning and a public legal fund for final storage had occasionally been proposed,[15] and this seemed at first to be the preferred option of the Economics Ministry. This model assumed that the dismantling costs are easy to estimate.

Then, starting October 2015, an expert commission[16] reviewed—on behalf of the government—the financing system and provided reform proposals to meet the before mentioned risks. Their recommendations and the new law published in December 2016 (BT 768/16) led to a fundamental change of the German funding system. Based on the reform proposals, a public legal fund was implemented in 2016, which will have to finance all aspects related to waste disposal (i.e., interim and final storage of all radioactive wastes). The fund was fed by the former provisions for these tasks totaling €23 bn including a risk premium. The utilities remain responsible for decommissioning and for the conditioning of waste, but all later tasks as well as the operation of the interim storage facilities will be done by public companies and paid from the money set aside in the fund. The responsibility as well as risks, including the financial ones in the case of insufficient set-aside money, will have to be borne by the public, which infringes the polluter-pays-principle.[17]

However, as we have shown above, since both technical as well as procedural issues—and therefore the expected costs—are very uncertain, there is a risk that the fund will turn out to be too small and the remaining costs will have to be covered by the public sector. Therefore, it appears to us that the formation of a single public law fund would have been the appropriate solution: Since the business model of the traditional utility companies continues to be under threat, and further losses are foreseeable, this fund should be created as quickly as possible. The nuclear waste producers should be required to supplement this fund to cover the following: the

---

[14]See the statements of the Federal Administrative Court on the financing of waste landfills, as well as the discussion with Ziehm (2015): Endlagerung radioaktiver Abfälle. Zeitschrift für Neues Energierecht—ZNER, 19, (3).

[15]For example, there is a report on this recommendation commissioned by the BMWi, see: Däuper and Fouquet (2014): Finanzielle Vorsorge im Kernenergiebereich—Etwaige Risiken des Status quo und mögliche Reformoptionen. Studie im Auftrag des Bundesministeriums für Wirtschaft und Energie. Berlin, 10. Dezember 2014.

[16]KfK—Kommission zur Überprüfung der Finanzierung des Kernenergieausstiegs, Commission on the financing of the German nuclear phase-out.

[17]See Jänsch, et al. (2017): Wer soll die Zeche zahlen? Diskussion alternativer Organisationsmodelle zur Finanzierung von Rückbau und Endlagerung. GAIA, Jahrhundertprojekt Endlagerung, 26(2), pp. 118–120.

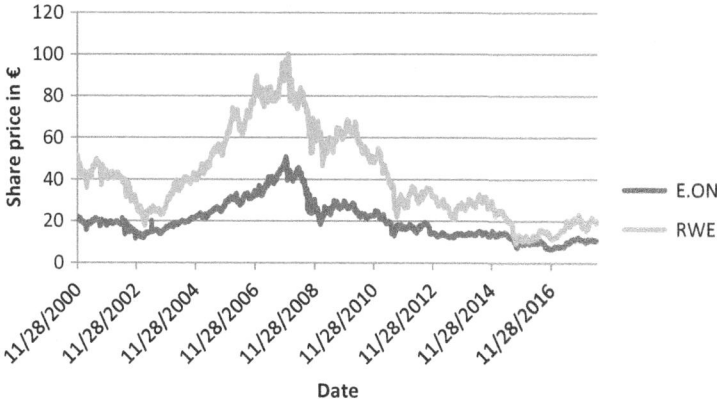

Fig. 5.9 Share price of big German utilities (E.on and RWE), 2001–2018. *Source:* Commodity System Yahoo

additionally required costs that were not covered by the provisions, including a realistic "cost increase factor"; an appropriate risk reserve; and the anticipated costs of the site selection process.

5.4.4 Implications for Corporate Strategies

As reported in Chap. 2, the energiewende and the decision on the phase-out of nuclear power in Germany (March 2011) marked the end of an era in which the big energy utilities were the Stackelberg leaders in energy policy making, with relatively little interference from policy makers or politicians. Since the 1970s, nuclear power production had been an extremely profitable activity, generating operational benefits of up to €1 mn per plant per day. The closure decisions and expected financial exposures forced the utilities to rethink their business models. Figure 5.9 shows the development of the shares of the two quoted companies, E.on and RWE, between 2011 and 2015. Although the utilities were initially prepared to fight another battle for lifetime extension in 2011, the political determination and massive public support for closure led to a reversal of the strategy, and they began attempting to divest as quickly as possible.

Over the past two decades, four large operators had dominated the German electricity market oligopolistically, and all four had significant nuclear assets. Their strategies for divesting themselves of their nuclear assets differed, but all four operators sought to separate these assets from the rest of their business and focus on potentially lucrative activities such as trading, grid operation, and renewables. Their strategies were also driven by a desire to shift responsibility for the additional costs of dismantling and waste storage elsewhere:

- Vattenfall was the first to divest its nuclear assets from Vattenfall Europe AG (*Aktiengesellschaft,* or joint-stock company) to the independent subsidiary company Vattenfall Europe Nuclear Energy GmbH (*Gesellschaft mit beschränkter Haftung,* or limited liability company) with a capital fund of only €500 million.
- The two stock companies tried a variety of methods to get rid of their nuclear activities: E.on did so by unbundling the nuclear assets into a "bad bank" company called Uniper in order to set up a new, renewables-based E. on without the old liabilities. But after encountering some government resistance, it accepted the obligation to keep the nuclear assets as a subsidiary (called Preussen Elektra) of the "New E.on".
- RWE also chose to split into two branches—one conventional and one for networks and renewables—but kept the two within one unified holding company (2015).
- EnBW, a large nuclear power group in southwest Germany considered outsourcing its nuclear liabilities to its majority owner, the State of Baden-Württemberg.

5.5 Conclusions

In this chapter, we have analyzed issues related to the decision to end the use of nuclear power in Germany after the Fukushima nuclear accident in March 2011. The decision represented the starting point of the energiewende. In fact, since the 1960s, all calls for an alternative "soft" path have been either supported or driven by the anti-nuclear movement. Chancellor Merkel's decision on March 14, 2011, which was set down in law later that summer, led to the closure of eight plants in spring of that year and the closure of all others in Germany by 2022 at the latest. The strong political support at all levels brought all debates over this decision to a swift end. Given the current state of the nuclear industry and the high liabilities, closures may even come earlier than 2022.

Model analysis carried out directly following the 2011 decision as well as ex-post analysis concur on the fact that the closure of seven old NPPs in 2011 had no significant effect on either the German or the Central Western European market. In March 2011, prices increased by a few €/MWh, only to continue their downward trend thereafter. Having become a net importer of electricity in 2011, Germany rapidly became a net exporter again and reached a record 54 TWh of exports in 2015. Alongside safety and cost arguments, there are also technical reasons for plant closure: nuclear power stations operate in a way that makes them very inflexible and prevents them from contributing to a flexible renewable-based energy system. Several studies also suggest that future closures of NPPs up to 2022 pose no threat to system adequacy either in Germany or in Central Western Europe; this is in line with current capacity forecasts of the European electricity industry (ENTSO-E), and also the calculations of the Pentalateral Forum (PLEF).

The real challenge after the closure of nuclear power stations both in Germany and in other countries will be the dismantling of plants and the search for long-term storage. In this respect, the German case provides ample evidence of the—widely underestimated—technical, institutional, and financial challenges that lie ahead. Dismantling is expensive and takes much longer than planned, as we have observed in monitoring the dismantling of the German NPPs. Operators now have incentives to further delay the process and seem to have reached a standstill, where no progress on dismantling can be observed. The search for a final storage site for high-level radioactive waste is extremely difficult and may last well into the late twenty-first century, if not beyond. Although €38 bn of provisions have been made by the companies of which €23 bn including a risk premium were transferred into a public legal fund for final storage these will be largely insufficient to cover the total costs, raising questions about the organizational models to be applied. We argue that a unified public fund for both, decommissioning and final disposal, is the best model to address future uncertainties.

Last but certainly not least, the energiewende also implied the end of the incumbent electricity utilities' conventional, fossil-nuclear business model. After an initial period of shock in 2011, they began to develop strategies to divest themselves of nuclear liabilities and shift the costs of waste storage, if not all liabilities, to the public sector. Faced with political determination and broad societal consensus, the nuclear industry gave up resisting the nuclear phase-out and changed its strategy: Its initial proposal to hand over the entire nuclear installations to the government, including open issues such as decommissioning and storage, is a clear sign that the industry has accepted a phase-out as inevitable and is focusing instead on reducing the costs or cutting its losses. After having fought against renewables for three decades, all of the restructured utilities now have business models that are focused on renewables (in addition to electricity trading and networks). Meanwhile, the costs of the nuclear legacy—in particular those of a high-level waste storage site—will likely end up being borne by the German government and taxpayers.

References

Bredberg, Ines, Johann Hutter, Kerstin Kühn, Katarzyna Niedzwiedz, Frank Philippczyk, and Frank Thömmes. 2017. *Statusbericht zur Kernenergienutzung in der Bundesrepublik Deutschland 2016*. Salzgitter: Bundesamt für kerntechnische Entsorgungssicherheit.

Bruninx, Kenneth, Darin Madzharov, Erik Delarue, and William D'haeseleer. 2013. Impact of the German Nuclear Phase-out on Europe's Electricity Generation – a comprehensive study. *Energy Policy* 60 (September): 251–261.

Brunnengräber, Achim, Maria Rosaria Di Nucci, Ana Maria Isidoro Losada, Lutz Mez, and Miranda A. Schreurs. 2015. *Nuclear waste governance: An international comparison*. Berlin, Germany: Springer.

Kunz, Friedrich, and Hannes Weigt. 2014. Germany s Nuclear Phase Out-A Survey of the Impact since 2011 and Outlook to 2023. *Economics of Energy & Environmental Policy* 3 (2): 13–28.

Kunz, Friedrich, Christian von Hirschhausen, Dominik Möst, and Hannes Weigt. 2011. Security of supply and electricity network flows after a phase-out of Germany's Nuclear Plants: any trouble ahead? IDEAS: RSCAS Working Papers, European University Institute RSCAS 2011/32. European University Institute. Robert Schuman Centre for Advanced Studies. Florence School of Regulation. Italy.

Leuthold, Florian, Hannes Weigt, and Christian von Hirschhausen. 2012. A large-scale spatial optimization model of the European Electricity Market. *Networks and Spatial Economics* 12 (1): 75–107.

Matthes, Felix Christian. 2012. Exit economics: The relatively low cost of Germany's nuclear phase-out. *Bulletin of the Atomic Scientists* 68 (6): 42–54.

Matthes, Felix Chr, Sabine Gores, and Hauke Hermann. 2011. *Zusatzerträge von ausgewählten deutschen Unternehmen und Branchen im Rahmen des EU-Emissionshandelssystems Analyse für den Zeitraum 2005–2012*. Berlin: WWF - Öko-Institut.

Nestle, Uwe. 2012. Does the use of nuclear power lead to lower electricity prices? An analysis of the debate in Germany with an international perspective. *Energy Policy* 41 (C): 152–160.

Schneider, Mycle, Antony Froggatt, Phil Johnstone, Andy Stirling, Tadahiro Katsuta, M.V. Ramana, Christian von Hirschhausen, Ben Wealer, Agnès Stienne, and Julie Hazemann. 2018. *World Nuclear Industry Status Report 2018*. Paris: Mycle Schneider Consulting.

Traber, Thure, and Claudia Kemfert. 2012. German Nuclear Phase-out Policy: Effects on European Electricity Wholesale Prices, Emission Prices, Conventional Power Plant Investments and Eletricity Trade. DIW Berlin, Discussion Paper 1219. Berlin, Germany: DIW.

von Hirschhausen, Christian, Clemens Gerbaulet, Claudia Kemfert, Felix Reitz, Dorothea Schäfer, and Ziehm Cornelia. 2015a. *Rückbau und Entsorgung in der deutschen Atomwirtschaft: Öffentlich-rechtlicher Atomfonds erforderlich*. Vol. 45. Berlin: DIW Wochenbericht.

von Hirschhausen, Christian, Clemens Gerbaulet, Claudia Kemfert, Felix Reitz, and Cornelia Ziehm. 2015b. German nuclear phase-out enters the next stage: Electricity supply remains secure – major challenges and high costs for dismantling and final waste disposal. *DIW Economic Bulletin* 5 (22/23): 293–301.

Wealer, Ben, Clemens Gerbaulet, Christian von Hirschhausen, and Jan Paul Seidel. 2015. Stand und Perspektiven des Rückbaus von Kernkraftwerken in Deutschland (»Rückbau-Monitoring 2015«). DIW Berlin, Data Documentation 81. Berlin, Germany: DIW Berlin, TU Berlin.

Wealer, Ben, Simon Bauer, and Christian von Hirschhausen. 2017. Decommissioning: Survey of International Experiences with Focus on Germany, France, Japan and South Korea and the U.S. Presented at the 2nd World Nuclear Decommissioning & Waste Management Congress (Europe) 2017, London, September 11.

Chapter 6
Renewable Energy Sources as the Cornerstone of the German Energiewende

Jonas Egerer, Pao-Yu Oei, and Casimir Lorenz

"I'd put my money on the sun and solar energy. What a source of power! I hope we don't have to wait till oil and coal run out before we tackle that."
Thomas Edison (late nineteenth century) (Quoted from James D. Newton (1987): Uncommon Friends: Life with Thomas Edison, Henry Ford, Harvey Firestone, Alexis Carrel, & Charles Lindbergh. San Diego: Harcourt Brace Jovanovich).

6.1 Introduction

At least since the 1980 study on the energiewende by Krause et al. (1980), renewable energies have been considered a viable alternative in Germany to conventional fossil fuels, and renewable energy technologies were seen as a "soft path" towards a more sustainable energy system. However, the energiewende of the 1980s focused solely on phasing out mineral oil and nuclear power and granting a stronger role to solar energy and energy efficiency, while maintaining a relatively high level of coal-based power generation. This perception has changed, and today there is a broad consensus on renewables being the very core of the energy mix. In fact, the German government's Energy Concept for 2050 declared the development of renewables as its

J. Egerer (✉)
Friedrich-Alexander-Universität Erlangen-Nürnberg, Erlangen, Germany
e-mail: jonas.egerer@fau.de

P.-Y. Oei
Junior Research Group "CoalExit", Berlin, Germany

TU Berlin, Berlin, Germany

DIW Berlin, Berlin, Germany

C. Lorenz
Aurora Energy Research, Berlin, Germany

© Springer Nature Switzerland AG 2018
C. von Hirschhausen et al. (eds.), *Energiewende "Made in Germany"*,
https://doi.org/10.1007/978-3-319-95126-3_6

number one energy priority.[1] The share of renewables in primary energy consumption was to rise to above 60% by 2050 (2020: 18%, 2030: 30%, 2040: 45%) and targets for the share of renewables in electricity consumption were set even higher: at least 80% by 2050 (2020: 35%, 2030: 50%, 2040: 65%) (BMWi and BMU 2010). These political objectives were formulated in detail in the law on renewable energies.[2] Renewables have thus become a cornerstone of the current energiewende.

This chapter discusses specific features of the German path toward a renewables-based electricity system and some challenges it is facing along the way. It also reports on the implications of a renewables-based electricity system for price formation and interrelations with conventional power plants. The next section recalls the development of renewables in Germany over the last 25 years from a niche source following the first feed-in law of 1990 to what has become Germany's number one electricity source since 2014, contributing over one third of the total supply and leaving lignite, coal, natural gas, and nuclear behind. Section 6.2 also sketches out government plans to reach its ambitious targets and the debate between the three main producers: solar, onshore wind, and offshore wind (plus to a certain extent bioenergy); while the renewable objectives for 2030 have already been clearly defined, opinions differ as to how to reach the 2050 objectives. We also survey the employment impacts of renewables. In Sect. 6.3, we argue that a renewables-based electricity system works very differently than the previous conventional system, for example, with respect to price formation, the dominant weight of fixed costs, the disappearing wedge between "peak" and "base" load, and the increasing role of flexibility. Section 6.4 takes a look at the issue of costs in the renewables transformation of the energy system, both from an aggregate perspective and from the perspective of individual technologies. The section also compares the costs of renewables with conventional generation (coal and nuclear), taking a public economics perspective, considering, for instance, the external (social) costs. We find that the renewables-based energiewende is welfare-enhancing compared to the high social costs of the previous fossil and nuclear-based energy system. Section 6.5 concludes.

6.2   Renewables as the Core of the Electricity System

6.2.1   1990–2015: From a Niche Player to the Main Electricity Supplier

As reported in Chap. 2, Germany introduced the first legal initiative to develop renewable energies in 1990 following similar activities at the European level. The law on feeding in electricity into the grid (*Stromeinspeisungsgesetz*, StrEG) of

---

[1]"Renewable energies as a cornerstone of future energy supply" (BMWi and BMU 2010, 7).

[2]Introductory paragraphs 1 and 2 of the EEG 2005; according to the Energy law (EnWG 2005), the share of renewables should be "continuously rising" (§ 1).

December 7, 1990, provided for fixed feed-in tariffs (FiTs) to integrate renewable sources, at the time mainly small local hydropower, into the energy system. In a time of purely monopolistic concession owners, the law obliged utilities to compensate producers of renewables and then to pass these costs on to consumers in the final price. A cap of 5% was set for renewables, and the law included the possibility of an exemption for utilities that were particularly affected by the feed-in of renewables.

From 1990 until today, the legislation on renewables has been spread among a number of specific laws, and has not yet been integrated into the more general energy law. One might explain this sector-specificity by the strong lobbying power that proponents of renewable energies had since the first specific legislation was proposed, and they have resisted any integration of this legislation into the more general energy law to this day. In fact, between 1998 and 2014, the responsibility for policies on renewables was with the Ministry of Environment, which worked to some extent in competition with the Energy Department of the Economics Ministry, traditionally more inclined towards conventional energies. However, the sector-specific legislation even survived the merger of the two departments into the Federal Ministry for Economic Affairs and Energy in 2014.

Major reforms of the renewables legislation took place in the EEG 2000,[3] thanks to a red-green initiative pushed by Hans-Joachim Fell (Greens) and Herrmann Scheer (SPD). Subsequently, legislation was extended by the EEG 2004,[4] the EEG 2009,[5] the EEG 2012,[6] the EEG 2014,[7] and the EEG 2017.[8] The EEG 2000 banned the 5% cap and provided for a substantial increase of the use of renewable energies (to 15% by 2015), in order to attract private capital into the sector and allow economies of scale. The 2004 revision of the law adapted the feed-in tariffs and introduced particularly favorable conditions for bioenergy. Whereas all previous laws focused on a fixed feed-in tariff, the EEG 2009 contained the first provision for the direct marketing of renewables by producers (*Direktvermarktung*). The EEG 2012 included specific provisions for offshore wind parks and for geothermal energy, a source that has remained marginal until today.

---

[3]Gesetz für den Vorrang Erneuerbarer Energien, March 29, 2000 (EEG 2000), Bundesgesetzblatt 2000, 13, p. 305.

[4]Gesetz für den Vorrang Erneuerbarer Energien, July 21, 2004 (EEG 2004), Bundesgesetzblatt 2004, p. 1918.

[5]Gesetz zur Neuregelung des Rechts der Erneuerbaren Energien im Strombereich und zur Änderung damit zusammenhängender Vorschriften, October 25, 2008 (EEG 2009), Bundesgesetzblatt 2008, 49, p. 2074.

[6]Gesetz zur Neuregelung des Rechtsrahmens für die Förderung der Stromerzeugung aus erneuerbaren Energien, July 28, 2011 (EEG 2012), Bundesgesetzblatt 2011, 42, p. 1634, amended by the law of August 17, 2012, Bundesgesetzblatt 2012, 38, p. 1754.

[7]Gesetz zur grundlegenden Reform des Erneuerbare-Energien-Gesetzes und zur Änderung weiterer Bestimmungen des Energiewirtschaftsrechts, July 21, 2014 (EEG 2014), Bundesgesetzblatt 2014, Part I, 2014, 33, p. 1066.

[8]Gesetz zur Einführung von Ausschreibungen für Strom aus erneuerbaren Energien und zu weiteren Änderungen des Rechts der erneuerbaren Energien, October 13, 2016 (EEG 2017), Bundesgesetzblatt 2016, 49, p. 2258.

From the EEG 2014 onwards, the European Commission took a stronger position vis-à-vis the renewables legislation in Germany (and other Member States), launching a legal debate over whether guaranteed feed-in payments were to be considered as state aid. Subsequently, the renewables laws were revised to include more "market-based" elements: the EEG 2014 started a trial period for contracting 400 MW of large photovoltaic projects with an auctions mechanism and imposed a "market premium", that is, an uplift on the regular wholesale market price, and direct marketing by the producers of renewable energies.

With the EEG 2017, auctions replaced the feed-in tariff for wind power and large-scale photovoltaic projects (>750 kW). The annual auction budgets until 2022 cover about 2800 MW in onshore wind, 600 MW in large-scale photovoltaic systems (the annual target is 2500 MW including small solar systems), and 400 MW in technology-neutral auctions. Thereby, the auction mechanism applies regional limitations for the share of onshore wind power in the northern coastal regions and of photovoltaics in the south. The process of implementing auctions for offshore wind power is more complex and still ongoing for several years due to the specific characteristics of offshore projects. In general, the practical implementation of "pilot auctions" has proven complex due to increased transaction costs. It is therefore unclear, whether the number of participants will remain high and auctioning is really an appropriate way forward for low-cost renewables supply in the longer term. In 2017–2018, auctions on onshore wind and photovoltaic were oversubscribed several times due to a large project pipeline. This has resulted, so far, in very competitive bids for onshore wind and in particular for large-scale photovoltaic. In fact, the first technology-neutral auction has contracted only photovoltaic projects, bidding lower prices than onshore wind projects.

The institutional framework in Germany has been effective in pushing renewables to become a major source of electricity. In fact, it has led to a boom in renewable electricity, particularly from wind but also to a lesser degree from solar installations and bioenergy. Between 1990 and 2017, the share of renewables in gross electricity production increased from 3.6% to 33.3%, the number one source of electricity, ahead of lignite (22.5%), hard coal (14.1%), natural gas (13.2%), and nuclear energy (11.7%). In absolute terms, production has risen from 20 TWh (mostly hydropower) to 218 TWh.[9]

Figure 6.1 provides an account of the growth of renewables during this period: onshore wind clearly took the pole position from hydropower in the early 2000s, whereas solar power has been showing the highest absolute growth rates for some years since the late 2000s. Still, as large-scale photovoltaic in Germany has become cost competitive to wind power, it might see a comeback in the next years with higher growth rates than currently predicted. Offshore wind has growing numbers

---

[9]Source: Arbeitsgemeinschaft Energiebilanzen (AGEB). 2018. "Bruttostromerzeugung in Deutschland ab 1990 nach Energieträgern." Arbeitsgemeinschaft Energiebilanzen e.V. February 2018. In parallel, a 2009 law on heat from renewables aimed at a share of 14% of renewables in energy consumption for heat (space heating and cooling, process heat, warm water).

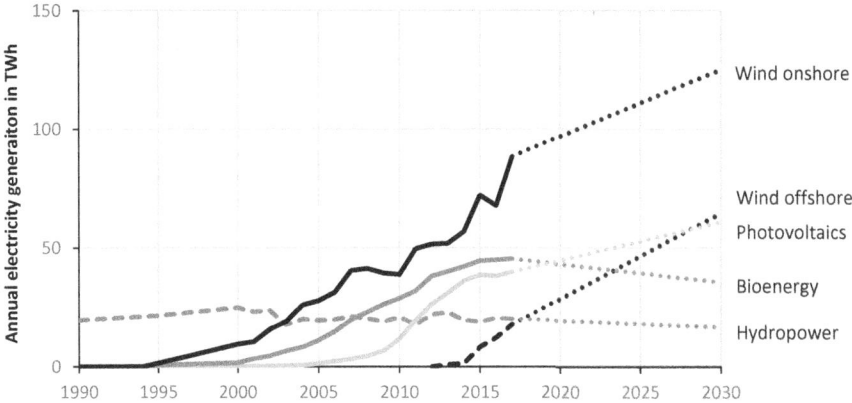

Fig. 6.1 Electricity production from renewable sources (1990–2017) and projection until 2030. Source: Own depiction based on AGEB (2018) and BNetzA (2017) (AGEB. 2018. "Bruttostromerzeugung in Deutschland ab 1990 nach Energieträgern." Arbeitsgemeinschaft Energiebilanzen e.V. February 2018; BNetzA. 2017. "Bestätigung des Netzentwicklungsplans Strom für das Zieljahr 2030." Bonn, Germany)

since 2014, but its future role is still unclear and will depend upon whether the significant cost disadvantage can be reduced. The projections in Fig. 6.1 reflect the main scenario of the German Grid Development Plan[10] which reaches a renewable share of about 55% in electricity generation for the year 2030.

6.2.2   Significant Employment Effects

In macroeconomic terms, too, renewables have grown from a niche segment to center stage of the German energy sector. This is particularly the case for employment, where renewables have outpaced the traditional sectors by far. In the mid-2010s, about 300,000 direct and indirect jobs had been created in the different segments of renewables. Figure 6.2 shows, in comparing the employment effects of renewable energies, that Germany still outweighs other European countries with most jobs being in the field of wind power and bioenergy. The overall number of renewable jobs in Germany has been somewhat higher around 2010, before the photovoltaic business declined and related jobs decreased from over 100,000 to 27,000 in 2016. On the second place follows Italy (180,000), followed by France and Spain (140,000), and the United Kingdom (110,000). The distribution of jobs also differs among the countries, according to the type of renewable energies used. Whereas most jobs in Italy, France, and Spain relate to bioenergy and heat pumps,

---

[10]BNetzA. 2017. "Bestätigung des Netzentwicklungsplans Strom für das Zieljahr 2030." Bonn, Germany.

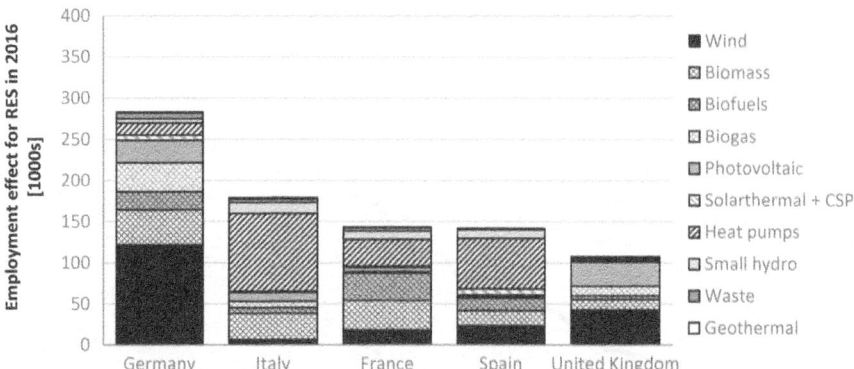

Fig. 6.2 Employment distribution for renewable energies in selected European countries (2016). Source: EurObserv'ER 2017 (2017) [EurObserv'ER. 2017. "The State of Renewable Energies in Europe." 17th EurObserv'ER Report. Paris, France: Observ'ER (FR), ECN (NL), RENAC (DE), Frankfurt School of Finance and Management (DE), Fraunhofer ISI (DE) and Statistics Netherlands (NL)]

the United Kingdom doubled its renewable jobs mainly in wind power and photovoltaics between 2012 and 2016.

The regional distribution of the employment effects within Germany follows mainly the availability of renewable resources. Figure 6.3 shows the distribution of jobs in renewable energy sources (RES) to the federal states of Germany, and its distribution by subsector: jobs in the wind industry focus on northern Germany, solar has become more important in central and southern Germany, and jobs in bioenergy are well distributed and correlated to the size of federal states. Fewer jobs exist in hydro and geothermal energy, most of which are in the South. Estimations on the net employment effect for Germany, including all positive and negative factors, predict the highest rise in employment in the construction sector, where job growth clearly outweighs the job losses in the mining and service sectors. In a study by Dehnen et al. (2015), the authors estimate the future annual net effect up to 2020 to be on average 18,000 jobs.[11]

The broad regional dispersion of employment in the renewables sector (with the exception of the city-states Berlin, Bremen, and Hamburg, which have less space per capita than the other federal states) is also an advantage, when compared with the local clustering of the former, fossil-nuclear energy system. This becomes particularly evident when comparing the employment of renewables with those of the lignite and nuclear sector, which is highly concentrated in a small number of regions. The shift from conventional to renewable capacities, therefore, has different positive and negative effects on the various regions. Both past and remaining jobs in these conventional sectors are mostly concentrated in the lignite mining regions of Brandenburg, NRW, and Saxony. The number of jobs in renewables, however, has

---

[11]Dehnen, Nicola, Anselm Mattes, and Thure Traber. 2015. "Die Beschäftigungseffekte der Energiewende." Berlin, Deutschland: DIW Econ.

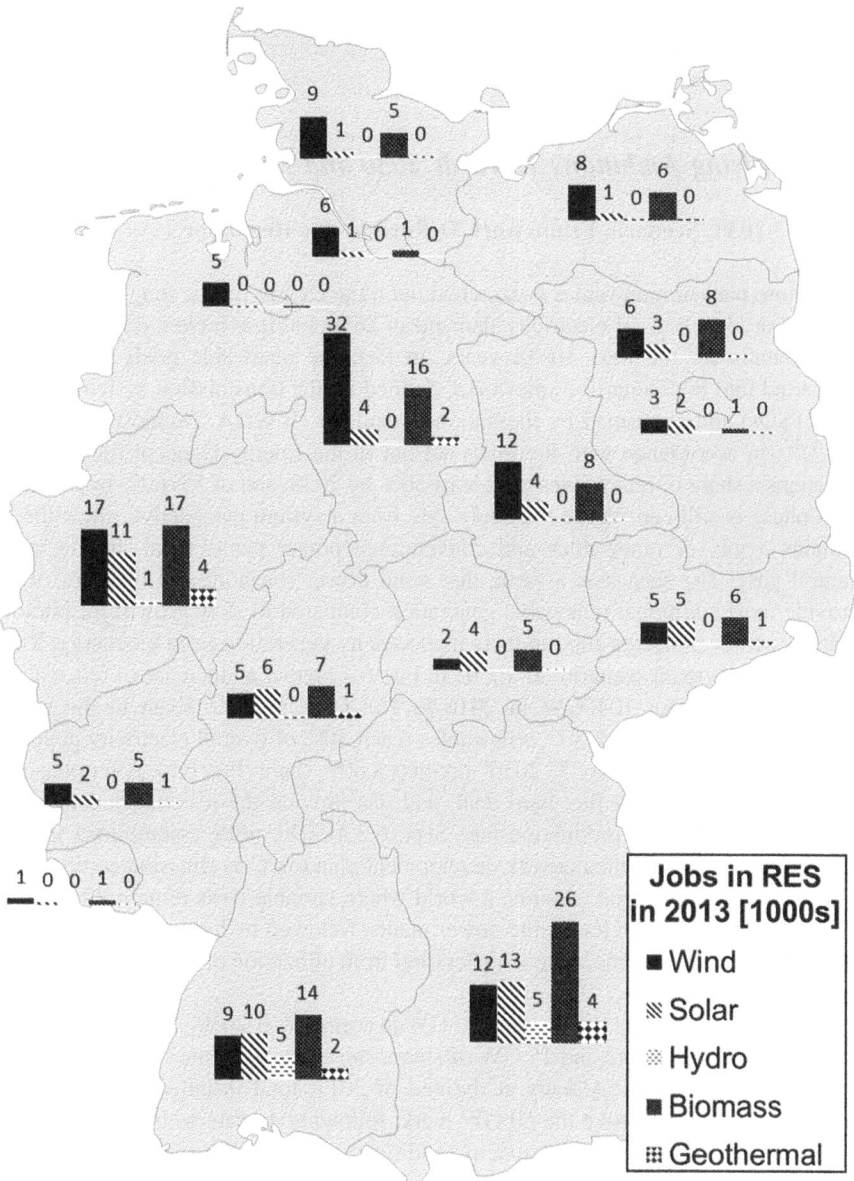

Fig. 6.3 Employment in renewable energies in Germany in 2013. Source: Own illustration based on data from Ulrich and Lehr (2014) (Ulrich, P., Lehr, U.—GWS mbH (2014): Erneuerbar beschäftigt in den Bundesländern: Bericht zur aktualisierten Abschätzung der Bruttobeschäftigung 2013 in den Bundesländern; Osnabrück)

outnumbered by now the remaining jobs in the coal business even in those states (see Chap. 3 on hard coal and lignite).

6.2.3   Rising Ambitions Towards 2030 and 2050

6.2.3.1   2030: Scenario Framework Defined by the Regulator

The future path of renewables is sketched out quite clearly for the long term with a share of at least 80% of electricity demand in 2050 and it is broken down in quite some details for the next 10–15 years. In fact, the renewable goals have been converted into the scenario frameworks, defined by the transmission system operators (TSOs) and confirmed by the national regulator (BNetzA, Bundesnetzagentur 2017)[12]: in accordance with the goals set out in the Energy Concept for 2050, it envisions a share of renewables of at least 50% by 2030, and of 55–60% by 2035.[13] This phase is still considered relatively safe from a system perspective, since there remains a mix of renewables and conventional power plants (coal, lignite, and natural gas). The scenarios assume that wind power (onshore and offshore) will provide most additional renewable generation compared to slow growth for photovoltaics and even decreasing numbers in electricity generation from bioenergy. The numbers in the main scenario "B 2030" in Fig. 6.4 predict an increase of renewable capacity from about 104 GW in 2016 to 153 GW in 2030. Even in the most pessimistic scenario "A 2030", renewables reach 50% of overall electricity generation in 2030 while scenario "C 2030" predicts a 60% share (Fig. 6.5). After phasing out nuclear in 2022, lignite, hard coal, and gas provide the residual demand not covered by renewable production (see Sect. 6.3.4). The price assumptions in the modelling exercise for the network development plan (on CO_2 emission certificates, hard coal, and natural gas) assume a world where variable costs remain the lowest (and utilization highest) for lignite power plants, followed by hard coal, and natural gas which also gains some share in generation from utilization of gas-fired combined heat-and-power plants.

The example of wind power, with 74 GW in scenario "B 2030", of which 59 GW is expected to be onshore and 15 GW offshore, shows the dynamic of the renewable transformation process. Already at the end of 2017, total installed onshore wind capacity in Germany broke the 50 GW mark, following 4 years with an unforeseen 4.6 GW in average annual capacity additions. Consequently, the preliminary

---

[12]BNetzA. 2017. "Bestätigung des Netzentwicklungsplans Strom für das Zieljahr 2030." Bonn, Germany.

[13]The scenario framework of the four TSOs, produced every 1–2 years in the context of the network development plan, provides a firm corridor for future developments. The exercise produces an outlook with three 2030 scenarios and one 2035 scenario calibrated to governmental objectives that establishes a "roadmap" not only for the subsequent network development plan but also for all of the stakeholders involved in the process.

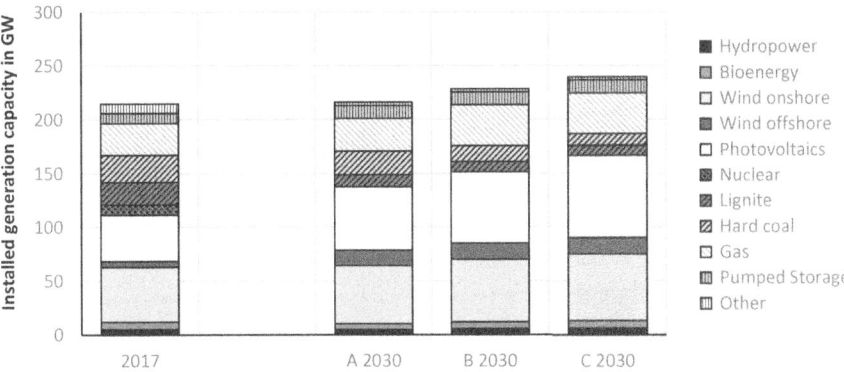

Fig. 6.4 Generation capacity in Germany (2017 and official scenarios for 2030). Source: Scenario framework in the Grid Development Plan (BNetzA 2017)

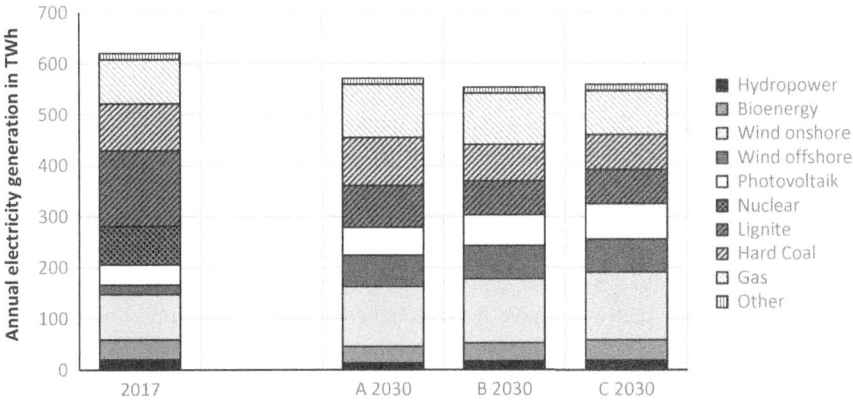

Fig. 6.5 Electricity generation in Germany (2017 and official scenarios for 2030). Source: Scenario framework in the Grid Development Plan (BNetzA 2017)

scenario framework has been adjusted in the 2019 version of the network development plan, increasing mainly onshore wind capacity (+11 GW) and some photovoltaics (+2 GW) for 2030.

6.2.3.2   2050: Pathway Beyond 80% Renewable Electricity

By contrast, the path from 2030 to 2050, the date by which renewables have to cover at least an 80% share of demand, is more uncertain. Higher renewable share challenges the role of conventional power plants. Due to the ongoing decline in costs for renewable electricity generation, onshore wind and large-scale photovoltaic have become competitive to new fossil-fired power plants in Germany. On the contrary, fossil-fired power plants see raising generation costs by lower utilization

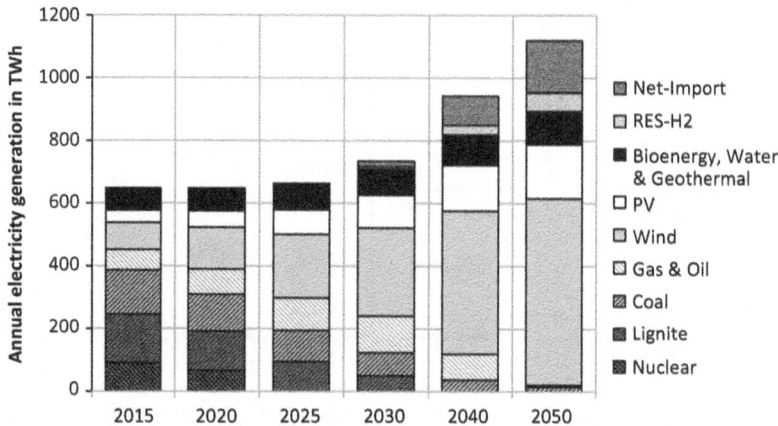

Fig. 6.6 Pathway towards a low-carbon energy system in Germany in 2050. Source: Nitsch (2016, 33)

rates and higher prices for CO_2 emission certificates in a carbon-constrained world. Between 2030 and 2050, large fossil-fired power plants, serving as "base load" capacities, will widely have been phased out and the role of storage and other flexibility options will have to increase. The scenarios of the grid development plan describe the first phase of this transformation towards sector coupling with assumptions on heat pumps, e-mobility, power-to-gas, power-to-heat, small battery storages for photovoltaic systems, and demand-side management of large electricity consumers. Several pathways have been sketched out for that future, and they all converge that, while the technology mix cannot be predicted with certainty, there is no doubt about the technical feasibility of scenarios reaching the target of 80% of renewables or even close to 100%, by 2050.[14]

The German government has regularly relied on a team of economists and engineers to produce bi-annual scenarios for the electricity system in 2050, called "lead study" (*Leitstudie*). Figure 6.6 shows a scenario leading to 2050, designed by the main author of the team, (Nitsch 2016, 21), that is in line with the climate targets for 2050. The scenario calculations show that fossil-fuel generation is largely phased out after 2030, and the future system largely relies on wind, photovoltaic, bioenergy, and some hydrogen. The scenario foresees a significant increase in electricity consumption, from currently 600 TWh to over 1100 TWh, due to the large-scale electrification in the transport and heating sectors. Imports of renewables from other countries might also play a significant role, but rather in the medium- or long-run.

---

[14]For example, the vision of a 100% RES-based system sketched out by SRU (2011) relies on extensive exchanges with the neighbouring countries, mainly Scandinavia.

6.3   A New Era for the Electricity System

Clearly the focus on renewable energies in the electricity system constitutes a major break with the conventional, fossil-nuclear system. At the beginning of the energiewende, the fundamental character of this transformation was not widely understood, and some observer still consider it as a short-term phenomenon, believing that conventional power will remain the pillar of the system, requiring only marginal modifications to the regulatory framework; this vision has been expressed, for instance, in the European Roadmaps of the European Commission (EC 2011). However, when considering the technical, economic, and institutional implications of a system based on a very high share of renewables, such as Denmark or Germany, there are clear indications of disruption with the old system, in which many of the traditional features of the past are modified, such as the "energy-only market", the differentiation into "base" load and "peak" load, etc. This subsection describes some of the elements of this disruption, and we also report on a similar line of argumentation presented by the think tank Agora Energiewende, summarized in Box 6.1: "12 Insights on Germany's energiewende".

6.3.1   The Merit Order Effect

The rise of variable renewables from a small niche market to the center stage will bring with it significant modifications to the electricity system. While some of the effects are still ongoing and many other changes will affect the functioning of the German and the European electricity markets, it is already evident that the conventional electricity market is no longer working as it used to, and business models for energy companies are undergoing substantial change as well. An important change has been introduced with the cost structure of the variable renewable technologies wind and solar: both are capital-intensive but have negligible incremental costs in contrast to classical conventional energy sources, which feature relatively high incremental costs and comparable low capital costs.

An increasing supply of renewable electricity at almost zero marginal costs changes the hourly wholesale electricity market price by shifting the supply curve to the right, in particular in hours with high availability of renewable generation (see Fig. 6.7). In a fully competitive market setting, the marginal power plant that sets the hourly price (intersection between supply and demand curve) will have lower costs than before the energiewende. The difference between the two prices is called the "merit order effect", which reduces profits of power plants in the short-term and in the medium-term requires adjustments of the power plant portfolio. Assuming that the slope of marginal generation costs increases with supply, the merit order effect will be stronger in hours of high demand (peak) and weaker in hours with low demand (off-peak). The strength of the effect depends on whether demand is assumed to be elastic or inelastic and the level of time disaggregation (number of

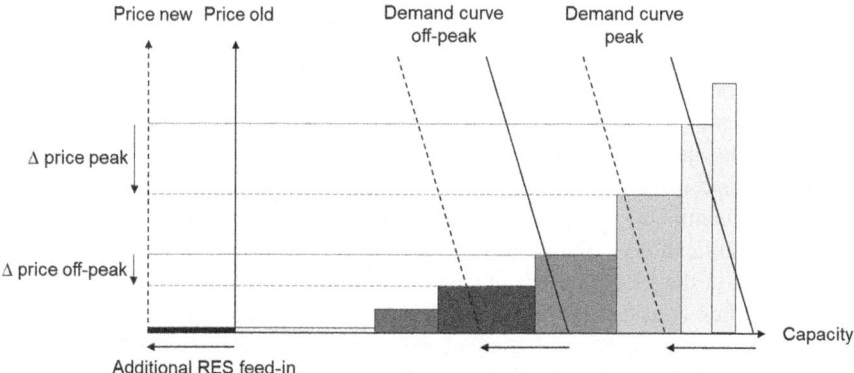

Fig. 6.7 The merit order effect in off-peak and peak demand hours. Source: own depiction

time slices): the higher the level of aggregation (averaging), the lower the merit order effect will be.

Different studies and papers have produced different estimates of the merit order effect attributable to renewable electricity. Table 6.1 shows a solution of ex-post estimations of the merit order effect in Germany in the years 2006–2011. While there is some variance due to different methodological approaches and data used, the merit order effect appears to be significant, at an average of ~0.7 cent/kWh; this corresponds to 15% of the wholesale electricity price (averaged over the entire period).[15]

6.3.2 No More Distinction Between "Peak Load" and "Base Load"

The conventional electricity system has long relied on a clear distinction between "peak load," defined by high prices and eventual price spikes, and a "base load," with relatively modest prices. Power plants have all been calibrated to this structure, consisting of base load plants, such as lignite and nuclear, "mid load" by hard coal and combined cycle gas turbines (covering higher demand during the day), and "peakers," such as gas turbines. Figure 6.8 shows the development of hourly average electricity prices in Germany, in 2011 and 2014, which allows the identification of different effects. Overall, spot prices significantly decreased, by almost 20 €/MWh, mainly due to the merit order effect, lower fuel prices, lower CO_2 prices, and also due to reduced demand. In addition, one also identifies the dampening effect of midday solar electricity (9–16 h), which further reduces prices, by about 4.50 €/MWh during that time. A similar, somewhat weaker effect is triggered by wind in the early evening

---

[15]Assuming inelastic demand, the renewable electricity has thus reduced the annual electricity bill of wholesale consumers by 3.5 billion € (500 TWh × 0.7 cents/kWh).

Table 6.1 Ex-post estimations of the merit order effect in the years 2006–2011

	2006	2007	2008	2009	2010	2011	2012	2015	2020
Cludius et al. (2013)				−0.52	−0.72	−1.14			
Sensfuß and Ragwitz (2007)	−0.78								
Sensfuß (2011)		−0.58	−0.53	−0.6	−0.52	−0.87	−0.89		
Traber et al. (2011)									−0.32
Weigt (2009)	−0.62	−1.04	−1.3						
EWI (2012)								−0.2	−0.5
Speth and Warzecha (2012)					−0.56	−0.56			
Speth and Klein (2012)						−0.748			
Vereinigung der Bayerischen Wirtschaft e.V. (2011)	Average −0.8 ct/kWh (2006–2010)								

Source: Bundesministerium für Wirtschaft und Energie (BMWi). 2014. Zweiter Monitoring-Bericht "Energie der Zukunft". p. 38. https://www.bmwi.de/Redaktion/DE/Publikationen/Energie/zweiter-monitoring-bericht-energie-der-zukunft.html

hours, when a lot of wind blows. With a rising share of renewables, and higher flexibility on the demand side, but also on the supply side, the traditional differentiation between "peak" and "off-peak" prices is strongly reduced.

Conceptually, the previous concept of "base load" disappears in a renewables-based electricity system. This has significant economic consequences as conventional power plants are unable to cover their fixed costs through high inframarginal rents obtained in "peak" hours. It has also consequences for the cost coverage of younger (recently built) power plants and for decisions on future investments, which become less attractive; it also affects the additional rents to be gained from old, amortized power plants, and the decision when to retire them.

6.3.3   Wholesale Electricity Prices in Germany

While the "merit order effect" has certainly contributed to a reduction in the overall wholesale price, it is not the only factor driving the decline in electricity prices observed over the past years. Figure 6.9 shows the general trend of the electricity wholesale price at the EEX-Energy Exchange, by annual averages, from 2007 to

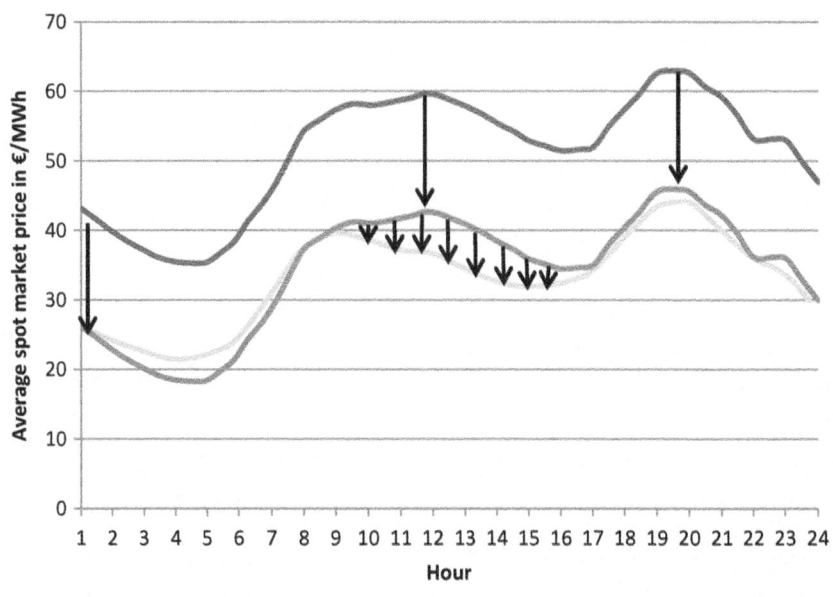

Fig. 6.8 The disappearing difference between "peak" load and "base" load in a renewables dominated system. Source: Own depiction, based on BDEW (2015) (BDEW—Bundesverband der Energie- und Wasserwirtschaft e.V.—BDEW. 2015. Erneuerbare Energien und das EEG: Zahlen, Fakten, Grafiken; Energie-Info, Anlagen, installierte Leistung, Stromerzeugung, EEG— Auszahlungen , Marktintegration der Erneu erbaren Energien und regionale Verteilung der EEG—induzierten Zahlungsströme, Berlin)

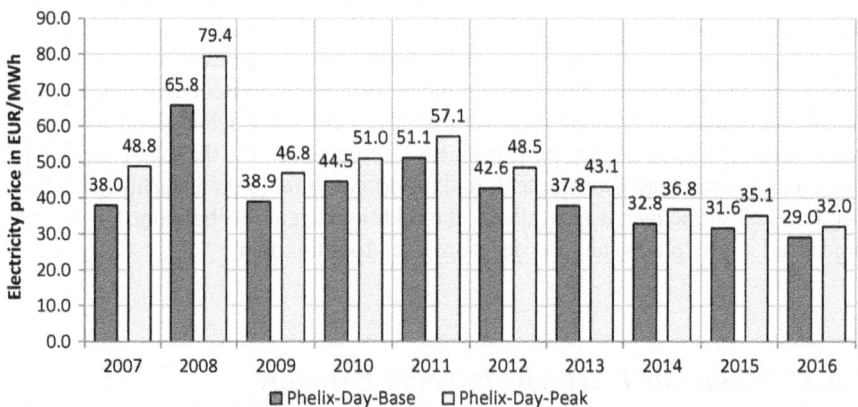

Fig. 6.9 Average electricity wholesale price in Germany (2007–2016). Source: BNetzA (2015, 2017) (BNetzA. 2015. "Monitoringbericht 2015." Bonn, Germany; BNetzA. 2017. "Monitoringbericht 2017." Bonn, Germany)

2016.[16] Apart from the price peak in 2008, driven by high coal and natural gas prices, the level has remained surprisingly stable, hovering around 50 €/MWh up to 2011. Surprisingly, the nuclear moratorium on seven plants in Germany in March 2011 did not have a lasting effect on electricity prices. In fact, after the first nuclear phase out decision in 2001, the large utilities decided to invest in eight new hard coal power plants, adding 6.2 GW in new generation capacity after 2012. All in all, one observes a general trend of decreasing wholesale prices since the beginning of the energiewende, as well as converging day-peak and day-base prices. The price decline indicates, that in many hours with some renewable generation, the price setting marginal generator becomes lignite or the most efficient hard coal generators. In result, the low wholesale prices have caused major disturbances for short-term operation and high write offs for new investment projects by the incumbent energy industry.

6.3.4 Simulations of a Renewable System in Germany in 2030

The effects of renewables on the energy system also modify the way that conventional sources are dispatched, both in the yearly aggregate and in an hourly cycle. An analysis of 15-min load, wind, and photovoltaic data (published by ENTSO-E and the TSOs) can be used to demonstrate the effects of a dominant share of electricity from renewables. Figure 6.10 shows the "residual load", i.e. the part of load not covered by renewable electricity, for the years 2017 and 2030 (with projections based on the latest B 2030 scenario, cf. Sect. 6.2.3), respectively. The peak load is only modestly reduced, as the calculation does not consider demand-side flexibility and increased flexibility in generation from bioenergy; however, both the shape of the curve, and the aggregate electricity produced by conventional sources are substantially modified, as renewables enter the sector at scale. For example, total electricity provided by conventional power plants decreases by about 40%, from 415 TWh (2017) to 250 TWh (2030). Depending on CO_2 and fuel prices in 2030, one can estimate the operational hours by technology.[17] In 2030, renewables will see excess supply in about 1200 h. Assuming no other must-run generation, the remaining 9.5 GW in lignite capacity will have between 6350 and 6950 full-load hours. If combined cycle gas turbines become cheaper than firing hard coal, they might operate in 4900–6350 h, while the remaining 14.8 GW in hard coal capacity only run in 2150–4900 h. Clearly, this raises the question how these firm capacities can be financed, which we address in the next section.

---

[16]Since 2016, EEX prices have somewhat recovered to about 40 EUR/MWh after the mothballing and shut-down of several conventional power plants.

[17]The analysis of the residual load neglects on the one side possible trade with neighbouring countries, which might allow higher operational hours for lignite power plants. On the other side, must-run CHP generation and the variable character of wind and photovoltaics might favor more flexible conventional power plants.

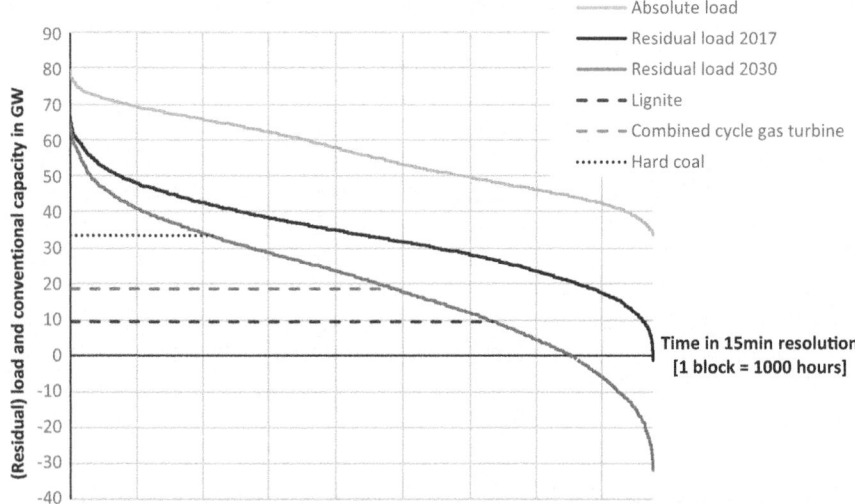

Fig. 6.10 Residual load for conventional electricity (2017 and 2030). Source: Historic TSO data and modeling results, based on scenario framework defined by BNetzA

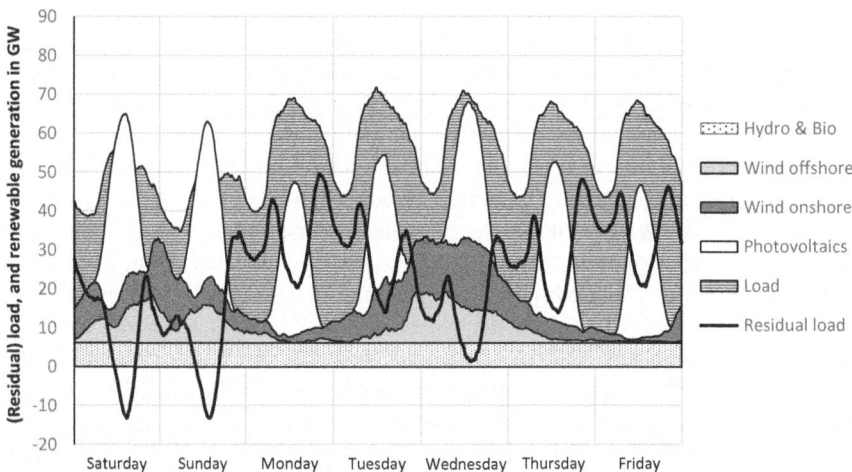

Fig. 6.11 Renewable supply and load for a solar-intensive week (week 22) in 2030. Source: Own calculation based on ENTSO-E and TSO data

Figures 6.11, 6.12, and 6.13 provide examples of the potential effect of renewables at certain specific hours of the year in Germany (based on 2017 weather data): Fig. 6.11 shows a representative week in spring (calendar week 22, month of May), with significant hours of solar penetration. One observes a morning and evening peak in residual load as photovoltaic generation cuts into the peak demand at noon

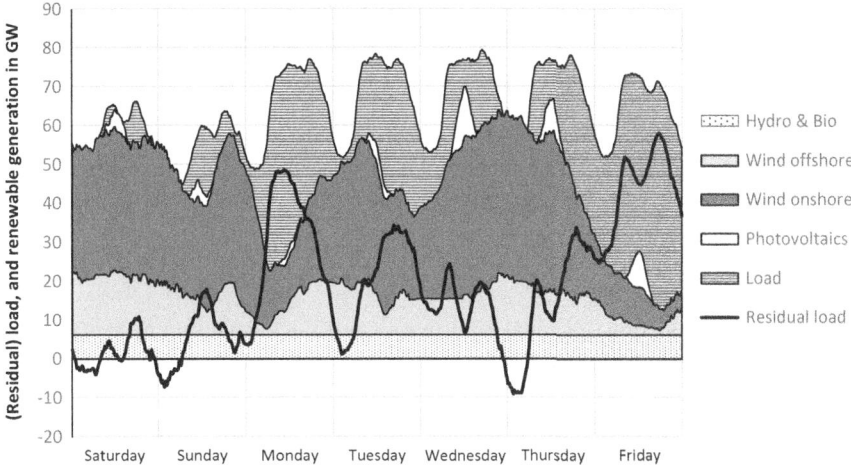

Fig. 6.12 Renewable supply and load for a wind-intensive week (week 50) in 2030. Source: Own calculation based on ENTSO-E and TSO data

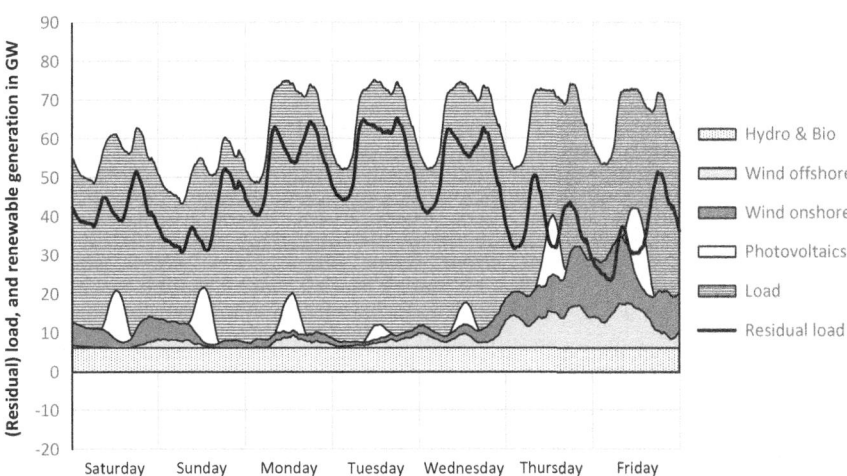

Fig. 6.13 Electricity supply and load for a week with little renewables (week 4) in 2030. Source: Own calculation based on ENTSO-E and TSO data

and steep residual load changes of almost 50 GW within few hours. Figure 6.12 depicts a similar situation in a week with high wind (here: calendar week 50, in December): while the electricity from solar capacities is quite modest, onshore and offshore wind provide sufficient electricity to cover the entire demand during the weekend but show similar residual load ramps as photovoltaics. Figure 6.13 shows a situation where neither wind nor photovoltaics provide significant generation for several days. In this case, backup capacity is required to assure load with both natural gas and coal power plants. This leads us to the question of how these could be financed.

6.3.5   Is a New Market Design Required?

6.3.5.1   Conceptual Issues ...

The massive introduction of capital-intensive renewables changes the way prices are set, and suggests that the conventional, "energy-only" market design may not be well suited for the new system. In fact, as in other European countries, the discussion about the design of appropriate capacity instruments in Germany is vivid and controversial. In particular, after the introduction of the energy concept for 2050 in September 2010 and the nuclear moratorium of March 2011, the German energy industry and policy makers have been engaged in intensive discussion over the advantages and disadvantages of capacity instruments to guarantee the proper functioning of the electricity sector and to ensure supply security as well as resource adequacy. On the one hand, the "carbon power push" of the 2000s led to high overcapacities and low prices in the German electricity system, making it difficult to see the need for capacity instruments. On the other hand, a relatively strong merit order effect and low wholesale electricity prices throughout the 2010s, in combination with the planned closure of nuclear power plants in the near future, lead some observers to conclude a need for capacity instruments.

The conceptual discussion about capacity instruments is broad and unlikely to lead to a consensual assessment. One approach is to find a theoretically optimal structure, e.g. welfare-optimal, that designs an „optimal" market independently of time and space. Such a discussion has emerged, e.g., between proponents of an „energy-only" market design (such as Professor Hogan from the Harvard Electricity Policy Group, HEPG) and a capacity-based market design (Cramton and Ockenfels 2011; Cramton and Stoft 2005). It was largely conducted on theoretical grounds and with an assumed objective of welfare maximization; likewise, Cramton and Ockenfels (2011) suggestion of reliability contracts abstracted from the concrete country or region under consideration. Another stream of literature insisting on the institutional aspects focusses more on the transaction costs of implementing different instruments, and combining different objective functions, e.g., supply security, consumer interests, climate impact, etc. (Beckers and Hoffrichter 2014). The main argument here is that recovering high fixed costs through random price variations implies significant risks for the investor, and thus high capital costs.

6.3.5.2 ... And a Pragmatic Solution for Germany

The German response to the challenges of new market designs was very pragmatic and politically sensible at the same time: on the one hand, several capacity instruments were introduced, quite ad-hoc, after 2011; but on the other hand, political attempts were made to contain the capacity debate to certain market segments, and to maintain the "energy-only" design as long as possible (see Neuhoff et al. 2013).

In reality, capacity instruments were gradually introduced into the German market, of different sized and institutional design:

- As early as summer 2011, the German regulator (*Bundesnetzagentur*, BNetzA) put in place a capacity instrument (without calling it such): It was a small strategic reserve, negotiated bilaterally between the BNetzA and potential providers, focusing primarily on a balance between supply and demand in South Germany. Sometimes called a "winter reserve", this strategic reserve was calibrated to the winter peak demand, and a market design that emulated a "copperplate" in Germany, that is, an electricity system without any network congestion (see the Chap. 8 for a critical discussion of this assumption). The strategic reserve was formalized in an ordinance (*Kraftwerksreserveverordnung*) in 2012. Originally expected to expire in 2017, it was later extended to 2021.[18]
- A second capacity instrument was introduced in 2014: a full-fledged "strategic reserve". Old capacities that utilities have nominated for closure, can enter the strategic reserve, where they bid for electricity delivery in cases of particular capacity shortage; however, once they entered the strategic reserve, they are not allowed to participate in the ordinary wholesale market.[19]
- Last but not least, a very peculiar capacity instrument was later introduced to compensate some lignite power plants, the so-called "lignite reserve", in 2015 (Oei et al. 2015; SRU 2017) In fact, 2.7 GW of rather old lignite plants in East and West Germany were placed into this reserve, and obtained some fixed payments (1.6 billion € in total) for being "on reserve" for another 4 years.

The German utilities pursued no clear strategy, but increasingly moved away from the energy-only market concept, to embrace different forms of capacity instruments. In particular, after Germany's two large European neighbors, the UK and France chose to pursue a strong national strategic reserve, the mood changed and the German industry, too, demanded a comprehensive capacity instrument to assure

---

[18]In 2015, the reserve for the winter 2016/2017 was about 4 GW, contracted both in South Germany and in neighboring countries, mainly Austria.

[19]In 2016, the strategic reserve contained about 5 GW of capacity, plus an additional 2 GW of capacity allocated explicitly to South Germany.

system stability.[20] An alternative instrument was developed by Matthes et al. (2012), targeting a "low-carbon" capacity market, essentially a selective capacity instrument open only for flexible natural gas-fired capacity. This specific instrument became particularly popular in Southern Germany, where the general preference is for local and (relatively) clean electricity from natural gas plants rather than coal electricity "imported" mainly from North German coal power plants.

The de facto establishment of different segments of capacity markets was accompanied by an "energy-only" rhetoric by the government. In fact, the government's energy strategy of 2015 included a strong statement in favour of an energy-only market, in connection with a small strategic reserve. With hindsight, the pragmatic compromise, i.e., a sufficiently large strategic reserve but no general instruments favoring CO_2-intensive power plants, appears as appropriate for the first phase of the energiewende; in particular, it avoided a "watering can approach" for CO_2-intensive power plants. This approach may have to be revisited, though, as the energiewende enters into the next phase, in the mid-2020s, with nuclear and coal plants leaving the market, and renewables becoming not only the leading, but also by far the dominant source of supply.

Box 6.1 12 Insights on Germany's energiewende by Agora Energiewende

Along similar lines to the arguments presented in this chapter, Agora Energiewende (2013), a think-tank on technical and economic reforms of the German energy system, also argue that the functioning of a renewables-based electricity market will substantially change in the near future. They describe their findings as follows:

Insight 1: It's all about wind and solar

Two winners have emerged from the technology competition initiated by the German Renewable Energy Act: wind power and photovoltaics, the most cost-effective technologies with the greatest potential in the foreseeable future. All other renewable technologies are either significantly more expensive or have limited potential for further expansion (water, biomass/biogas, geothermal energy) and/or are still in the research stage (wave power, energy from osmosis processes, etc.).

Insight 2: "Base-load" power plants will disappear altogether, and natural gas and coal will operate only part-time

(continued)

---

[20]Insiders have reported that the rather liberal position favoring an energy-only market by RWE, the largest German utility, was eliminated with the decision adopted by the French Parliament ("Assemblée Nationale") on December 18, 2012, to introduce a national capacity instrument ("tradable certificates") that was originally supposed to benefit mainly the French incumbent EdF. As one of the most powerful companies within the energy industry, RWE contributed to the shift of the association toward a strong capacity instrument.

Box 6.1 (continued)

Wind and PV will form the basis of the power supply, with the rest of the power system being optimized around them; most fossil-fueled power plants will be needed only at those times when there is little sun and wind, they will run fewer hours, and thus their total production will fall: "Base load" power plants will be a thing of the past. Rapid changes in feed-in from renewables as well as forecasting uncertainties will create new requirements for both short- and long-term flexibility. Over the medium term, combined heat-and-power as well as biomass plants will need to be operated according to the demand for electricity. Demand-side management and storage contribute to maintaining system balance.

Insight 3: There's plenty of flexibility—but so far it has no value

In the future, fluctuations in wind and PV production will demand significantly greater flexibility from the power system. Technical solutions to provide sufficient flexibility readily exist today. The challenge is not about technology and control, but rather about incentives. Leveraging small-scale flexibility options at the household level by using smart meters is currently too expensive.

Insight 4: Grids are cheaper than storage facilities

Grids decrease the need for flexibility: fluctuations in generation (wind and PV) and demand are equilibrated across large distances. Grids enable access to cost-effective flexibility options in Germany and Europe. Transmission grids reduce overall system costs with relatively small investment costs. Expanding and upgrading distribution grids is also less expensive than local storage facilities. New storage technologies will only become necessary as the share of renewable energy exceeds 70%. Local PV battery systems may provide a business case for individual investors sooner because of savings in taxes and fees.

Insight 5: Securing supply in times of peak load does not cost much

At certain times (e.g., during windless days in the winter), wind and PV are not sufficient to cover peak loads, and, for this reason, controllable resources will be required in the same order of magnitude as today. Peak load can be met reliably by firm generation capacity, or be reduced through demand-side measures; almost a quarter of the demand (approx. 15–25 GW in Germany) arises in only very few hours of the year (<200). Gas turbines can meet this demand quite cheaply (35–70 million € per year per GW), controllable loads or retired power plants might be even cheaper. European cooperation reduces the cost and simplifies securing supply in times of peak loads.

Insight 6: Integration of the heat sector makes sense

The heat sector offers enormous potential for increasing system flexibility. CHP plants already provide a link between the electricity and heat sectors; in the medium term, dual-mode heating systems, capable of using either fuel or

(continued)

Box 6.1 (continued)

electricity will be deployed; over the longer term, the systems will be integrated through use of a common fuel: natural gas, biogas, or power-to-gas.

Insight 7: Today's electricity market is about trading kilowatt hours—it does not guarantee system reliability

Today's electricity market handles energy quantities (Energy-Only). The energy-only market may not provide sufficient incentives for new and existing resources to continuously ensure system reliability. The energiewende brings this issue to the forefront because power production from wind and PV will reduce the average market price of electricity and with it the operating times of fossil-fueled power stations.

Insight 8: Wind and PV cannot be refinanced through marginal-cost based markets

Wind and solar power have operating costs close to zero. Wind and PV produce electricity when the wind blows and the sun shines, regardless of electricity price. Therefore, in principle, wind and PV cannot be refinanced in a marginal-cost based market, even when their total costs are below those of coal and gas. High CO_2 prices do not fundamentally change this effect.

Insight 9: A new energiewende market is required

The future energiewende market must fulfill two functions: i/ steer the installation of capacity in order to achieve an efficient balance between demand and supply; ii/ send investment signals for renewable energy as well as for conventional facilities and make energy demand and storage (longer term) more flexible. The new market will create two sources of revenue: i/ revenue (as before) from the sale of electricity quantity (MWh) in the marginal-cost based Energy-Only Market, and ii/ revenue from a new investment market for megawatts (MW). In addition, fossil-fueled power plants, renewable energy, demand-side resources, and storage systems will compete to provide ancillary services (e.g., balancing energy). Installing a new mechanism instead of the current feed-in tariffs for renewables is only justified if it results in increased efficiency.

Insight 10: The energiewende market must actively engage the demand side

Greater demand-side flexibility is fundamental to increasing the use of wind and PV. Demand response is usually cheaper than electricity storage or supply-side options. Current regulations for grid tariffs and ancillary services often work at cross purposes with demand response, and should therefore be reformed. The new market for investments in firm capacity must be designed such that demand-side resources able to shift loads can actively participate.

Insight 11: The energiewende market must be considered in the European context

The ongoing integration of the German power system into the European system makes the energiewende simpler and more affordable because i/ the

(continued)

> **Box 6.1** (continued)
> fluctuations of wind and PV energy production become less pronounced over a larger geographic region, ii/ firm capacity can be collectively shared, and iii/ low-cost flexibility options in Europe can be more fully utilized (e.g., energy storage resources in Scandinavia and Alpine countries). European electricity trading stabilizes market prices.
>   Insight 12: <u>Efficiency: A saved kilowatt hour is the most cost-effective kilowatt hour</u>
>   Energy efficiency decreases total costs; increased energy productivity enables the decoupling of economic growth from energy consumption. Every kilowatt saved means less burning of natural gas and coal and lower investments in new power plants (fossil and renewable). The challenge lies less in technology and more in creating the right incentives.
>   *Source:* Agora Energiewende (2013).

6.4   What About Costs?

One of the main points of arguments both for and against the renewables-based energiewende in Germany relates to the costs involved. Since different cost concepts are used in this debate, one can make arguments <u>against</u> the renewables-based energiewende ("too expensive") as well as arguments <u>for</u> it ("economically efficient"). In this subsection, we apply different approaches to assess the "costs" of the renewable targets. Needless to say, results differ depending on the concept of "costs" used, the observed time horizon, and the alternatives in the comparison. This subsection presents three different cost analyses: i) a short-term analysis of additional costs of renewables compared to the existing fossil-nuclear electricity system, the so-called EEG surcharge ("EEG-Umlage"); ii) a dynamic perspective on the private costs and benefits of renewables in the context of total system costs within a more and more carbon constrained world; and iii) a public economics perspective taking into account the external, environmental costs of the competing fuels.

6.4.1   Short-Term Private Costs of Renewables: The EEG Surcharge

One commonly cited measure of cost is the renewables surcharge ("EEG-Umlage") designed to pass on the additional costs of the renewables feed-in law to consumers of electricity. The surcharge is calculated as the differential costs between the feed-in payments and the spot market revenues of renewable electricity generation. Figure 6.14 shows the development of the renewables surcharge from 2003 until 2018. In the years 2010 and 2011, one can observe a steep increase following high annual

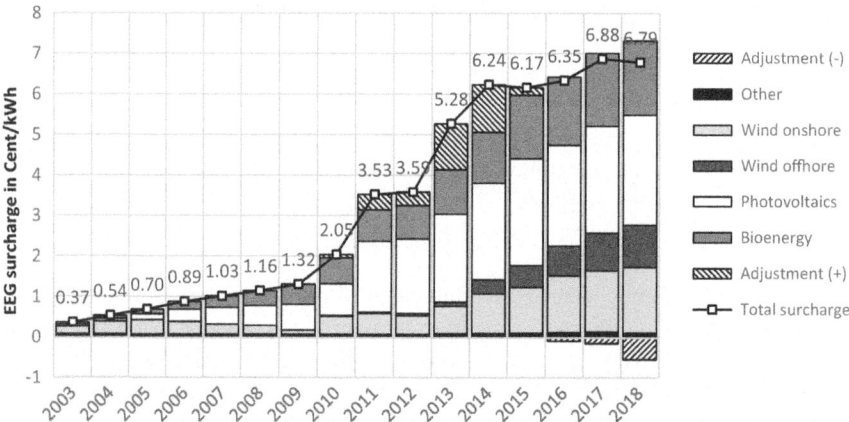

Fig. 6.14 The renewables surcharge to private households in Germany. Source: Own depiction based on 50Hertz (2018) (50Hertz. 2018. "EEG-Umlage." www.netztransparenz.de; BMWi. 2014. "Wie hat sich die EEG-Umlage über die Jahre entwickelt?" www.bmwi-energiewende.de)

investment levels in photovoltaics at a time when feed-in tariffs have still been in the range of 20–40 €cents/kWh. In 2013 and 2014, too low projection of EEG costs in previous years and resulting negative levels of the EEG account required another steep increase. Since then these additional payments could be reduced, resulting in only small increases of the EEG surcharge due to large investments in mainly onshore and offshore wind capacity.

The mid-term private cost of renewables will depend on the combined development of the renewable surcharge and the wholesale electricity price. Scenarios with stable wholesale prices result in a surcharge, which plateaus in the mid-2020s around 7.6 €cents/kWh and decrease thereafter to about 4.4 €cents/kWh in 2035 for a system with more than 60% in renewables supply (Oeko-Institut 2016). Main drivers are the phasing out of historically higher subsidies and the decreasing technology cost for new capacity investments in wind and solar. The winners of the 2017 renewable auctions for new onshore wind and large-scale photovoltaic projects receive only a guaranteed feed-in tariff of 3–5 cent/kWh. Overall, the projections for the surcharge are very sensitive to changes in the cost allocation amongst consumers and related exemptions, changes in overall electricity demand levels, and to higher wholesale prices.

From a political economy perspective, the distribution of the 24.2 billion € (2016) in renewables surcharges is very interesting. An even allocation per kWh would result in a renewable surcharge of less than 4 cent/kWh for a gross electricity demand of about 595 TWh per year in Germany. However, 4% of all German companies, mainly energy intensive industries, are to some extend exempt from these charges, i.e., paying less than 1.38 cents/kWh instead of the full fee of 6.88 cents/kWh in the year 2017. As these companies stand for more than half the industrial demand (Fig. 6.15), the majority of industrial electricity consumption pays no or a significantly lower EEG surcharge, resulting in higher surcharges for consumers that are

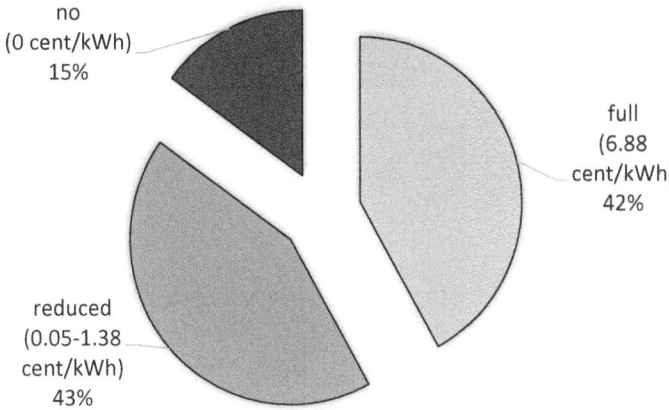

Fig. 6.15 Renewable surcharge (*EEG-Umlage*) levels for the industry in 2017. Source: BDEW (2017) (BDEW. 2017. "Erneuerbare Energien und das EEG: Zahlen, Fakten, Grafiken." Berlin, Germany: Bundesverband der Energie- und Wasserwirtschaft e.V.—BDEW)

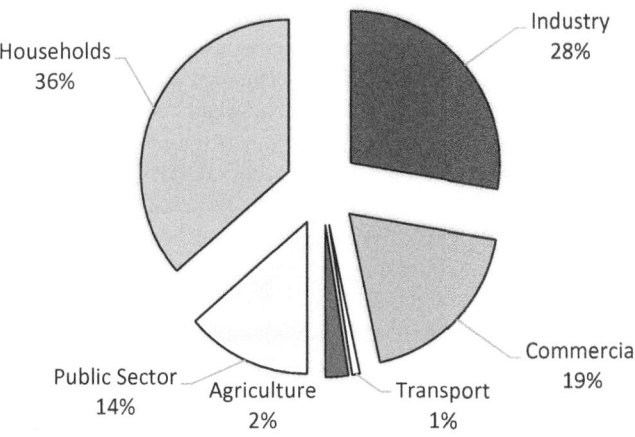

Fig. 6.16 Distribution of renewable surcharge (*EEG-Umlage*) costs to consumers in 2016. Source: BDEW (2017)

not exempt. Small household consumers contributed most (36%), compared to industry consumers (28%), the rest being distributed between the commercial sector (19%), the public sector (14%), agriculture (2%), and transportation (1%) (Fig. 6.16). Therefore, large energy intensive companies in Germany may be seen as strongest short-term beneficiaries of the energiewende, paying no or reduced renewable surcharges and one of the lowest electricity price in Western Europe because of the merit order effect.

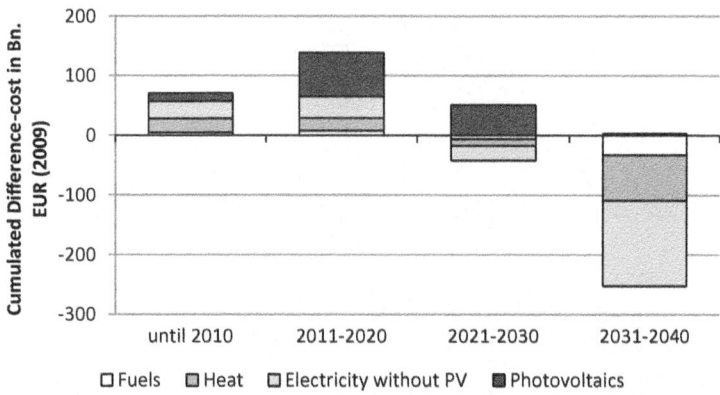

Fig. 6.17 Energy system costs of the energiewende, as compared to a conventional "business-as-usual" (BAU) case. Source: BMU (2012)

6.4.2 Dynamic Perspective on Private Costs: Renewables as a Sound Long-Term Investment

The dynamic analysis of long-term costs and benefits of the energiewende towards renewables provides another perspective, which considers future cost reductions of renewable technologies. A study (so-called *Leitstudie*) for the Ministry of Environment (BMU 2012) conducted a comparison between the cumulative renewables costs of this renewable transformation (RES scenario) relative to a benchmark, that is, to the costs incurred under a business-as-usual (BAU) framework. Figure 6.17 shows the difference between the two scenarios in terms of private production costs (i.e., excluding the social costs). Private production costs include capital costs, variable fuel costs, and costs for carbon emission certificates. While results indicate that the decade 2011–2020 of the RES scenario is particularly expensive in the electricity sector, due to the high feed-in still guaranteed to solar (for 20 years), the differences disappear in the subsequent decade (2021–2030). Payments to solar start to decrease and other renewable technologies already reduce costs, in both, the remaining electricity system and the heat sector, compared to the BAU framework. The cost difference vanishes completely in 2026, meaning that renewables start to stabilize and even decrease energy costs for consumers in the RES scenario. After 2030, the trend is fully reversed, and total energy system costs are significantly lower than in the BAU framework. In 2040, initial higher costs for the RES scenario have been fully compensated.

6.4.3 Public Economics Perspective

Yet another perspective emerges when adopting a public economics perspective, that is, when taking into account not only the private costs to the consumer, but also the social costs incurred by society through different forms of electricity provision. The simplest analysis compares the *social* costs of the three pillars of the old electricity system: nuclear, coal, and renewables. For better comparison, we discuss levelized cost of electricity (LCOE), a common measure for the cost of electricity provision per kWh. In addition to private costs, which have fixed price cost components (investment cost, lifetime, interest rate, full-load hours per year, and fixed operation & maintenance (O&M) costs) and variable price components (fuel cost and efficiency), this measure can also include additional social costs specific to the generation technology.

From the social perspective, nuclear appears to be the most expensive of all electricity sources, in particular when accounting for all types of costs. With respect to LCOE, Toke (2012) and Boccard (2014) indicate that nuclear has no cost advantage over other sources of electricity generation, in particular due to its high capital costs;[21] the capital costs of nuclear power plants have risen continuously since the 1970s, and initial investment costs for ongoing projects are likely to be in the range of 6000 €/kW.[22] In addition, significant costs incurred in R&D and the development of new reactors are all being paid by the public sector as will be the largest share of costs for disposing spent fuel which are still largely unknown. Even after six decades of nuclear energy use there are no permanent disposal sites anywhere in the world that guarantee safe storage of nuclear fuel rods for tens of thousands of years. Another important cost factor is insurance against potential major accidents. The high costs of major accidents at nuclear power plants are difficult to quantify; currently, society is bearing the majority of these costs because nuclear power plant operators are subject to very few insurance requirements. Irrespective of what form or combination of insurance forms (public, private, or a mix) proves most economically advantageous, the costs must be included in the cost calculation. The economic viability of nuclear power will also be diminished due to reduced full-load hours and higher flexibility requirements in a renewables-based electricity system and further tightening of safety regulations currently being developed at the pan-European level.[23] Depending on the assumptions, total social costs of nuclear energy range between 20 and 40 cents/kWh.

---

[21]Boccard (2014) concludes "the future cost of nuclear power in France to be at least 76 €/MWh and possibly 117 €/MWh."

[22]See discussion in Hirschhausen (2017), and the survey paper by Wealer et al.(2018).

[23]After the Fukushima nuclear disaster, EU Energy Commissioner Günther Oettinger recommended mandatory stress testing of European nuclear power plants. The results pointed to the urgent need for retrofits at some plants. A draft regulation will form the basis for the binding rules on liability and compulsory inspection routines to be introduced in all countries. See European Commission, Draft proposal for a Directive amending Nuclear Safety Directive IP/13/532, June 13, 2013. Francois Lévèque (2013, Nucléarie On/Off. Paris, Dunod, p. 171) provides the most intuitive explanation of why the civil use of nuclear power cannot be considered an economical energy alternative: "Nuclear power is the child of science and the military" ("L'énergie atomique est la fille de la science et de la guerre"), own translation.

The social costs of fossil fuel based electricity includes the greenhouse gas externalities, the effects of sulphurdioxydes (SO_2), nitrogene oxides (NO_x), mercury (H_g), and groundwater contamination. Local negative externalities are fine dust particles and noise, the displacement of local populations to make way for new opencast lignite mines as well as long-term costs for later subsidence damages of underground coal mining. Estimates of the social costs of lignite, the most CO_2-intensive fuel, are in the range of 80–120 €/MWh, or about two to three times the current wholesale price of electricity (see Chap. 4). Clearly, from a social perspective, burning coal (and other fossil fuels) reduces welfare and there is no serious progress in making fossil-fired generation a component of the future energy system by reducing negative externalities. System-wide carbon capture, transport, and sequestration (CCTS), required for reducing carbon emissions, seems very unlikely and even small steps towards tighter regulation in the Industrial Emissions Directive at European level (e.g., for SO_2 and NO_x) have almost been blocked by Germany and its eastern neighbors in 2017.

Compared to nuclear and fossil fuel based electricity generation, the costs of externalities of renewable electricity generation are significantly lower (e.g., 0.2 cent/kWh for wind and 1.3 cents/kWh for photovoltaic, (Küchler and Wronski 2015). However, being less centralized and smaller than conventional power plants, they are more numerous and therefore more visible to the public. In consequence, especially onshore wind power can be exposed to regional "not in my backyard" (NIMBY) opposition. In terms of LCOE, the costs for the EEG surcharge indicate that renewables have been more costly than operation of the existing and written-off nuclear and fossil-fired power plants, when social costs are not internalized. However, already today, cost reductions in the last years allows LCOE for new onshore wind and large-scale photovoltaics to be lower than those of new nuclear or coal-fired power stations in Germany. As a result of expected cost reduction for new investments in wind and photovoltaics and increasing costs for burning fossil fuels in a carbon-constrained world with a significant price for carbon emissions, renewables are likely to replace most of fossil fuel generation in the electricity sector in Germany years before 2050.

In addition to the LCOE, a technical-economic analysis of renewables also has to take into account the costs of system security: that is, it must balance the intermittency of the renewable supply. Different approaches have been developed in the literature to calculate the costs of system integration. However, these approaches often hinge on a series of assumptions on the costs of transmission allocated to the renewables, the social costs of firm capacity, such as the use of natural gas plants as a backup, etc. However, recent technological trends have led to a situation in which the intermittency of renewables is mitigated by low-cost storage technologies and renewables are still competitive with the social costs of other generation technologies. Deutsch and Graichen (2015) calculate scenarios in which the combination of solar and storage (e.g., Lithium-Ion) costs less than 10 cents/kWh, with further cost reductions in the future; this is below the social costs of any electricity generated from fossil fuels, let alone nuclear power. Thus, from a social-welfare policy perspective, it is clear that renewable generates electricity at the lowest cost independent of the intermittency issue.

6.5 Conclusions

Variable renewable energy sources such as wind and solar are the cornerstone of the energiewende. Because of this focus, the German energiewende has become a unique case of low-carbon transformation worldwide. This chapter has provided a survey of the evolution of renewables from a niche player to the dominant market player, implying a fundamental change in the functioning of the electricity market. In general, we find that although there are new technical, legal, and economic challenges related to the large-scale use of renewables in Germany, the process is unfolding as it should, and is turning out to be less difficult than it is sometimes considered to be from an outside perspective. From the viewpoint of public economics, the focus on renewables is socially efficient since the costs are lower than the two main alternatives, coal and nuclear.

Since the first specific law on renewables in 1990, the use of renewables in Germany has grown exponentially, based mainly on onshore wind resources with contributions from bioenergy, photovoltaics, and offshore wind. Meanwhile, extensive support for photovoltaics has benefitted the global breakthrough of the most freely available of all technologies. Several updates of the law on renewables (EEG 2000, 2009, 2012) have not changed the overall approach (Morris and Pehnt 2016). Systematic reductions to tariffs have brought technical progress and cost reductions in the manufacturing industry, and industry has been flourishing as a result (Weigt and Leuthold 2010). Since 2014, electricity from renewables has been surpassing all other conventional energies in terms of market share, and expections forsee more than 50% in 2030 and beyond 80% by 2050. At present, no technical nor political obstacles can be identified that would prevent these targets from being reached.

Clearly, the focus on variable renewables has had a disruptive effect on the electricity sector and led to significant changes not only in market design and price formation but also in corporate strategies. The conventional thinking in terms of regular base load and high-price peak load no longer holds sway; the merit order effect is lowering variable prices and the clear need to rethink the financing of capital cost-intensive renewables with negligible variable costs has become patently clear. The German government has so far resisted creating a formal capacity market for conventional generation, although the instruments have been put in place that provide fixed-cost support to some selected conventional generators.

We have also revisited three different concepts to judge the costs of the focus on renewables. The renewables surcharge has risen from 1 €cent/kWh in 2009 to almost 7 €cents/kWh in 2018, to large extend driven by overpayment for solar installations; this conceptual error has since been corrected so that the surcharge will decrease over time. A more dynamic analysis of the differences between a renewables-based and a conventional system shows clear advantages of the former: the lion's share of investment will be made in the period 2011–2030, and large benefits are to be expected beyond 2030. Last but not least, renewables clearly have an advantage over coal, and fossil fuels in general, and all the more over nuclear, in terms of "social costs"; this effect is intensified as low-cost storage technologies (such as batteries) and cross-sectoral integration of the energy system become more established, which reduces concerns about intermittency and integration costs. The public economics perspective therefore suggests that the renewables-based strategy is clearly economically efficient; that renewable energies have lower social costs than coal, burdened by carbon emissions, and nuclear energy, burdened by high capital costs, long-term costs for nuclear fuel disposal, and uninsured risks. From a dynamic perspective, the German approach is also beneficial in reducing outlays for imported energy fuels (coal, gas, and uranium). In sum, although the initial costs were high (and unevenly distributed) renewables deployment in the German energiewende has been so far a success, indicating the feasibility of this approach to low-carbon transformation.

References

Agora Energiewende. 2013. *12 Insights on Germany's Energiewende.* A discussion paper exploring key challenges for the power sector, Berlin.

Thorsten Beckers, and Albert Hoffrichter. 2014. *Grundsätzliche und aktuelle Fragen des institutionellen Stromsektordesigns – Eine institutionenökonomische Analyse zur Bereitstellung und Refinanzierung von Erzeugungsanlagen mit Fokus auf FEE.*

BMU. 2012. *Langfristszenarien und Strategien für den Ausbau der erneuerbaren Energien in Deutschland bei Berücksichtigung der Entwicklung in Europa und global.* Schlussbericht BMU-FKZ 03MAP146. Stuttgart: Deutsches Zentrum für Luft- und Raumfahrt (DLR), Stuttgart Institut für Technische Thermodynamik,Fraunhofer Institut (IWES), Kassel Ingenieurbüro für neue Energien (IFNE).

BMWi, and BMU. 2010. *Energy Concept – for an Environmentally Sound, Reliable and Affordable Energy Supply.* Berlin.

Boccard, Nicolas. 2014. The cost of nuclear electricity: France after Fukushima. *Energy Policy* 66 (March): 450–461.

Cramton, Peter, and Axel Ockenfels. 2011. *Economics and Design of Capacity Markets for the Power Sector.* Maryland: University of Maryland, University of Cologne.

Cramton, Peter, and Steven Stoft. 2005. A Capacity Market that Makes Sense. *The Electricity Journal* 18 (7): 43–54.

Deutsch, Matthias, and Patrick Graichen. 2015. *What If. . . There Were a Nationwide Rollout of PV Battery Systems?* Berlin: Agora Energiewende.

EC. 2011. *Energy Roadmap 2050.* Communication from the Commission to the European Parliament, the Council, the European Economic and Social Committee and the Committee of the Regions. Brussels: European Commission.

Energiewirtschaftsgesetz vom 7. 2005 Juli. (BGBl. I S. 1970, 3621), das zuletzt durch Artikel 2 Absatz 6 desGesetzes vom 20. Juli 2017 (BGBl. I S. 2808, 2018 I 472) geändert worden ist. https://www.gesetze-im-internet.de/enwg_2005/EnWG.pdf

EWI. 2012. Der Merit-Order-Effekt der erneuerbaren Energien – Analyse der kurzen und langen Frist. Energiewirtschaftliches Institut an der Universität Köln

Johanna Cludius, Hauke Hermann, and Felix Chr. Matthes. 2013. *The merit order effect of wind and photovoltaic electricity generation in Germany 2008-2012*. CEEM Working Paper 3-2013. http://ceem.unsw.edu.au/sites/default/files/documents/CEEM%20%282013%29%20-%20MeritOrderEffect_GER_20082012_FINAL.pdf

Krause, Florentin, Hartmut Bossel, and Karl-Friedrich Müller-Reissmann. 1980. In *Energie-Wende: Wachstum und Wohlstand ohne Erdöl und Uran*, ed. Öko-Institut Freiburg. Frankfurt am Main: S. Fischer.

Küchler, Swantje, and Rupert Wronski. 2015. *Was Strom wirklich kostet: Vergleich der staatlichen Förderungen und gesamtgesellschaftlichen Kosten von konventionellen und erneuerbaren Energien*. Berlin, Germany: Forum Ökologisch-Soziale Marktwirtschaft e.V.

Matthes, Felix, Ben Schlemmermeier, Carsten Diermann, Hauke Hermann, and Christian von Hammerstein. 2012. *Fokussierte Kapazitätsmärkte. Ein neues Marktdesign für den Übergang zu einem neuen Energiesystem*. Studie für die Umweltstiftung WWF Deutschland. Berlin: Öko-Institut e.V. - LBD-Beratungsgesellschaft mbH - RAUE LLP.

Morris, Craig, and Martin Pehnt. 2016. *Energy Transition: The German Energiewende*. An initiative of the Heinrich Böll Foundation. First Released in November 2012, Revised in July 2016, Berlin.

Neuhoff, Karsten, Jochen Diekmann, Clemens Gerbaulet, et al. 2013. Energiewende und Versorgungssicherheit: Deutschland braucht keinen Kapazitätsmarkt. *DIW Wochenbericht* 80 (48): 3–4.

Nitsch, Joachim. 2016. *Die Energiewende nach COP 21 – Aktuelle Szenarien der deutschen Energieversorgung*. Kurzstudie für den Bundesverband Erneuerbare Energien e.V. Langversion, Stuttgart.

Oei, Pao-Yu, Clemens Gerbaulet, Claudia Kemfert, Friedrich Kunz, Felix Reitz, and Christian von Hirschhausen. 2015. *Effektive CO_2-Minderung im Stromsektor: Klima-, Preis- und Beschäftigungseffekte des Klimabeitrags und alternativer Instrumente. 98. Politikberatung kompakt*. Berlin: DIW.

Oeko-Institut. 2016. *Projected EEG Costs up to 2035: Impacts of Expanding Renewable Energy According to the Long-Term Targets of the Energiewende*. Berlin: Study for Agora Energiewende.

Sensfuß, F, and Ragwitz, M. 2007. Analyse des Preiseffektes der Stromerzeugung aus erneuerbaren Energien auf die Börsenpreise im deutschen Stromhandel – Analyse für das Jahr 2006/ Gutachten des Frauenhofer Instituts für System- und Innovationsforschung für das Bundesministerium für Umwelt, Naturschutz und Reaktorsicherheit. (Forschungsbericht)

Sensfuß, F. 2011. Analysen zum Merit-Order Effekt erneuerbarer Energien Update für das Jahr 2010 Karlsruhe, 4. November 2011 Frauenhofer ISI. https://www.erneuerbare-energien.de/EE/Redaktion/DE/Downloads/Gutachten/analysen-merit-order-effekt.pdf?__blob=publicationFile&v=2

Speth, V, and Klein, A. 2012. *The impact of different wind and solar portfolios on spot market prices – a market model*. Proceedings of 11th International Workshop on Large-Scale Integration of Wind Power into Power Systems as well as on Transmission Networks for Offshore Wind Power Plants, Lisbon, Portugal, 13–15 November 2012.

Speth, V, and Warzecha, J. 2012. *The impact of wind and solar on peak and off-peak prices – evidence from two year price analysis*. 12th IAEE European Energy Conference, Venice.

SRU. 2011. *Pathways towards a 100% Renewable Electricity System*. Special report. Berlin: Sachverständigenrat für Umweltfragen.

———. 2017. *Start Coal Phaseout Now*. Berlin: German Advisory Council on the Environment.

Toke, David. 2012. *Nuclear Power: How Competitive Is It under Electricity Market Reform?* Presentation given at the HEEDnet seminar presented at the HEEDnet Seminar, London, July 17.

Traber, T. Kemfert, C. Diekmann, J. 2011. Strompreise: Künftig nur noch geringe Erhöhung durch erneuerbare Energien DIW Wochenbericht 6/2011. https://www.diw.de/sixcms/detail.php?id=diw_01.c.455270.de

Hirschhausen, Christian von. 2017. *Nuclear Power in the 21st Century – An Assessment* (Part I). DIW discussion paper 1700, Berlin.

Vereinigung der Bayerischen Wirtschaft e.V. Kosten des Ausbaus der erneuerbaren Energien. 2011. http://www.baypapier.com/fileadmin/user_upload/Downloads/Standpunkte/Studie_Kosten_Erneuerbare_Energien.pdf

Wealer, Ben, Clemens Gerbaulet, Claudia Kemfert, and Christian von Hirschhausen. 2018. *Cost Estimates and Economics of Nuclear Power Plant Newbuild: Literature Survey and Some Modelling Analysis.* Presented at the 41 st IAEE International Conference, Groningen, NL, June 11.

Weigt, Hannes. 2009. Germany's wind energy: The potential for fossil capacity replacement and cost saving. *Applied Energy* 86: 1857–1863. https://doi.org/10.1016/j.apenergy.2008.11.031.

Weigt, Hannes, and Florian Leuthold. 2010. Experience with renewable energy policy in Germany. In *Harnessing Renewable Energy in Electric Power Systems: Theory, Practice, Policy*, ed. Boaz Moselle, Jorge Padilla, and Richard Schmalensee. Washington, DC: RFF Press.

Chapter 7
Energy Efficiency: A Key Challenge of the Energiewende

Claudia Kemfert, Casimir Lorenz, Thure Traber, and Petra Opitz

> *"A saved kilowatt hour is the most cost-effective kilowatt hour."*
> *Agora Energiewende (2013, p. 2): Lesson 12.*

7.1 Introduction

A significant improvement in energy efficiency is crucial for the success of the energiewende. Energy efficiency plays an important role in reducing primary energy demand and fuel costs, and in many cases, it constitutes the least-cost option for GHG emissions reduction. Other benefits arise from its positive impact on local air quality, human health, and productivity. By making the use of fossil fuels more

---

This chapter summarizes work by the authors and their respective organizations, DIW Berlin and DIW Econ, see references, we draw particularly on a study on deep decarbonization coordinated by Claudia Kemfert et al. (2015); the usual disclaimer applies.

C. Kemfert (✉)
DIW Berlin, Berlin, Germany

Hertie School of Governance, Berlin, Germany

German Advisory Council on the Environment (SRU), Berlin, Germany
e-mail: sekretariat-evu@diw.de

C. Lorenz
DIW Berlin, Berlin, Germany

TU Berlin, Berlin, Germany

T. Traber
DTU Denmark, Copenhagen, Denmark

P. Opitz
DIW Econ., Berlin, Germany

© Springer Nature Switzerland AG 2018
C. von Hirschhausen et al. (eds.), *Energiewende "Made in Germany"*,
https://doi.org/10.1007/978-3-319-95126-3_7

energy-efficient, the external costs of energy provision can be reduced. And as fossil fuel combustion is reduced, black carbon as well as sulfur dioxide emissions are projected to decline significantly. Demand for fossil fuels is therefore expected to decrease as well, which will help to reduce exposure to volatile international prices. This will correspond directly to increased energy security and will also reduce Germany's dependence on energy imports. In addition, energy efficiency investments may trigger positive employment effects through growth in the the building sector, and in the longer term, household energy savings will boost spending in other sectors.[1] Furthermore, fossil fuels are limited resources, and their efficient use increases the potential for their non-energy use in sectors where substitution is more costly or impossible.

The important role of energy efficiency in the energiewende is clear. This chapter summarizes the German approach to energy efficiency, both in general and at the sectoral level in areas such as building, industry, and transport. It also discusses obstacles that could prevent an increase in energy efficiency from a general viewpoint and in the specific case of Germany. In the following section, we provide a survey of the energy efficiency targets identified in Germany's Energy Concept 2050 and other documents, and discuss the results of this policy, which have been mixed to date. Section 7.3 then provides a sectoral analysis, looking specifically at the building, industry, and transport sectors. Section 7.4 looks at specific energy efficiency policies going forward, in particular the National Action Plan on Energy Efficiency (NAPE) and some key future challenges. Section 7.5 concludes.

7.2 Energy Efficiency Targets and Achievements

7.2.1 Issues Related to Energy Efficiency

Energy productivity, defined as GDP per unit of domestic energy consumption, is the key indicator of progress in energy efficiency. It measures the use of energy in relation to the performance of the economy as a whole. The measure may focus on either primary or final energy productivity. For assessing improvements in energy efficiency at consumption level, final energy productivity is the more appropriate indicator, as it disregards the fuel mix and the efficiency of the energy transformation. These parameters are of more interest when analyzing the total primary consumption as needed for e.g., the analysis of international energy security and climate policy.

Energy demand is influenced by a variety of factors besides the energy efficiency improvements that are intended to decrease demand. Historically, growth of population, GDP, and the sectoral share of industry contributed to increases of energy

---

[1]Blazejczak, J., Edler, D., Schill, W.-P. (2014): Steigerung der Energieeffizienz: ein Muss für die Energiewende, ein Wachstumsimpuls für die Wirtschaft. DIW Wochenbericht Nr. 4/2014, pp. 47–60.

demand, while growing prices and a growing GDP-share of the service sector dampens demand growth. In addition, climatic conditions play a substantial role. Besides these factors the stock of buildings and the stock of electric appliances are shaping the energy demand of a country.

In addition to the multitude of variables affecting energy demand, there are also numerous uncertainties about the extent to which demand will decline with energy efficiency improvements as a result of what is known as "rebound effect." The rebound effect describes the response of consumers to the lower per unit cost of energy after an investment. A direct rebound effect is the increase in consumption of a good, in this case energy, resulting from the lower cost of use. An indirect rebound effect occurs when lower consumption has reduced the price of energy, and the lower cost in turn enables increased household consumption. Hence, the rebound effect describes an adjustment and optimization of household consumption following changes in prices (Borenstein 2013).

7.2.2 Efficiency Objectives for Germany

Efficiency targets have a long history in German energy policy, and they were a cornerstone of the Energy Concept 2050 announced in 2010 (BMWi and BMU 2010). Table 7.1 summarizes the initial efficiency targets of the Energiewende. These include a reduction of primary energy consumption of 20% by 2020 and of 50% by 2050 and an energy productivity increase of 2.1% annually (compared to average growth of 1.7% for the period 1990–2012). In the building sector, a nearly carbon-neutral building stock should be in place by 2020, with a thermal refurbishment rate for residential buildings of 2% annually. Gross electricity consumption should be decreased by 10% (2020) and by up to 25% (2050, base year: 2008), and final energy consumption of the transport sector should be reduced by 10% (2020) and by 40% (2050, base year: 2005).[2] Moreover, gross electricity generation by combined heat and power (CHP) plants should increase from 101 TWh in 2014 to 110 TWh in 2020 and 120 TWh in 2025.[3] Compared to 1990, primary energy consumption should decrease slightly and overall energy productivity should improve by 2020.

---

[2]The indicators assume a macroeconomic growth rate of 0.8% annually.

[3]§1 (1) *Gesetz für die Erhaltung, die Modernisierung und den Ausbau der Kraft-Wärme-Kopplung* (Kraft-Wärme-Kopplungsgesetz - KWKG), Kraft-Wärme-Kopplungsgesetz vom 21. Dezember 2015 (BGBl. I S. 2498).

Table 7.1 Energy efficiency objectives of the energiewende

Energy efficiency					
	2014	2020	2030	2040	2050
Primary energy consumption (compared with 2008)	−6.5%	−0.2			−0.5
Energy productivity (final energy consumption)	1.2% p.a. (av. 2008–2014)	2.1% p.a.			
Gross electricity consumption (compared with 2008)	−4.6%	−0.1			−0.25
Thermal refurbishment of residential buildings	~1% p.a. (2012 value)	2% p.a.			
Final energy consumption of transport sector (compared with 2005)	+0.2%	−0.1			−0.4

Source: BMWi (2015), Löschel et al. (2015)

In addition to these initial targets, the government added three additional targets (BMWi and BMU 2012, 16):

- 20% reduction of heat demand in the existing building stock by 2020
- reduction of primary energy consumption of buildings by about 80% up to 2050[4]
- rise in the thermal refurbishment rate of the existing building stock to 2% per year

7.2.3 Achievements to Date: Aggregate Results

However, a closer look at the results achieved to date shows that significant challenges still lie ahead in adjusting the measures and instruments currently in place. Energy productivity in Germany is increasing steadily, although the current rate is insufficient to reach the efficiency targets set as part of the energiewende. Figure 7.1 shows developments in Germany compared to other major European economies. Interestingly, the GDP per unit of energy consumption [in kilogram of oil equivalent (KGOE)] is quite similar across Western Europe, between 8 and 10 €. Poland, by contrast, has only half of that level of energy productivity. Together with the UK, Germany has achieved relatively high energy productivity gains by European comparison, slightly higher than the EU average.[5]

---

[4]Both heat demand and primary energy consumption in this case do not include renewable energies (Löschel et al. 2014b, 17). The use of renewable energy will therefore help in improving both indicators, which is reasonable from the point of view of GHG emissions reductions.

[5]Eurostat (2018): Data Explorer, last accessed 05.05.2018 at http://appsso.eurostat.ec.europa.eu/nui/show.do?dataset=nama_r_e3gdp&lang=en

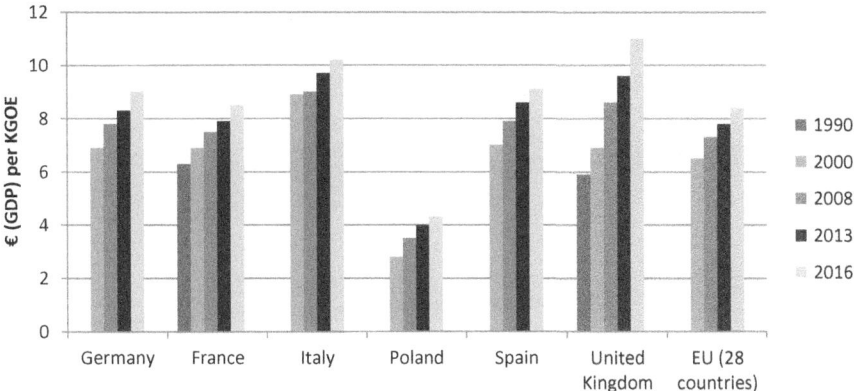

Fig. 7.1 Energy productivity 1990–2016: Germany in the European context. Source: Eurostat (2018) Data Explorer, last accessed 05.05.2018 at http://appsso.eurostat.ec.europa.eu/nui/show.do? dataset=nama_r_e3gdp&lang=en

Although Germany has achieved some success in reducing primary energy consumption and gross electricity consumption, it has still failed to reach its targets for energy productivity improvements. Up to 2013, primary energy consumption fell by just 4.0% compared to the base year 2008 (Löschel et al. 2014b, 6). Recent estimations show that depending on the assumed growth rates and primary energy productivity, a gap of between 9.9% and 12.8% (or 1.445 and 1.751 PJ)[6] will still have to be closed to achieve compliance with the 20% primary energy consumption target for 2020 (Fraunhofer ISI 2014, 10). Final energy consumption increased by 1.19% between 2008 and 2013 (see Fig. 7.2). To reach the envisaged targets, it is crucial that GDP growth be further decoupled from energy consumption and that energy productivity increases more rapidly.

Final energy productivity has varied across different time periods. While average energy productivity increased by 2% from 1990 to 2000 and by 1.3% from 2000 to 2004, the increase was even more substantial between 2004 and 2008 (as much as 2.6%). However, from 2008 to 2012, the rate of energy productivity improvement slowed to an average rate of 1.1%. To achieve the 2050 target, an annual average energy productivity increase of about 2.6% from 2012 to 2020 would be required (Löschel et al. 2014a).

Structural processes underway in the economy, including price fluctuations and the changing sectoral and sub-sectoral composition of GDP, have a significant impact on both GDP and energy consumption. These factors affect the aggregate indicator as well. This makes it difficult not only to interpret the overall indicator but

---

[6]The basis of calculation is temperature-adjusted primary energy consumption of 14,594 PJ in 2008. Fraunhofer ISI / Fraunhofer IFAM/ Prognos/ Ifeu (2014): Ausarbeitung von Instrumenten zur Realisierung von Endenergieeinsparungen in Deutschland auf Grundlage einer Kosten-/Nutzen-Analyse. Wissenschaftliche Unterstützung bei der Erarbeitung des Nationalen Aktionsplans Energieeffizienz (NAPE). Zusammenfassung.

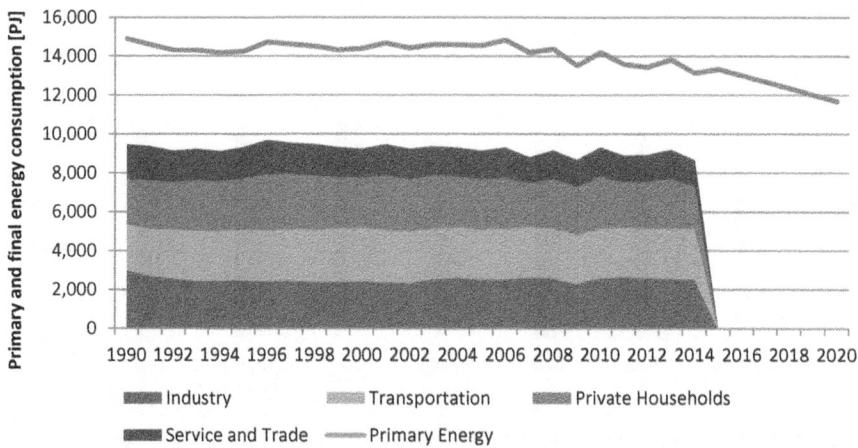

Fig. 7.2 Primary and final energy consumption development (in PJ). Source: Own depiction based on data from BMWi (2014) *Energy Data*: Complete Edition. Berlin, Germany

in particular to disentangle the success of policy measures aiming at energy efficiency increases from other factors. It is therefore necessary to monitor energy efficiency improvements at a disaggregated sectoral level.

7.2.4 Disaggregated Results

7.2.4.1 Commercial Sector, Buildings, Transport

The commercial and service sector performed best (3% energy productivity improvement annually since 1991), followed by road freight transportation (2.3%), individual road transportation (1.5%), industry (1.3%), and private households (1%) (Löschel et al. 2014a, 50). For industry, which accounts for 29% of total final energy demand in Germany, a future annual increase of 1.3% in energy productivity was agreed upon in negotiations between government and industry over CO_2 tax relief. Whereas the less energy-intensive service sector, which includes the public sector, is less affected by business cycles, the industrial sector is more sensitive to international market developments. Energy productivity in industry decreased in 2003 and 2009, years with low market demand for industrial goods and low utilization of existing capacities. Overall, energy productivity has improved since 1991. Technological processes and increasing use of CHP plants have contributed to this development. The growing importance of less energy-intensive sectors has also played a role.

Moreover, building construction is central to improving energy efficiency, and produces positive results. Although the building stock grew over the period 2000–2012 with the rise in living standards (measured in terms of living space in m^2/

person), heat consumption in residential buildings decreased by around 450 PJ or 20% (Schlomann et al. 2014, 23). Temperature-adjusted specific heat demand per m^2 declined by 10.8% from 2013 to 2008.[7] If these trends continue, achieving the 2020 targets for residential heat consumption will be possible if the refurbishment rate of buildings increases to 2% annually, as targeted (Löschel et al. 2014b, 13). However, it is important to note that further efficiency improvements in buildings will be more difficult to achieve in the years to come and will also become ever more costly as effiency levels rise. The even more challenging target for 2050 of reducing primary energy demand in the building sector by about 80% will call for a substantial increase in investment in this sector.

With regard to envisaged energy savings in the transport sector, the target appears to be quite ambitious: a 10% reduction by 2020 compared to 2005. In 2012, the achieved reduction was only 0.6% compared to the reference year. Over the same period, transport services[8] for passenger and freight traffic increased by 4 and 9%, respectively. Energy consumption in passenger and freight traffic, which decreased between 2005 and 2012 by 2.9% per year, was therefore more than compensated for by the overall increase in transport services. Energy consumption in the various subsectors of transport differed substantially:

- In road transport, energy consumption declined by almost 2% from 2005 to 2012 and in rail transport it decreased even more rapidly (4.6%).
- In shipping (2% increase) and aviation (including fuel tank contents on international flights), energy consumption increased by almost 8%.

7.2.4.2 Electricity (TT)

A remarkable development was observed in the electricity sector (see Fig. 7.3). Electricity consumption peaked at around 620 TWh between 2006 and 2008 due to the rapid economic growth in Germany during this period. However, since 2007, electricity consumption has decreased slightly, dropping to 581 TWh gross electricity consumption in 2009 during the financial crisis. This represents a reversal of the continuous increase in gross annual electricity consumption that was observed up to 2007, and a slight decrease in gross annual electricity consumption since 2008.

Although the average annual decrease from the base year 2008 to 2013 was about 0.55%, this rate needs to be doubled to 1.1% annually to reach the 2020 reduction target of 10%. This, too, is a daunting task. Private electricity consumption is currently decreasing, but at the same time, electricity consumption in sectors like transport is still increasing compared to 1999.

Assuming the kind of economic development projected in the German government's energy scenarios, growth in electricity productivity needs to increase to 1.6%

---

[7]BMWi (2014): Die Energie der Zukunft. Erster Fortschrittsbericht zur Energiewende. December 2014, page 32.

[8]Services are measured in passengers/km and tons/km.

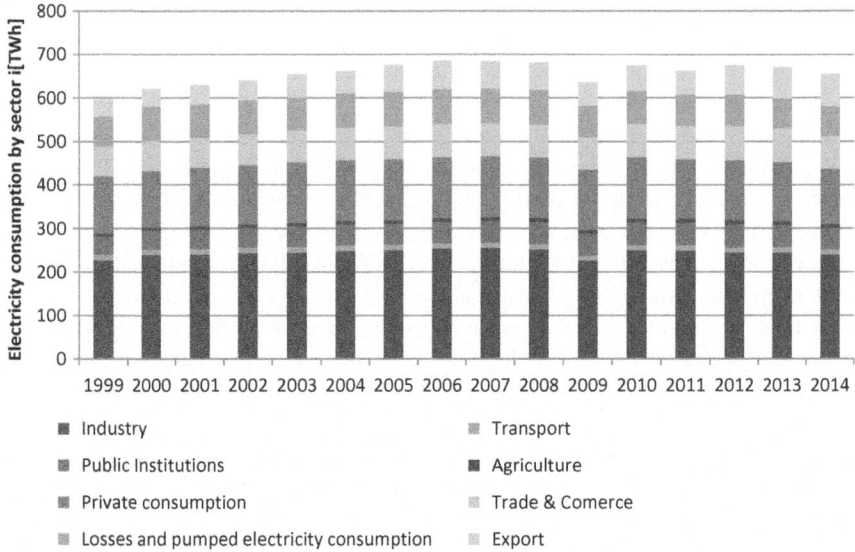

Fig. 7.3 Total gross electricity consumption from (1999 to 2014). Source: Own graph based on data from BMWi (2014) Energy Data: Complete Edition. Berlin, Germany

annually. This corresponds to an increase of 0.2 percentage points compared to the average annual productivity increase of about 1.4% in Germany from 2008 to 2013 (Löschel et al. 2014a, 46). Although considerable progress in energy efficiency has been achieved in recent years, present results indicate the need for efforts to strengthen this trend and increase energy productivity to achieve the 2020 goals.

7.3   Sectoral Policy Analyses

7.3.1   Policies for the Building Sector

Energy consumption in residential and commercial buildings represents more than 40% of total final energy use and is the main source of energy savings potential identified on the demand side (Öko-Institut and Fraunhofer ISI 2014, 14). The building sector therefore needs to contribute substantially to achieving the targets of reducing primary energy demand and of increasing final energy productivity. Buildings differ widely by type, age, owner, and user. Their final energy demand is driven mainly by heating and cooling, hot water supply, and lighting. These drivers depend heavily on heat losses through the building envelope (roof, walls, cellar, windows, and ventilation) and the energy standard of the equipment in place. Measures to reduce energy consumption therefore need to focus on all these elements.

Market failures, information problems, and human behavioral issues (bounded rationality of behavior) can threaten the success of measures to reduce energy consumption, and a variety of appropriate policies should be put in place to overcome these potential obstacles. Homeowners' and experts' assessments of the economic viability of refurbishment often differ, leading to differences between the estimated "optimal" rate of refurbishment and the actual rate of practical implementation. A viable approach for this sector would integrate different demand- and supply-side policies—for instance, raising the standards for insulation and heat production while providing financial assistance and working to raise public awareness for energy efficiency issues. For new buildings, the use of renewable energy for heat should be obligatory, and mandatory efficiency standards should be established.

Currently existing policies can be grouped into the following broad categories:

Administrative law

- policies creating minimum standards for the energy performance of new and existing buildings and their energy equipment combined with gradual tightening of these standards over time through additional ordinances and laws. This approach also includes the use of renewable heat in accordance with technical availability and economic viability.
- regular inspection of heating and air-conditioning systems.
- changes in the principal-agent relationship (building owners and tenants). In May 2012, the German government amended the tenancy law to split financial benefits and costs between landlords and tenants to facilitate refurbishment. Landlords are now allowed to increase rents up to 11% per year to cover the costs of energy renovation.
- obligatory use of renewable heat in new buildings.

Economic incentives

- public financial assistance to building owners through soft loans, investment grants, and subsidies that reduce liquidity constraints.

"Soft" instruments

- provision of information and advice on energy consumption reduction in buildings through energy consulting and building labelling (energy performance certificates) to reduce informational market barriers and to identify potential for savings based on high energy efficiency standards.
- Information and awareness-raising campaigns.

These instruments should lead, on the one hand, to 25% lower energy consumption for new buildings after 2016. On the other hand, building insulation should be improved by 20% from 2016 on. This requires a doubling of the refurbishment rate. The aim is to reach a 20% reduction of primary energy consumption between 2008 and 2020.

7.3.2   Policies for Industry

The majority of German industry is export-oriented and, thus, exposed to international competition. Not surprisingly, this sector is oriented primarily toward international regulations, and energy efficiency policies in industry are defined mainly by European legislation. Apart from the EU-ETS, which covers a large part of German industry, EU policies for industries have two main areas of focus:

Technology-driven activities Energy efficiency standards were introduced for energy-related products (ErP) through the 2009 EU directive establishing a framework for the setting of eco-design requirements for energy-related products. The directive defines minimum standards for energy-using products used in all sectors. It sets implementing regulations for various types of products. Examples include power transformers, water pumps, industrial fans, and electric motors. The regulations applying to electric motors serve as an example:

• The EU Motor Regulation (640/2009) defines requirements for the environmentally compatible design of electric motors and the use of electronic variable-speed drive control and creates four international efficiency classes for induction motors.
• The European Energy-related Product Standard EN 50598 focuses on the drive system as a whole and defines requirements for energy-related products (energy efficiency, eco balancing) for drive systems in electrically-driven machines.

Process-driven activities Energy management tools such as voluntary and obligatory energy audits are being used at regular intervals for all non-SME companies, and energy management systems are being installed in accordance with ISO 50001 standards. These systems provide a means for companies and organizations to create the necessary systems and processes for operational control and continued improvements in energy performance. Public funding is provided to SMEs for energy consulting.

In addition to the EU rules, the German government concluded an agreement with the German business community in 2012 on energy efficiency improvements up to 2022.[9] The agreement is in response to a decision formulated in the Energy Concept 2010 to extend the exemption of energy-intensive industries from the eco-tax, in place since 1999, under certain conditions. The conditionality was linked to verifiable implementation of energy management systems in accordance with the ISO (International Organization for Standardization) 50001 starting in 2013, combined with the agreement to set increased energy efficiency targets that would become binding in 2015. The following efficiency targets have been agreed upon:

• 1.3% energy efficiency increase in 2013 as a condition to apply for the eco-tax exemption in 2015.

---

[9]BMWi (2012): Vereinbarung zwischen der Regierung der Bundesrepublik Deutschland und der deutschen Wirtschaft zur Steigerung der Energieeffizienz vom 28. September 2012.

- For subsequent years, the targets are 2.6% in 2014, 3.9% in 2015, and 5.25% in 2016 to apply for the respective tax exemptions.
- Monitoring will be conducted by an independent economic institute and targets for tax exemptions in 2019–2022 will be set by 2017.

   The agreement is designed to provide tax exemptions as a financial incentive to ensure that after having slowed to a 1% annual rate of improvement between 2008 and 2012, future energy efficiency improvements stay on track with the overall target.

7.3.3 Policies for the Transport Sector

The transport sector is the second most important sector of the German economy, consuming almost 29% of total final energy. Currently, it is not on track to meet targets for 2020. Due to the diversified structure of the sector, a package of policies is needed to improve the sector's climate performance. The Mobility and Fuels Strategy of the German Government in place since June 2013 takes the difficulties of the sector into account. It is intended not as an overarching mobility strategy but as an initial, concrete contribution to achieving the targets in the transport sector. Since the majority of energy used in the sector is consumed by road transportation (82% in 2012), most policies aim at reducing fuel consumption of vehicles, for instance, by promoting fuel switch and technology improvements. Incentives have also been created through the vehicle tax and emission standards for new cars and light commercial vehicles, as formulated in the corresponding EU directives.

7.4 Energy Efficiency Policies Going Forward

7.4.1 The National Action Plan on Energy Efficiency (NAPE)

To address the slow progress achieved to date on energy efficiency targets, the German government introduced its National Action Plan on Energy Efficiency (NAPE) in 2014 (see BMWi 2014). The Action Plan, summarized in Table 7.2, is estimated to lead to an additional 390–460 PJ of primary energy savings, although this will still not be sufficient to close the existing GHG emissions reduction gap.

Since abatement costs are comparatively low and the savings potential is high, a special focus of this policy is on energy efficiency in buildings. The 2020 targets for residential heat consumption can be reached if the current decline in heat demand continues and if the 2% refurbishment rate is achieved. However, although the NAPE has improved financial support for refurbishment of buildings and tax deductions are planned, more ambitious instruments are necessary to achieve the 2020 targets and the even more stringent 2050 targets. This is due in particular to the following challenges:

Table 7.2 Key measures of the National Action Plan on Energy Efficiency (NAPE)

Measure	Predicted reductions by 2020	
	Volume of reduction in PJ	GHG in Mt CO_2-equivalent
Immediate measures		
Quality assurance and optimization of the existing energy consultations	4.0	0.2
Tax encouragement of energy-saving redevelopment	40.0	2.1
Further development of the CO_2 Building Renovation Program	12.5	0.7
Introduction of a competitive tendering scheme	26–51.5	1.5–3.1
Promotion of contracting (incl. deficiency guarantee)	5.5–10	0.3–0.5
Further development of the KfW Energy-efficiency Program	29.5	2.0
Energy efficiency networks initiative	74.5	5.0
Top-Runner-Strategy—on national and EU-level	85.0	5.1
Obligation for large-scale enterprises to conduct energy audits	50.5	3.4
National efficiency label for old heating systems	10.0	0.7
Further immediate measures of the NAPE	about 10	about 0.5
Sum of immediate measures	**350–380**	**21.5–23.3**
Further measures		
Measures starting in October 2012	43,0	2,5
Provisional estimator for the effect of the additional operating process	up to 40	up to 4
Total	**390–460**	**ca. 25–30**
Measures in the transport sector	110–162	7–10

Values in bold are intermediate and total sum. Source: BMWi (2014, 21)

- Studies have estimated necessary additional annual investment of about 26.4 billion € to achieve the 2020 target. That would mean the current level of about 100 billion € of annual investment in building construction would need to rise to about 126 billion €.[10] The question is how to create incentives for private investors to generate this level of additional investment. Policies and instruments (KfW soft loan programs) currently in place and even the planned increase in financial support for these programs appear insufficient.
- How can higher energy savings be achieved today, for instance, through "deep renovation," to avoid lock-in effects that could raise mitigation costs in the future?
- Opportunities for refurbishment are currently not being fully utilized. Although the refurbishment rate is about 1% p.a., 3% p.a. of the building stock is subject to

---

[10]BMVBS (2013): "Maßnahmen zur Umsetzung der Ziele des Energiekonzepts im Gebäudebereich – Zielerreichungsszenario." BMVBS-Online-Publikation 03/2013. Berlin, Germany.

some non-energy renovation. This indicates potentially missed opportunities for energy efficiency improvement (BPIE 2014, 41). Similarly, what can be done to ensure that these opportunities are utilized without creating problems for non-energy renovation? Such problems may occur, for instance, through the imposition of deep energy efficiency refurbishment mandates that require much higher up-front investment than non-energy renovation.

• By 2020, all refurbished and new heat systems should be on track to meet 2050 targets since no additional refurbishments are expected for buildings already refurbished between 2020 and 2050. The challenge is to create incentives for replacement of outdated heating systems with modern and innovative systems that can help to avoid lock-in effects.

7.4.2 Conflicts Between Targets and Policy Options

7.4.2.1 Heating

The solution to challenges in the area of power generation for heating depends on a number of key issues. There is substantial evidence of fundamental conflicts between the refurbishment rate and projected heat demand reduction as sub-targets for increased energy efficiency in the existing building stock. This will require a rethinking of the relationship between these sub-targets and the overarching target of climate change mitigation through reduction of GHG emissions. The refurbishment rate and the reduction of heat demand are two sub-targets that have been set to encourage carbon-neutral development in the building sector through the implementation of special policy measures. However, one of these two sub-targets alone would be sufficient. The refurbishment rate can serve only as a rough indicator of progress in reducing heat demand given that there is no precise definition of the scope and quality of refurbishments required to achieve energy savings objectives. For instance, refurbishments carried out as part of government-subsidized energy incentive programs may differ significantly in terms of efficiency from renovations carried out by a building owner independently.

In addition, there is also a problem with the definition of heat demand. This term does not distinguish between renewable and fossil heat. However, if heat demand is met completely by renewable energies, it satisfies the overarching national GHG emission reduction target, and any heat reduction impact on climate change is nullified. Therefore, the introduction of the concept of "net heat demand" considers renewable heat at "0" emissions, as the reduction of total heat demand is the more appropriate criterion.[11] Alternatively, targets for the building sector should be re-formulated in terms of primary energy demand, in line with the Energy Efficiency

---

[11]BMVBS (2013, p. 58): "Maßnahmen zur Umsetzung der Ziele des Energiekonzepts im Gebäudebereich – Zielerreichungsszenario." BMVBS-Online-Publikation 03/2013. Berlin, Germany.

Ordinance (EnEv), which defines primary energy demand for buildings as "non-renewable primary energy demand" (Löschel et al. 2014a, Z-13). The positive effects of such an approach would be:

- providing building owners more freedom to make decisions on least-cost options for GHG emissions reductions, for instance, on the level of insulation vs. utilization of renewable heat.
- reducing the necessary targets for CO_2-neutral refurbishment of existing stock and thereby lowering the required investment.

Such an approach coincides with ongoing discussions over the application of the Renewable Heat Law to the existing building stock. If renewable heat in existing buildings became obligatory, or if obligatory refurbishment measures could be offset by renewable heat, the choice of measures would be left more open to building or flat owners. This could trigger lower-cost solutions, which would become even more important at later stages in the transition of the building sector, when additional, more advanced insulation technologies become more costly.

Scenario analysis has found that the "net heat demand" approach would help to achieve the 2020 targets at lower refurbishment rates, and that in some cases, it would even lead to overcompliance with the implicit CO_2 reduction target in the building sector. Setting a CO_2 target for the building sector would certainly be an appropriate target adjustment, and it would by no means render ongoing efforts irrelevant. Rather, it would make it possible to reduce the heat demand in buildings, thereby increasing the attraction of higher private investment in refurbishment of buildings, and it would thereby help to overcome the respective obstacles. Merging the two main laws in the building sector—the Energy Efficiency Ordinance (EnEv) and the Renewable Heat Act, which is mentioned as an opportunity in the NAPE—appears to be a viable strategy for increasing the effectiveness of the legal framework.

Apart from improved regulation in 2012, the principle-agent problem in which energy bills are not paid by the purchasing party (building or apartment owner) but by the tenant has still not been dealt with adequately. This issue is of key importance in Germany, since about half of all flats are not owner-occupied. Balancing interests between building or flat owner and tenant is crucial to future investments in energy efficiency. The magnitude of refurbishment costs borne by tenants is an important political issue, and a number of proposals have been made to address this problem. One is the proposal to adjust the rent index, a basic tool for determining rents in new rental contracts, to criteria determining the energy efficiency of the flat in question. Given that the rent index does not include costs of heating and hot water, rents would rise on flats with higher energy efficiency standards, but tenants would save energy costs. The case is the reverse for non-energy-efficient flats. However, such an approach would not solve the problem of social affordability of energy-efficient homes for low-income households, which is also the subject of policy debate. For tenants, the energy savings often only offset the rent increases after a substantial period of time. In addition, the current reform of the housing allowance for the poor needs to take into consideration adjustments for affordability of flats refurbished in line with energy-efficiency standards (BMUB 2014, 33)

A further important and unresolved barrier to energy-efficient refurbishment of buildings is the tradeoff between lower life-cycle costs versus lower up-front costs (energy efficiency investment in buildings usually creates high up-front costs), which is also subject to high transaction costs. Demographic trends such as population aging have also led to a general unwillingness to invest in energy-efficient refurbishment. These issues even affect relatively affordable investments in energy efficiency. Moreover, fluctuations in funding for various support programs are adversely affecting market developments in this sector. Several proposals related to new policies and instruments are currently under discussion. The proposed measures aim to overcome specific barriers and should be viewed as supplementary to the already existing mix of policies and instruments.

Further tightening of the provisions of the EnEV is possible, but the potential impact of these changes is limited. The high up-front costs make it most cost-effective to carry out major renovations simultaneously—for instance, insulating the envelope of a building and simultaneously installing new heat supply systems. In some cases, however, homeowners prefer to carry out renovations individually rather than in combination. Other aspects, including the design of the building, its location, available technologies, etc., also affect the costs of renovations and homeowners' decisions. Administrative law is generally limited and unable to take all these different aspects into account.

As far as the targets for 2050 are concerned, there is a great deal of uncertainty about basic elements that will shape the future structure of the building sector. Factors include the changes in the amount of available living space per capita, the number of new buildings being built, progress in the refurbishment of buildings of different ages, and the technologies used for heating, cooling, and ventilation. Some scenarios assume that about one third of the building stock in 2050 will be new, and that the remaining two thirds already exist today (Öko-Institut and Fraunhofer ISI 2014, xi). In such a scenario, the focus on transformation of the building stock will continue to be of primary importance. A simultaneous switch to renewable heating and cooling in the existing building stock would be imperative. Introducing the obligatory use of renewables in the building stock and in local and central heating systems is currently under discussion.

7.4.2.2 CHP Targets Versus Renewable Heat

The target for increasing electricity generation from CHP to 120 TWh in 2025 may or may not be in line with the planned RES targets for electricity generation and the heat savings targets for buildings up to 2050. The current scenarios for gross and net electricity consumption for 2020 and 2050, respectively, are summarized in Table 7.3; the target scenario for 2050 is about 20% lower than the reference scenario.

Table 7.3 Scenarios for electricity generation (2020 and 2050, in TWh)

	2020	2050
Gross electricity generation		
– Reference scenario	618	561
– Target scenario	576	459
Net electricity generation		
– Reference scenario	556	505
– Target scenario	518	413

Source: EWI, GWS, and Prognos (2014, 5) and own calculations

As CHP installations combined with the respective local heat grids require substantial investments and have an estimated life span of at least 30 years, it is assumed that the 120 TWh will also be achieved by 2050. Due to reduced electricity generation in 2050, CHP electricity will then amount to a share of almost 25% of net electricity generation. A potential conflict with the renewable energy target of 80% could arise unless a majority of CHP is provided by renewable sources.

Another aspect relates to heat demand: Heat generated by CHP plants is used for process heating in industries, for district heating, and for local (decentralized) heating of buildings. According to the forecasts, the share of process heating is expected to increase and to become the main driver of the overall increase in the heat supply by CHP. The share as well as the overall amount of district heating will decline (from 70 to 35%, and from 110 TWh to 64 TWh, respectively) in the period from 2020 to 2050. Financial assistance is also currently being provided to build local heat grids for residential buildings in densely populated areas. In combination with renewable heat and process heat, low-temperature local heat grids could provide flexible heating options. The overall increase of local heating, however, is estimated to increase from 2.5 TWh (2020) to 5.8 TWh by 2050 (EWI, GWS, and Prognos 2014, 219). Taking into consideration the declining residual heat demand due to increased energy efficiency of buildings (insulation) and additional competition from highly-efficient heating technologies (for example, condensing boilers) and renewable energies (heat pumps and solar heating), the CHP target is not likely to be met. If the majority of CHP plants rely on fossil fuels (natural gas) given that the potential for biomass CHP has almost been exhausted, rising CO_2 prices will lead to declining economic advantages of CHP (EWI, GWS, and Prognos 2014, 218). In order to avoid CHP investments that result in stranded assets or lock-in effects, the approach to the building sector needs to be adjusted.

7.4.2.3 Electricity Consumption Reduction Versus New Applications

Although reduction of electricity consumption is one of the targets set in the Energy Concept 2050, it may contradict targets and solutions in other sectors. The target does not distinguish between electricity generated from fossil fuels and from renewables, but focuses mainly on cost savings through efficiency improvements. In terms of absolute numbers, however, it may need to be adjusted. For example, the fuel switch envisaged for the transport sector and the heat sector will increase electricity demand substantially. In the transport sector, increased demand is expected due to the replacement of petrol and diesel with electric mobility and the use of power-to-gas technologies in the production of fuels (mainly hydrogen). The replacement of fossil heat with heat pumps in the building sector will further heighten electricity demand. Depending on scenario assumptions, electricity demand from such "new" applications may even overcompensate for reductions in traditional spheres of electricity consumption. Recent climate change scenarios propose that electricity consumption be examined from a variety of different viewpoints, distinguishing between "classic electricity consumers" (in line with present applications) and "new electricity consumers" in order to avoid conflicting targets (Öko-Institut and Fraunhofer ISI 2014). The additional demand from these new applications is calculated to add up to 300–400 TWh in 2050.[12]

Additionally, for some of the traditional electricity uses, demand may increase in absolute numbers. A modal shift of freight transportation is envisaged from road to rail, which would require not only an increase in railroad investment but would also imply an increase in electricity consumption by railroads. This will not threaten the overarching goal of GHG emissions reduction if the additional electricity is generated from renewables, which is indeed assumed in the scenarios. From this point of view, refocusing the current gross electricity consumption target on non-renewable electricity would be appropriate, but this target should also be accompanied by electricity productivity targets in order to achieve cost effectiveness and spur technological change.

7.5 Conclusion

Energy efficiency is an important pillar of the energiewende, and ambitious efficiency targets of all kinds have been defined. Yet progress is slow, and in some cases, the targets even appear to be contradictory—for instance, reducing electricity consumption while at the same time decarbonizing the transport sector through the use of electrical power. This chapter has provided an overview of the current situation in this sector, the perspectives and potential of energy efficiency policies in Germany, and the challenges going forward.

One of the most pressing tasks for the near future is the significant transformation of the large and multifaceted heating sector. The heating sector, which also includes cooling, hot water supply, and process heat, is responsible for roughly half of final energy demand; most of this sector still relies directly on fossil fuel combustion. According to the regular Monitoring Reports on the Energiewende, current energy

---

[12]Agora Energiewende. 2015. Wie hoch ist der Stromverbrauch in der Energiewende? - Energiepolitische Zielszenarien 2050—Rückwirkungen auf den Ausbaubedarf von Windenergie und Photovoltaik. Berlin.

efficiency policy has contributed to Germany's failing to meet its overall GHG emissions reduction targets for 2020 (Löschel et al. 2014b, 23, 2015). Scenario results indicate that the current underperformance in efficiency improvements cannot fully be compensated for by additional renewable electricity generation, as this would mean doubling of the amount of renewable electricity generation between 2014 and 2020. However, renewable heat is making an increasingly important contribution to reducing fossil fuel consumption and CO_2 emissions, which will take some of the pressure off energy savings. Central to this transformation is the adjustment and further elaboration of an integrated concept for a carbon-neutral building sector up to 2050, combining efficiency improvement (heat demand reduction) with renewable heat in a coherent manner and including renewable CHP. Heat demand reduction and renewables can be treated as substitutes for achieving the goal of carbon-neutral buildings and, thus, to the overarching GHG emission reduction goal.

The transport sector also poses major challenges and is currently not adequately regulated to reach the desired energy efficiency objectives. It is going through a slow learning process with new fuels and traction systems. Since the abatement costs are relatively high, transport has been low on the climate policy agenda. With the emergence of competitive renewable energies and increasing acceptance of new modes of transportation, this picture may change in the near future. New demand patterns may reduce overall transport demand, allowing the sector to move to a larger share of renewables such as renewable electricity and biofuels.

References

Agora Energiewende. 2013. *12 Insights on Germany's Energiewende.* A Discussion Paper Exploring Key Challenges for the Power Sector, Berlin, Germany.

BMUB. 2014. *The German Government's Climate Action Programme 2020.* Berlin.

BMWi. 2014. *Making More out of Energy – National Action Plan on Energy Efficiency.* Berlin.

———. 2015. *The Energy of the Future, Fourth 'Energy Transition' Monitoring Report – Summary.* Berlin.

BMWi, and BMU. 2010. *Energy Concept – for an Environmentally Sound, Reliable and Affordable Energy Supply.* Berlin.

BMWi, and BMU. 2012. *First Monitoring Report 'Energy of the Future' – Summary.* Berlin.

Borenstein, Severin. 2013. *A Microeconomic Framework for Evaluating Energy Efficiency Rebound and some Implications.* NBER Working Paper 19044. Cambridge, MA: National Bureau of Economic Research

BPIE. 2014. *Renovation Strategies of Selected EU Countries: A Status Report on Compliance with Article 4 of the Energy Efficiency Directive.* Brussels.

EWI, GWS, and Prognos. 2014. *Entwicklung der Energiemärkte – Energiereferenzprognose.* Endbericht 57/12. Studie im Auftrag des Bundesministeriums für Wirtschaft und Technologie. Basel, Köln/Osnabrück.

Kemfert, Claudia, Petra Opitz, Thure Traber, and Lars Handrich. 2015. *Deep Decarbonization in Germany - A Macro-Analysis of Economic and Political Challenges of the 'Energiewende' (Energy Transition),* DIW Berlin, Politikberatung kompakt 93. Berlin: DIW Berlin, German Institute for Economic Research.

Löschel, Andreas, Georg Erdmann, Frithjof Staiß, and Hans-Joachim Ziesing. 2014a. *Statement on the Second Monitoring Report by the German Government for 2012*, Summary. Berlin, Mannheim, Stuttgart: Federal Ministry for Economic Affairs and Energy.
———. 2014b. *Statement on the First Progress Report by the German Government for 2013*. Summary. Berlin, Münster and Stuttgart.
———. 2015. *Statement on the Fourth Monitoring Report of the Federal Government for 2014*, Summary. Berlin, Münster, Stuttgart: Federal Ministry for Economic Affairs and Energy.
Öko-Institut, and Fraunhofer ISI. 2014. *Klimaschutzszenario 2050 – 1. Modellierungsrunde*. Studie im Auftrag des Bundesministeriums für Umwelt, Naturschutz, Bau und Reaktorsicherheit. Berlin.
Schlomann, Barbara, Matthias Reuter, Bruno Lapillonne, Karine Pollier, and Jan Rosenow. 2014. *Monitoring of the 'Energiewende': Energy Efficiency Indicators for Germany*. Working paper Sustainability and Innovation S 10/2014. Karlsruhe.

Chapter 8
The Role of Electricity Transmission Infrastructure

Clemens Gerbaulet

In the simplest—and most recommended—regulatory approach, a plan for transmission network expansion would be prepared by the System Operator ...The transmission facilities that are included in the plan will be built ... under some kind of cost-of-service remuneration. ... The simple idea behind this simple scheme that is in use in several countries is just to make the business of transmission investment as "unexciting" ("boring" or "uneventful") as possible. Sophistication and complexity in transmission planning—"leaving it to the market," for instance—only cause indecision by investors, higher capital costs and—most frequently—lack of investment.
Luis Olmos, and Ignacio Perez-Arriaga (2009, 5286): A comprehensive approach for computation and implementation of efficient electricity transmission network charges.

This chapter is based on a series of papers and studies carried out in the Masmie and EE-Netze projects supported by the Stiftung Mercator, and complemented by other research papers (see references); it has been updated from it has been updated from Chap. 2 of my dissertation (Gerbaulet 2017). The modeling work was carried out using various versions of ELMOD, a European-wide electricity market and network model developed by Leuthold et al. (2012) and updated by Egerer et al. (2014). Thanks to Friedrich Kunz, Alexander Weber, and the entire ELMOD team for support and advice; the usual disclaimer applies.

C. Gerbaulet (✉)
TU Berlin, Berlin, Germany

DIW Berlin, Berlin, Germany
e-mail: cfg@wip.tu-berlin.de

© Springer Nature Switzerland AG 2018
C. von Hirschhausen et al. (eds.), *Energiewende "Made in Germany"*,
https://doi.org/10.1007/978-3-319-95126-3_8

8.1   Introduction

In addition to the core objectives of the energiewende analyzed in previous chapters, the infrastructure required to assure a reliable, clean, and economic electricity system is among the crucial conditions that have to be established for the energiewende to succeed. In this context, the electricity transmission infrastructure is a particularly important ingredient of the energiewende given the changing geographic distribution of electricity supply. Electricity transmission is a more controversial issue and also a stronger focus of attention because it involves important trade-offs between ambitious expansion projects by transmission system operators, decisive fuel choices overseen by regulators (e.g., decisions not to favor coal electrification through network extension), and the public debate about the appropriate siting of transmission corridors.

This chapter summarizes issues surrounding electricity transmission in the context of the energiewende. Even though infrastructure is an important ingredient of the energiewende, its importance has been exaggerated in the policy debate and in the public debate as well. Often hailed as a "critical factor" in the energiewende—and sometimes as the final nail in its coffin—transmission infrastructure has not been a demonstrable obstacle to the energiewende thus far, thanks to the highly developed network inherited from the old system and its continuous improvement over the last decade. Even in the medium term—that is, into the 2020s—no serious roadblocks for the energiewende are to be expected, provided that the transmission system operators (TSOs) and regulatory agencies stick to the path of transmission expansion that has proven reliable so far.[1]

The next Sects. 8.2 and 8.3 describe network planning and development from its inception in the 2000s until today. Over this period, a new method of transmission planning has been implemented, creating more transparency for transmission policies, which had not been open to public scrutiny under the old system. We also discuss two critical regulations that distort transmission planning: One is the algorithm created to identify expansion needs based on the assumption of a "copper plate" that by design ignores potential transmission constraints in the establishment of the dispatch merit order.[2] The other is the high equity remuneration of TSOs (9.05% or as of 2019 6.91%), incentivizing these companies to engage in high levels of investment. Section 8.4 then traces a decade of network development in Germany. As elaborated in detail, rates of transmission investment remain consistent over the years, and important connections, such as links between the former GDR and West Germany have been completed. Section 8.5 discusses the current debate of introducing multiple price zones in Germany. Moreover, it summarizes results of a study on the effects of establishing multiple price zones in Germany suggesting that there

---

[1]This chapter does not consider electricity distribution infrastructure.

[2]In 2015, the possibility of 3% curtailment of renewable electricity feed-in was introduced into the grid planning process, which reduced the need for investments that are needed only a very few hours per year to accommodate both high wind and photovoltaic (PV) feed-in.

is no need to split the German electricity market into zones. Section 8.6 details an interesting recent development: the explicit integration of carbon dioxide (CO_2) constraints into network planning. Finally, Sect. 8.7 concludes.

8.2 Transmission Planning and Incentives

8.2.1 Network Planning Before the Energiewende

Electricity transmission, that is, long-distance transport of electricity at voltage levels of 200 kV and above, has been a major point of discussion in the energiewende and has received extensive attention from technical professionals, policy makers, and economic analysts. Even though investment expenses in electricity distribution infrastructure are significantly higher (with about five times the level of average annual investments), transmission is considered particularly challenging because of the changing structure of electricity generation, away from centralized conventional sources towards more decentralized renewable sources. In this context, a general belief prevailed in the 2000s that a major shift in grid architecture might occur, not only in Germany but also across Europe, towards the diffusion of "Supergrids," European-wide high-voltage corridors ("Stromautobahnen") that were expected to accompany the low-carbon energy transformation at the European level.

Before the advent of the energiewende, investments were made into the development of electricity transmission infrastructure with the intention of a congestion-free connection between mostly fossil and nuclear generators and industrial as well as household consumers. Transmission planning was a technical exercise, and a congestion-free infrastructure was considered to be normal; the costs of maintaining a "copper plate" were passed on to the electricity prices. Within the eight (later: four) vertically integrated energy companies, transmission was carried out alongside generation and sales, with the distribution grid following based on local requirements. The transmission operators coordinated their activities somewhat in the "German Network Association" Deutsche Verbundgesellschaft (DVG), see for a historical overview Boll (1969).

Vertical unbundling was first introduced with the European Electricity Directive of 1996[3] and reinforced by the Acceleration Directive of 2003,[4] along with an increasingly heated policy debate. It increased pressure to modernize the institutional setting for transmission planning, which had proven to be intransparent and not open to either public or administrative oversight. The first network planning

---

[3]See EC (1996). Directive 96/92/EC of the European Parliament and of the Council of 19 December 1996 concerning common rules for the internal market in electricity.

[4]EC (2003). Directive 2003/54/EC of the European Parliament and of the Council of 26 June 2003 concerning common rules for the internal market in electricity and repealing Directive 96/92/EC.

exercise that involved some public consultation was coordinated by the German Energy Agency (*deutsche energieagentur,* dena), but the monopoly of data and network calculations remained with the TSOs (a situation that prevails to this day): the dena I-network study (dena 2005) concluded that 850 km of new-built lines and 392 km of line upgrades were needed by 2015. To accelerate network development, the "law on developing electricity transmission infrastructure" of 2009 (*Energieleitungsausbaugesetz,* EnLAG) was passed covering upgrades of 20 lines that were considered particularly important, totaling 1855 km in length.[5] For four of these lines, the law foresaw the possibility of laying some sections underground (Ganderkessee to Wehrendorf, Lauchstädt to Redwitz, Diele to Niederrhein, and Wahle to Mecklar). In the context of the Energy Concept 2010 and the first coordinated effort to create a European-wide Ten-Year Network Development Plan (TYNDP) (ENTSO-E 2010), the dena I network study (dena 2005) was followed by the dena II network study (dena 2010), once again conducted by the four TSOs with the participation of two academic reviewers. The results suggested the need for upgrading and new builds far beyond any historical averages: with 1500–3600 km of newly built lines and up to 5700 km of upgrades by 2020. A critical analysis of the dena II network study[6] pointed out the discrepancy between the results and a reasonable economic level of network extensions, as well as the TSO's incentives to overinvest (high rate of return, no congestion clause, etc.).

8.2.2   2011: A Renewed Institutional Framework for Network Planning

After the two dena network studies, a further need to modify the institutional framework of transmission planning was perceived, and the responsibility for network planning was shifted to the electricity sector regulator, the Federal Network Agency (*Bundesnetzagentur,* BNetzA) to streamline and accelerate the planning process. Following the mechanisms of structured network planning defined at the European level, prescribed by the third directive on the internal European Electricity Market,[7] transmission planning was completely reorganized by the new German Energy Law (Energiewirtschaftsgesetz, EnWG) of 2011.[8] In §12, the new law

---

[5]EnLAG (2009). Gesetz zum Ausbau von Energieleitungen (Energieleitungsausbaugesetz—EnLAG).

[6]Hirschhausen, Christian von, Robert Wand, and Christina Beestermöller. 2010. "Bewertung der dena-Netzstudie II und des europäischen Infrastrukturprogramms." Gutachten im Auftrag des WWF Deutschland. Berlin, Germany: TU Berlin.

[7]See EC (2009). Directive 2009/72/EC of the European Parliament and of the Council of 13 July 2009 concerning common rules for the internal market in electricity and repealing Directive 2003/54/EC.

[8]The German Energy Law of 2005 (BGBl 2005, Part I, p. 1970 http://www.bgbl.de/xaver/bgbl/start.xav?startbk=Bundesanzeiger_BGBl&jumpTo=bgbl105s1970.pdf), which implemented the

prescribed a more open and interactive, stakeholder-oriented process, organized and controlled by the regulator, that included a series of public consultations and ended with a law passed by parliament. The TSOs were given the task of setting up a 10-year and 20-year network development plan for measures considered "necessary and urgent" (*vordringlicher Bedarf*):

- In a first step, the TSOs are asked to develop a scenario framework (*Szenariorahmen*), including three scenarios for the energy mix for the next two decades. There are essentially three scenarios for a time frame of ten years (called A, B, and C), with Scenario B updated to cover 20 years as well. These scenarios also contain assumptions on electricity demand, fuel prices, etc. Once the scenario framework is handed over to the regulator (BNetzA), it undergoes public consultation, and is then approved, possibly with amendments by the regulator.
- Subsequently, the TSOs use the scenario framework to develop a long-term network development plan (*Netzentwicklungsplan*, NEP), which is first consulted on publically, then handed over to the regulator, who undertakes a second consultation. The regulator has to approve, or reject, the individual lines that are proposed. At the same time, the regulator carries out an environmental assessment (*Umweltprüfung*).
- Finally, the approved network development plan is passed to both chambers of parliament, the federal chamber (*Bundestag*) and the representatives of the 16 federal states (*Bundesrat*). Once approved by both chambers, it is published as the Federal Requirements Plan Law (*BBPlG , Bundesbedarfsplangesetz*).

Parallel to this new provision in the energy law, the German parliament also passed a law in 2011 that streamlined administrative responsibility within the regulatory agency, the BNetzA: The "transmission network development accelera-tion law" (*Netzausbaubeschleunigungsgesetz Übertragungsnetz*, NABEG) assigned responsibility for tracing the individual routes (*Bundesbedarfsplanung*) to the remit of the national regulator, alongside the decision-making process on the (local) siting of the corridors (*Planfeststellungsverfahren*). By concentrating activities at the federal level that had previously been carried out by the 16 federal states individu-ally, it was hoped that the implementation of the network development plan could be streamlined and accelerated.

The new procedure was applied for the first time in the network development exercise "NEP 2013," started in 2011. It has raised both public awareness and the participation of a broad range of stakeholders. After the first completion of this sequential procedure, the first Federal Requirements Plan Law was adopted and put

---

Directive 2003/54/EC (EC, 2003b), was adapted in the Gesetz zur Neuregelung en ergiewirtschaftsrechtlicher Vorschriften (BGBl 2011, Part I, p. 1554 http://www.bgbl.de/xaver/bgbl/start.xav?startbk=Bundesanzeiger_BGBl&jumpTo=bgbl111041.pdf) implementing the directive 2009/72/EC (see previous footnote). Since 2011, the planning process has been adapted, especially regarding the frequency of the planning procedures, see BGBl 2015, Part I, p. 2200 http://www.bgbl.de/xaver/bgbl/start.xav?startbk=Bundesanzeiger_BGBl&jumpTo=bgbl115s2194.pdf

into effect in 2013. For the following two years, the process of scenario framework, network planning, and consultation has been carried out annually with only small changes regarding pilot projects for underground alternating current (AC) cables in 2014 and a larger iteration in 2015. Due to changes in the energy-political framework, the procedure has been ceased in the year 2016 and no NEP has been conducted. Moreover, as of 2015, following the new energy law 2015, it is foreseen to perform the process on a biannual level to account for overlaps of the annual cycles.

8.2.3   Remaining Inefficiencies and Investment Incentives

8.2.3.1   Methodology Guarantees Congestion-Free Electricity Feed-in

It is always better to have slight overinvestment in infrastructures than to have too little investment. In addition, the procedure of transmission investments, including regulation and oversight, should be as simple as possible to avoid "sophistication and complexity in network planning" (Olmos and Pérez-Arriaga 2009, 5286). However, while the new procedure of network planning was a step in the right direction, it had two major drawbacks: First, even though some underlying assumptions were stated explicitly in the scenario framework, the monopoly power of the TSOs in setting up the network development plan was maintained. And second, the modified institutional framework of transmission planning did not significantly change the incentives for network development, leading to a situation where congestion-less electricity transmission has remained the point of reference, pushing TSOs toward ambitious expansion plans (for details, see Weber et al. 2013).

A point of criticism of the procedure used in the network development plan is that it adheres too closely the "old world" market design, which takes the geographical distribution of generation capacities as given so that neither constraints on feeding in electricity nor the costs of network expansion are taken into account. Effectively, this leads to a potentially over-dimensioned network, corresponding more or less to a "copper plate".[9] In fact, given the boundaries by the Szenariorahmen, the first step of the planning procedure to investigate the need for grid expansion, determines an "optimal market-based power plant dispatch" according to the merit order principle (with renewables benefitting from priority feed-in). A subsequent step determines the resulting network expansion needs using the pre-determined dispatch from step one. Any power producer (fossil, nuclear, or renewable) has the right to sell their electricity (once in the merit order), independently of their location in the network.

The disregard of the geographical component lead to an oversupply of electricity from CO_2-intensive coal plants in Northern Germany, where access to coal is less expensive, whereas the capacity requirements tend to be located in the South. As a

---

[9]Some of these points are laid out in detail in Jarass und Obermair (2012).

result, the methodology must accomodate the parallel feeding in of conventional and renewable electricity. This requires an oversized network where especially the North-South corridors would have to be oversized to fit situations of strong wind and full operation of hard coal and lignite power, a situation which tends to work against the climate objectives of the energiewende (Kemfert et al. 2016).

The arguments in favor of an integrated network planning algorithm seem obvious, particularly in light of international experience in countries like the UK or in the US, with its restructured systems.[10] Yet these experiences have not been taken into account in the network planning carried out as part of the German energiewende. It is unclear whether this strategy was chosen deliberately, perhaps to prevent major network congestion, following the example of Alberta, Canada, where the network operator is obliged to avoid congestion.[11] However, if this was indeed the far-sighted strategy, it should have been stated clearly by the regulator at the outset.

8.2.3.2 Equity Remuneration

Another driver of the levels of investment is the regulatory regime under which the TSOs operate. Eyre and Pollitt (2016) provide an extensive survey of regulatory regimes and the effects of incentives on the transmission planning process. Germany was definitely a latecomer in this process: After a long period of cost-based remuneration of network companies in the old system, what is known as "incentive regulation" was introduced in 2005. These "incentives" proved, however, to be quite favorable to the TSOs: In addition to a large share of the costs that were exempted from benchmarking because they were declared "not modifiable" by the TSO, all new investment projects received a generous return on equity of 9.05%[12]; at an average rate of inflation of 0–2%, this corresponds to a real rate of return of 6–7%, for an almost risk-free activity. Even the downward adjustment to a return on equity to 6.91%, which will be in place in the next regulatory period beginning in 2019, is significantly above the risk-free return. Comparing this to the average return of a risk-free asset (in the range of 1–3%) reveals the high incentives for TSOs to invest heavily in new transmission infrastructure.

For grid planning, the so-called NOVA principle should be used to guide the network policy: It implies that the first priority is to optimize the use of the existing

---

[10]For a survey and another concrete application see Kemfert et al. (2016).

[11]The Brattle Group (2007): International Review of Transmission Planning Arrangements, p. 32.

[12]See BNetzA (2011). BK4-11-304 Beschluss hinsichtlich der Festlegung von Eigenkapitalzinssätzen für Betreiber von Elektrizitäts- und Gasversorgungsnetzen für die zweite Regulierungsperiode in der Anreizregulierung as well as BNetza(2016). Bundesnetzagentur legt Eigenkapitalrenditen für Strom- und Gasnetze fest. https://www.bundesnetzagentur.de/SharedDocs/Pressemitteilungen/DE/2016/161012_EKZ.html

grid; second, to upgrade existing lines, and only third, to build new lines.[13] Some technologies exist to increase the capacity of the lines, for instance, using high-temperature conductors in the existing grid, which potentially increases line capacity significantly or the installation of phase shifter transformers, which can somewhat direct the flows by influencing line impedances.[14] Also, the active monitoring of existing lines (*Leiterseilmonitoring*) has the potential to increase capacity in times of potential congestion, i.e., cold winter days, by indicating a surplus capacity of certain lines (due to the low temperatures, or times of high wind which cools the conductors). Even though the potential capacity upgrades are estimated in the range of 20–30%, this instrument was not pursued very actively by the TSOs (see Jarass and Obermair 2012; Jarass and Jarass 2016). Further options to conduct grid congestion management that are currently discussed could be grid-beneficial dispatch of storage systems, grid-beneficial dispatch of HVDC-systems, automatic redispatch, and further automation regarding the security assessment of the system.[15]

8.3 Overview of Network Development Plans

8.3.1 Projected Future Network Development

The TSOs in Germany responded quite rationally to the incentives and the mandate of a congestion-free grid: They planned substantial investments. It is difficult to quantify a degree of overinvestment as there is a very fine line between overinvesting and "staying on the safe side." Taking into account the current costs for congestion management (see Sect. 8.4.3) while many grid expansion measures are not completed show that the hypothesis of substantial overinvestment does not hold.

This applies to both the period of almost unregulated transmission planning in the two dena network studies, i.e., before 2011, and the subsequent period of more formalized network development procedures initiated with the energy law of 2011 (EnWG 2011). Network development plans developed in the second period envisaged an enormous increase in planned new builds and line upgrades relative to the average of 50–100 km of high-voltage transmission lines built over the previous decade. Table 8.1 shows the aggregate figures for network expansion and network upgrades as presented in the subsequent network planning exercises, beginning with the dena I network study (dena 2005).

---

[13]NOVA, in German, stands for NetzOptimierung, -Verstärkung und –Ausbau (network optimization, strengthening, and expansion).

[14]The pilot project of installing high-temperature conductors in 2013 was successful: the capacity of the 380 kV line between Remptendorf and Redwitz (East Germany to Bavaria) was increased by 400 MW or approx. 25%.

[15]Agora Energiewende, and Energynautics. 2018. "Toolbox für die Stromnetze—Für die künftige Integration von Erneuerbaren Energien und für das Engpassmanagement." Agora Energiewende.

Table 8.1 Aggregate transmission network expansion plans in Germany (2005–2017)

Planning document	Network newbuilt	Network upgrades
dena I network study (2005, 126)	851 km up to 2015	392 km up to 2015
dena II network study (2010, 364)	1500 km up to 2015, as well as 1700–3600 km from 2015 to 2020 depending on scenario	up to 5700 km depending on scenario up to 2020
Network development plan 2022 (50Hertz et al. 2012, p. 116; BNetzA 2012, 2013a, p. 418)[a]	AC: 1500 km "Startnetz" + 650 km newbuiltDC: 1600 km newbuilt + 300 km on existing AC corridors	AC: 400 km "Startnetz" + 2000 km upgrades
Network development plan 2023 (50Hertz et al. 2013, p. 87; BNetzA 2013a, p. 418)[b]	AC: 1400 km "Startnetz" + 600 km newbuiltDC: 1600 km newbuilt + 300 km on existing AC corridors	AC: 300 km "Startnetz" + 2500 km upgrades
Network development plan 2024 (50Hertz et al. 2014, p. 61; BNetzA 2015a)[c]	AC: 1400 km "Startnetz" + 648 km newbuiltDC: 1750 km newbuilt + 300 km on existing AC corridors	AC: 500 km "Startnetz" + 2750 km upgrades
Network development plan 2030 (50hertz et al 2017, p. 89, Bundesnetzagentur 2017, p. 325)	AC: 600 km "Startnetz" + 550 km newbuiltDC: 2150 km newbuilt + 300 km on existing AC corridors	AC: 900 km "Startnetz" + 3400 km upgrades

Sources: 50Hertz et al. (2017) and BNetzA (2017)[d]
[a]50Hertz, Amprion, TenneT, and TransnetBW. 2012. "Netzentwicklungsplan Strom 2012, 2. überarbeiteter Entwurf der Übertragungsnetzbetreiber"; BNetzA. 2012. "Bestätigung Netzentwicklungsplan Strom 2012 durch die Bundesnetzagentur für Elekt-rizität, Gas, Telekommunikation, Post und Eisenbahnen." Bonn, Germany: Bundesnetzagen-tur; BNetzA. 2013a. "Bestätigung des Netzentwicklungsplans Strom 2013 durch die Bundesnetzagentur für Elektrizität, Gas, Telekommunikation, Post und Eisenbahnen." Bonn, Germany
[b]50Hertz. 2013. "Netzentwicklungsplan Strom 2013, Zweiter Entwurf der Übertragungsnetzbetreiber"; BNetzA. 2013a. "Bestätigung des Netzentwicklungsplans Strom 2013 durch die Bundesnetzagentur für Elektrizität, Gas, Telekommunikation, Post und Eisenbahnen." Bonn, Germany
[c]50Hertz. 2014. "Netzentwicklungsplan Strom 2014, Zweiter Entwurf der Übertragungsnetzbetreiber"; BNetzA. 2015a. "Monitoringbericht 2015." Bonn, Germany. 2017. "Monitoringbericht 2017." Bonn, Germany
[d]50Hertz. 2017. "Netzentwicklungsplan Strom 2030, Version 2017, zweiter Entwurf der Übertragungsnetzbetreiber"; BNetzA. 2017a. "Bestätigung des Netzentwicklungsplans Strom für das Zieljahr 2030." Bonn, Germany

Figure 8.1 shows the number of kilometers planned in the various grid development plans by the TSOs. The grid reinforcements amount to about 4000–5500 km over time. Leaving the planned DC lines aside, most projects are newbuilt lines or line upgrades in existing corridors. New AC corridors amount to less than 2000 km, with a slightly declining trend over time. This trend can likely be attributed to increased use of the NOVA principle over time, as the various instances of public consultation of the grid development plans showed a clear tendency to not establish new corridors but to instead strengthen the already existing grid. The trend from opening new corridors towards reinforcements in the existing grid by using existing corridors can not only be attributed to increased importance of the NOVA principle

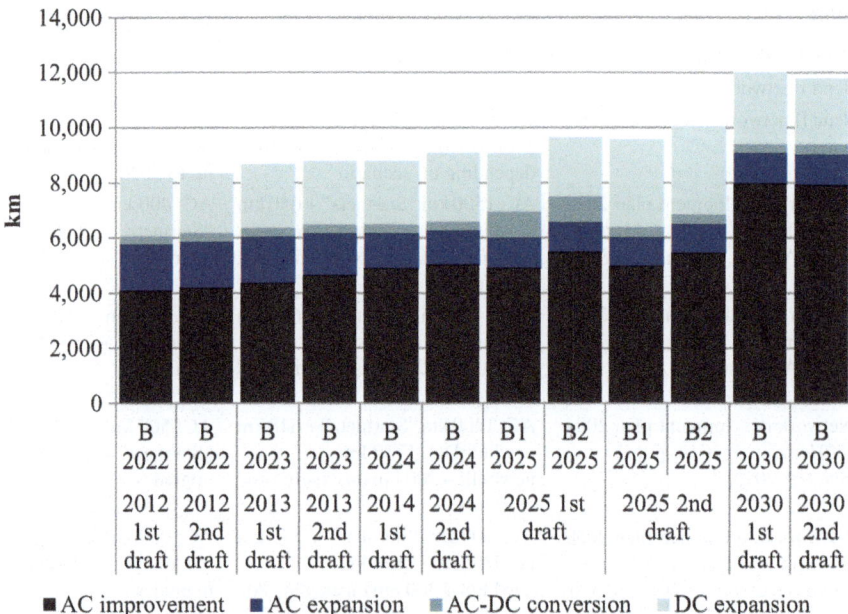

Fig. 8.1 High-voltage transmission expansion plans (in km) over time. Only main scenarios. Source: Adapted and updated form Gerbaulet (2017) based on 50Hertz et al. (2012, 2013, 2014, 2016, 2017) and BNetzA (2017) (50Hertz, Amprion, TenneT, and TransnetBW. 2012. "Netzentwicklungsplan Strom 2012, 2. überarbeiteter Entwurf der Übertragungsnetzbetreiber"; BNetzA. 2012. "Bestätigung Netzentwicklungsplan Strom 2012 durch die Bundesnetzagentur für Elekt-rizität, Gas, Telekommunikation, Post und Eisenbahnen." Bonn, Germany: Bundesnetzagen-tur; 50Hertz. 2013. "Netzentwicklungsplan Strom 2013, Zweiter Entwurf der Übertragungsnetzbetreiber"; 50Hertz. 2014. "Netzentwicklungsplan Strom 2014, Zweiter Entwurf der Übertragungsnetzbetreiber"; 50Hertz. 2016. "Szenariorahmen für die Netzentwicklungspläne Strom 2030—Entwurf der Übertra-gungsnetzbetreiber"; 50Hertz. 2017. "Netzentwicklungsplan Strom 2030, Version 2017, zweiter Entwurf der Übertragungsnetzbetreiber"; BNetzA. 2017a. "Bestätigung des Netzentwicklungsplans Strom für das Zieljahr 2030." Bonn, Germany)

over time. Also, public opposition against the TSOs expansion plans and potential new corridors has influenced the planning objectives. The so-called NIMBY (Not In My Back Yard) effect tends to become stronger in the German public, where decarbonization objectives and general favoring of renewable energy conflict with the increased need to transport electricity. This has not lead to a complete stop in the development of new corridors, but has shifted the objective slightly towards upgrading the already existing infrastructure, and also underground cables.

8.3.2   Onshore and Offshore Development Plans

In order to provide an impression of the physical realities of network development, we summarize the results of the 2012 network development exercise, including both onshore and offshore connections. The first draft of the TSOs' 2012 network development plan included aggregate network development measures of 6600 km onshore, corresponding to investments of about 20 billion €. The regulator accepted 5700 km of these lines, including 2800 km of new builds and 2900 km of upgrades on existing lines.

A new element of the 2012 planning exercise was the compilation of an offshore network development plan, called "O-NEP electricity 2013." Offshore connections were previously coordinated in a decentralized manner by the federal states (Lower Saxony and Schleswig Holstein for the North Sea, and Mecklenburg-Pomerania for the Baltic Sea). In the exercise, coordination was centralized in the national plan under the control of the regulator, BNetzA, to ensure closer adherence to the onshore development plan. The Offshore NEP-2013 includes the projects already being implemented, the so-called "starting offshore grid" ("Start-Offshorenetz"), as well as an additional 1135 km HVDC lines, and 595 km AC-lines in the North Sea, and 370 km AC-lines plus 60 km onshore connections in the Baltic Sea. Figure 8.2

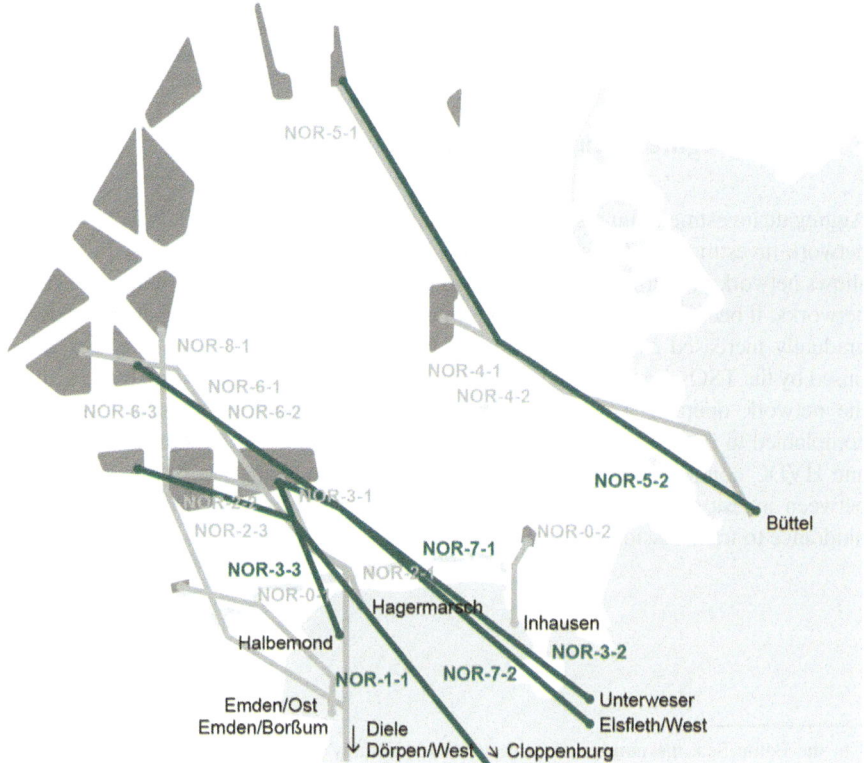

Fig. 8.2 Offshore Network Development Plan (NEP) for the German North Sea. The starting grid is gray; the projects of the scenario 2023B are in green. Source: Gerbaulet et al. (2013) (Gerbaulet, Clemens, Friedrich Kunz, Christian von Hirschhausen, and Alexander Zerrahn. 2013. "Netzsituation in Deutschland bleibt stabil." DIW Wochenbericht 80 (20): 3–12)

shows the offshore network development plan for the North Sea. While the basis of Fig. 8.2 is from 2013, the list of projects has remained constant over time. By 2018, some 1547 km had been realized.[16]

8.4 Network Expansion During the Energiewende

A look at the state of the current network investments in the German electricity sector appear to support the hypothesis that some delays of grid expansion currently experienced are not necessarily a binding constraint to the energiewende. Given an institutional design that favors the complete integration of the merit order, in addition to the high equity remuneration for TSOs, network extension has proceeded steadily over the last decade. This development appears likely to continue, since there are few issues in transmission network expansion that might create obstacles for the energiewende. This can be shown at two different levels: (1) the aggregate transmission network investments in Germany, and (2) the large number of line expansions completed.

8.4.1 Aggregate Network Investment in Germany

Aggregate investment figures show a constant trend toward slightly increasing level of network investment and operations & maintenance (O&M) expenditures. Figure 8.3 shows network investments in maintenance and new builds of high -voltage electricity networks. It becomes obvious that investments of TSOs (*Übertragungsnetzbetreiber*) gradually increased over time, at around 0.5 billion € annually. Some issues were raised by the TSOs regarding the continuity of the regulatory framework. In particular, the network operator TenneT, a subsidiary of the Dutch state-owned TSO, has complained to regulators of being overburdened by capital expenditures for offshore and HVDC connections.[17] However, these conversations are part of any discussion between investors and regulators; so far, the availability of capital has not been a hindrance to transmission network development.

---

[16]In the Baltic Sea, the number of projects is considerably lower. BNetzA. 2018b. "Offshore-Monitoring Stand des Ausbaus nach dem vierten Quartal 2017." Bonn, Germany: Bundesnetzagentur.

[17]See ZEIT ONLINE (2011): "Erneuerbare Energien: Stromnetzbetreiber sieht Ausbau von Windparks gefährdet." Die Zeit, November 16, sec. Wirtschaft. Last accessed September 19, 2016, at http://www.zeit.de/wirtschaft/2011-11/windparks-finanzierung

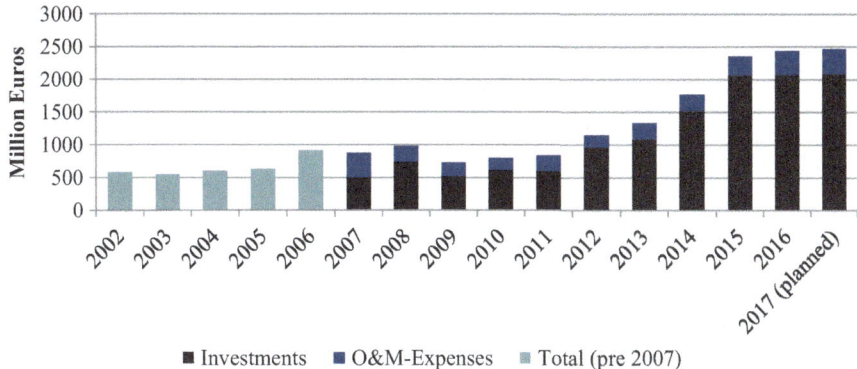

Fig. 8.3 High-voltage network expenditures in Germany (2002–2017). Sources: (BNetzA 2007, 2013, 2016, 2017) (BNetzA. 2007. "Monitoringbericht 2007." Bonn, Germany; BNetzA. 2013. "Monitoringbericht 2013." Bonn, Germany; BNetzA. 2016c. "Monitoringbericht 2016." Bonn, Germany; BNetzA. 2017. "Monitoringbericht 2017." Bonn, Germany)

8.4.2 Implementation of Network Development Plans

A second reassuring sign is the ongoing progress in network development seen throughout the energiewende to date. Both new builds and the expansion of existing lines have proceeded over the last decade. Figures 8.4 and 8.5 summarize developments in the transmission grid. Almost 500 km of line extensions or new builds were completed between 2009 and 2015, thus confirming the steady development of the aggregate figures shown above, albeit somewhat slower than anticipated.[18] Three lines connecting the former East and West Germany have been finished, and a large number of local extensions contributed to the development of the network.

The situation of the new HVDC corridors is quite different and more politically charged as well. As discussed in Chap. 11 on the trans-European energy infrastructure, the first years of the low-carbon transformation at the European level were characterized by a certain hype around trans-European HVDC South-North and East-West corridors, even extending to neighboring regions such as North Africa and Russia (Egerer et al. 2009). The inflated expectations have been brought back down significantly since then, but a similar hype has arisen around HVDC lines at the national level (Schröder et al. 2013). German TSOs have drawn up ambitious plans for HVDC corridors across the country, somewhat emulating the process at the European level.

---

[18]The dena I study had suggested a need for investment in 850 km of new-built lines, and 392 km of line upgrades.

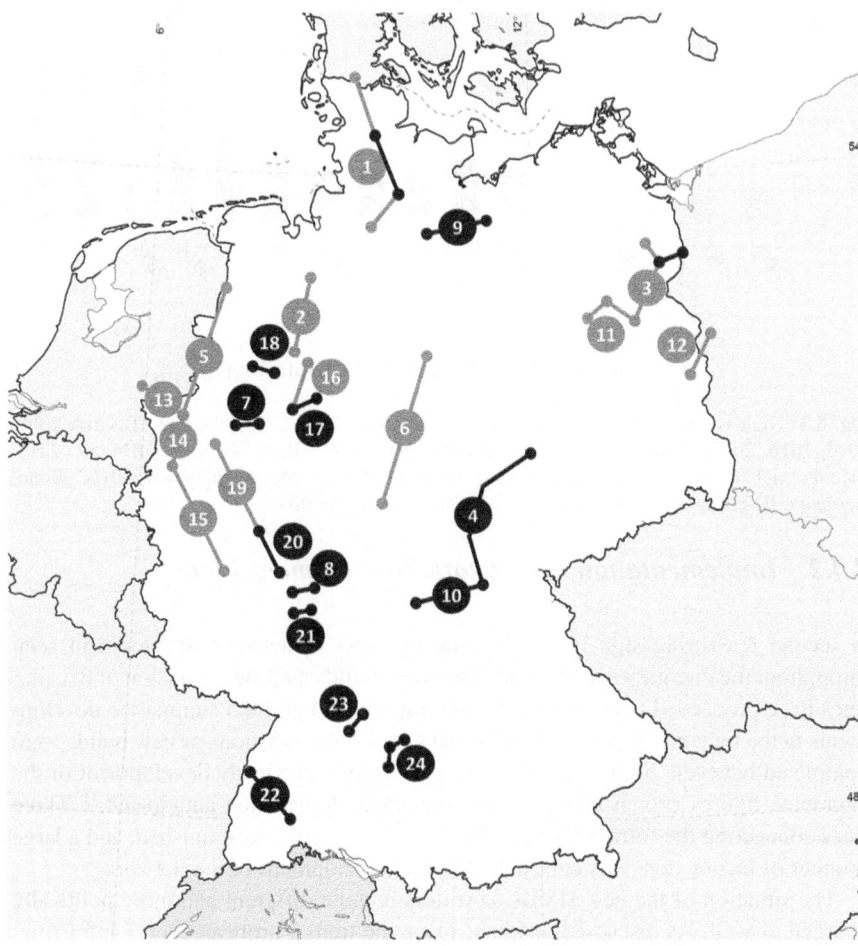

Finished projects are black, unfinished projects gray.

Fig. 8.4 Map of transmission expansion projects in Germany (2018). Finished projects are black, unfinished projects gray. Source: BNetzA (2018. "EnLAG-Monitoring Stand der Vorhaben aus dem Energieleitungsausbaugesetz (EnLAG) nach dem vierten Quartal 2017." Bonn, Germany)

Figure 8.6 shows the potential siting of four large HVDC corridors that appeared for the first time in the network planning exercise. Most of the four corridors run in a North-South direction. While the northern section of the corridors may have been better suited to renewables integration, the northern section of corridor A was not

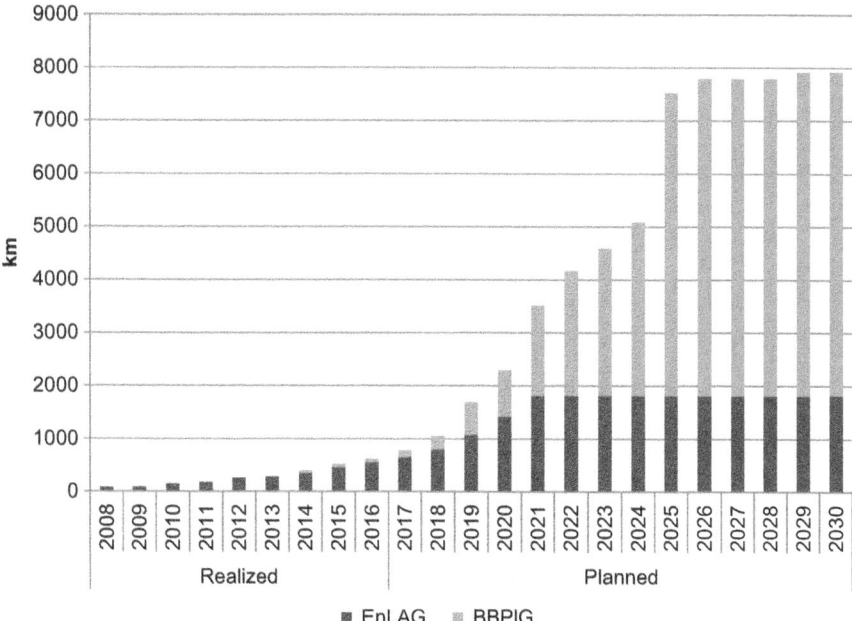

Fig. 8.5 Cumulative realized and planned transmission developments. Source: (Gerbaulet 2017) based on BNetzA (2016a, b) (BNetzA. 2016a. "BBPlG-Monitoring—Stand des Ausbaus nach dem Bundesbedarfsplangesetz (BBPlG) zum ersten Quartal 2016." Bonn, Germany; BNetzA. 2016b. "EnLAG-Monitoring—Stand des Ausbaus nach dem Energieleitungsausbaugesetz (EnLAG) zum ersten Quartal 2016." Bonn, Germany)

actively pursued during the first years of the energiewende. Since the 2025 iteration of the grid development plan, the northern connection point of corridor D is moved further north to Wolmirstedt, the southern connection point will be in Isar. The connection points of corridor C, which consist of two links, have been clarified in the 2030 version of the grid development plan. The first link connects Brunsbüttel and Großgartach, the second link Wilster and Bergrheinfeld, the location of the former nuclear power plant Grafenrheinfeld. Corridor B has not been considered as essential by BNetzA, thus taken out of the network development plan. This constitutes a strong sign that regulatory oversight can lead to adaptations of the NEP.

Fig. 8.6 Proposed high-voltage direct current (HVDC) transmission corridors in the 2013 network development plan and subsequent modifications. Source: Own depiction, based on BNetzA data

8.4.3   Levels of Congestion, Redispatch, and Ancillary Services

As a result of the steady network extension, the German electricity system continues to be managed well, which is a logical result of the institutional framework in place. Still, the costs for redispatch and infeed management have increased with the energiewende. Downward corrective measures by the TSOs (downward re-dispatch) were for 3.5 TWh, 2.5 TWh, 2.2 TWh,[19] 2.6 TWh, 8 TWh, 6.3 TWh, and 10.2 TWh in 2011 to 2017. The corresponding costs in 2015 were in the range of 402.5 million € for redispatch and 478 million € for renewable infeed management measures.[20] In 2016 the overall level of redispatch and infeed management decreased compared to the 2015 figures, and the level of downward redispatch reduced to 6.3 TWh and 220 million € for redispatch and 373 million € for infeed management. This development is mainly attributed to low wind infeed compared to 2015. In 2017 the values for redispatch and infeed management increased to an new high of 10.2 TWh redispatch and 5.5 TWh inveed management, summing up to about 1.4 billion €.

Model calculations on network congestion indicate that most redispatch occurs along one corridor running between Thuringia and Saxony in the former East Germany and Bavaria. This congestion occurs mainly in times of simultaneous lignite and wind feed-in, when Thuringia and Saxony are mainly exporting electricity (Mieth et al. 2015a, b). With the finishing of the one of the EnLAG pilot projects (EnLAG project No. 4), a 380 kV connection (with two circuits) between these two regions (Lauchstädt—Redwitz), this congestion has been reduced, but the neighboring connection Remptendof—Redwitz is still the corridor with the highest number of congested hours (1791 in 2017).

An additional indicator for the current network situation are the costs of system and ancillary services. These costs summarize efforts by the TSOs to keep the network system in balance at all times, and they include balancing reserves, redispatch and counter-trading, the provision of black-start capacity and reactive power, as well as infeed management, which refers to the costs of curtailing the feeding-in of renewables due to network congestion. Figure 8.7 indicates the absolute level of system costs (in the range of 1–1.4 billion € annually with the exception of 2015) but also a shift in cost structure over time. A large part of cost reduction is attributable to lower costs for reserve energy, which has become more competitive over time, whereas redispatch, infeed management, and grid reserve have shown rising costs over time. While numbers for the total system costs of 2017 have not been published at the time of the editorial deadline, the costs for redispatch and infeed management of 1.4 billion € alone will likely lead to a new peak.

---

[19]The source states 4.4 TWh but accounts for upward and downward redispatch measures.

[20]See the Monitoring Reports by BNetzA (2013, 2014, and 2016 for details).

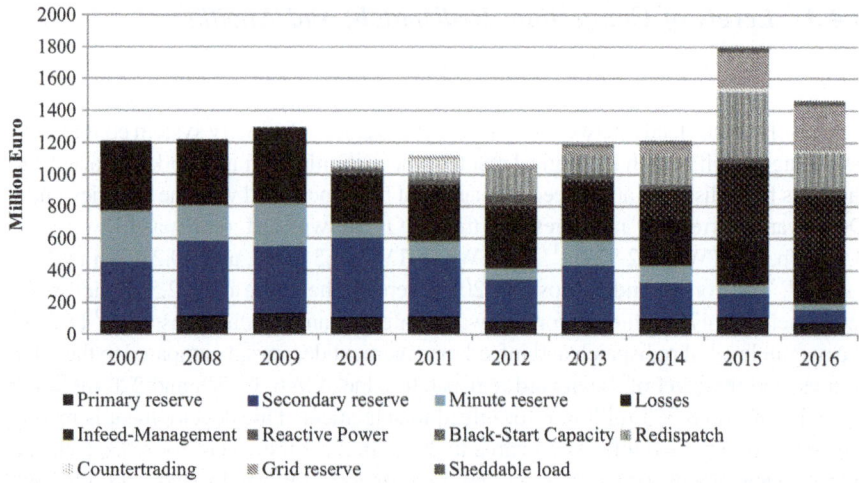

Fig. 8.7 Costs for system and ancillary services. Sources: BMWI (2015b) and BNetzA (2014, 2015, 2017) (BNetzA. 2014. "Monitoringbericht 2014." Bonn, Germany; BNetzA. 2015. "Monitoringbericht 2015." Bonn, Germany; BNetzA. 2017. "Monitoringbericht 2017." Bonn, Germany)

8.5 Effects from Splitting up Germany into Bidding Zones

8.5.1 A Controversial Debate

In the context of network development and Germany's future market design, the question of regionally differentiated electricity prices is raised frequently. Inter alia, the discussion covers the introduction of bidding zones within Germany as well as better coordination with neighboring countries to the North and East. Whereas the European Commission and the Agency for the Coordination of Energy Regulators (ACER) appeared to be in favor of internal bidding zones,[21] the German government has regularly voiced its opposition to such a move. An important study conducted on behalf of the network regulator found that the cons of bidding zones outweighed the pros: major disadvantages included the danger of abuse of market power and lower liquidity of the wholesale markets (Consentec and Frontier Economics 2011). Since then, all official governmental statements including the "White Book" of 2015 have argued in favor of one integrated bidding zone (BMWi 2015a). In 2018 ENTSOE has published the results of the first edition of the bidding zone review (ENTSO-E 2018) in which several bidding zone configurations for the central European region are analyzed. Apart from the split of the previously combined zone of Austria and Germany, which will be in place as of October 2018, other configurations included

---

[21] See European Commission (2014): Draft: Commission Regulation: Network Code for Capacity Allocation and Congestion Management. Title II, Chap. 2, Bidding zone configuration. BNetzA. 2015. "Monitoringbericht 2015." Bonn, Germany.

market splitting of large countries like Germany into two or three bidding zones. While smaller bidding zone configurations show operational improvements, the disadvantages of market splitting lead to the conclusion "no configuration is clearly classified as superior to any other." (ENTSO-E 2018, 12)

Several academic studies also analyzed potential configurations of bidding zones, but generally without a strong vote in favor of implementing them in practice (see among others Burstedde 2012; Breuer and Moser 2014). A study by the University of Duisburg-Essen suggests the introduction of bidding zones has rather modest effects (Trepper et al. 2013, 2015). In contrast to this, the Institute for Energy Economics of Cologne University (EWI) produced a study in favor of bidding zones.

8.5.2 Bidding Zones in Germany Would Have Minor Effects

A detailed study on the potential effect of bidding zones for the German electricity market by Egerer et al. (2016) provides additional quantitative evidence. Logically, because of the incentives given to TSOs explained above, their study reports a low level of congestion, and thus minor potential effects of bidding zones on dispatch and prices. Egerer et al. (2016) use a variant of the ELMOD electricity model to calculate the potential effects of bidding zones in a concrete setting that describes the German electricity market in 2012 and 2015. Figure 8.8 shows the distribution of

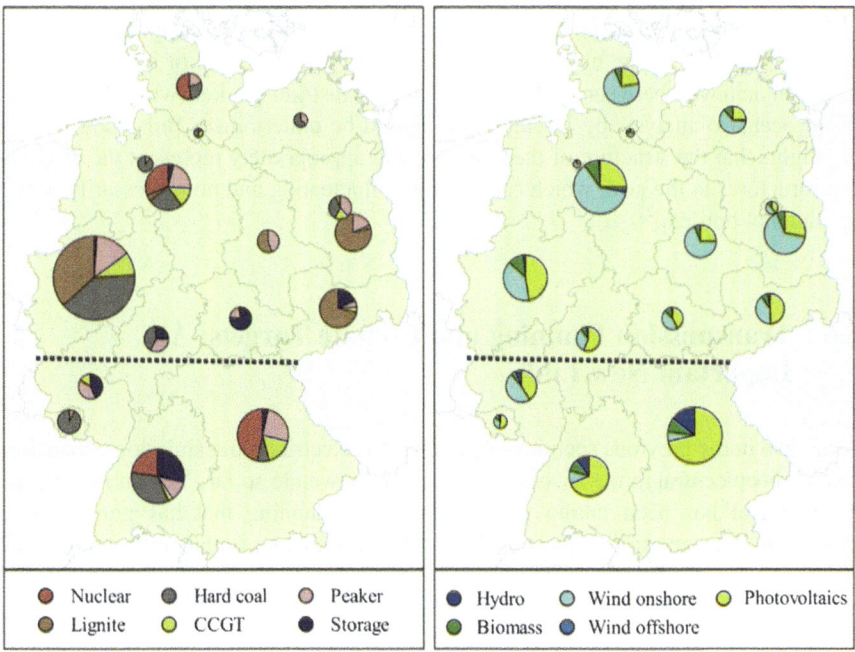

Fig. 8.8 Regional distribution of electricity generation capacities (2012). *Left:* Conventional capacities *Right:* renewable capacities. Source: Egerer et al. (2016)

conventional and renewable generation capacity, as well as the potential border between a Northern and a Southern zone: Whereas the conventional capacities are mainly located in the North, the distribution of renewables is somewhat more balanced. The model analysis yields only minor effects from the hypothetical introduction of bidding zones (which they analyzed for 2012 and 2015, respectively). The average price difference between the Northern and the Southern zone is 0.4 €/MWh in 2012 and 1.7 €/MWh in 2015, respectively; the latter corresponds to less than 5% of the average wholesale market price.

If a net transfer capacity (NTC) of 8 GW is established between the zones, redispatch is reduced by about 35%. Consumers in the North would have to pay 163 million € less, whereas those in the South pay 275 million € more. The effect on producers is the inverse: Producers in the North lose 199 million € (79 € of which are renewables), whereas producers in the South gain 201 million € (57 million € of which are renewables). The effects of setting up price zones were reduced when assuming that the high-voltage AC line between Thuringia and Bavaria, so-called South-West interconnector ("SüdWest-Kuppelleitung," Project No. 4) was built, which went online in 2017. The lines, two circuits of 380 kV each or almost 4 GW of capacity, relieved the congestion between Thuringia and Bavaria significantly. Consequently, the price difference narrows to a mere 0.4 €/MWh in this scenario.

The adverse effects of splitting up the German market more than outweigh these minor potential benefits. Not only would the transformation costs of splitting up an efficiently functioning market be high, it would also be difficult to define stable price zones. An important question would be how to create bidding zones: Two (North–South) or three, that is, a zone separating North-Rhine Westphalia from the rest. The liquidity of the market would suffer, and the high level of integration that has been achieved between the German and the Austrian market (which is not part of the scenario analysis by Egerer et al.) would be undermined. Burstedde (2012) highlights that the structure of the zones should appropriately represent the congestion structures in the grid, which can be highly fluctuating and not necessarily occur on national borders.[22]

8.6   Transmission Planning and Climate Targets: An Important New Link

As shown in the previous sections of this chapter, electricity transmission expansion has not been central to the successes of the energiewende so far. Yet, an interesting development has been taking place in network planning that has gone almost unnoticed: Transmission expansion has joined a number of other instruments that

---

[22]In 2018, the German regulator suggested that the one between South Germany and Austria be split, but details on the implementation were not provided.

are now treated as crucial for the attainment of climate goals. In the past, the only objective of transmission was to "connect" supply and demand, and the nature of the transported electricity was completely disregarded. However, the 2015 network planning exercise introduced climate goals into the planning process for the first time, making it clear that electricity networks also had to serve the objectives of the energiewende, including the greenhouse gas emission reduction targets (Gerbaulet et al. 2012).[23]

The first network development exercises of the 2000s ignored any GHG emission reduction targets set by the government, and it focused solely on facilitating market dispatch as implied by the merit order principle. However, it completely disregards negative climate externalities (except for the modest CO_2 price). Owing to the low prices for emission allowances in recent years, the Emissions Trading System, European Union's key tool for reducing CO_2 emissions, did not result in a shift away from lignite and hard coal toward the lower-carbon natural gas in Germany's energy sector. In fact, owing to their low power generating costs, lignite-fired power plants were almost always included in the dispatch. This resulted in very high CO_2 emissions, that even increased in 2012–2013, thus risking not to achieve the climate goals of the Energy Concept 2010 and the White Paper by the Federal Ministry for Economic Affairs from 2015 (BMWi 2015a) (see Chap. 4).

In 2014, after the German government announced its climate goals for 2020 and 2050, the electricity sector regulator, the Bundesnetzagentur, followed suit by imposing CO_2 emission targets on the modeling process for the first time ever in the history of German transmission planning. The 2014 scenario framework, approved by BNetzA in December 2014, contained mandatory emissions restrictions that were in line with the German government's emissions targets for the energy sector for 2020 and 2035. The constraint had to be applied by the TSOs and their consultants in the calculation of the energy mix, thus entering the scenario framework and the subsequent network development plan in form of a climate-friendly market result. Concretely, Scenario B2 prescribed a maximum of 187 and 134 million tons of CO_2 for 2025 and 2035, respectively. The current scenario[24] has an emission limit of 184 Mt and 127 Mt for 2030 and 2035, respectively. These numbers represent an 80% reduction pathway until 2050. While integration of climate goals is a welcome addition in the grid development process, the estimated CO_2-emission for 2030 and 2035 require a steeper emission reduction in the period following 2035 if the target of 95% decarbonization of the electricity sector is to be met.

In addition, future grid expansion planning in Germany will include the possibility to curtail up to 3% of the annual production of onshore wind farms and solar

---

[23]This section is based on Mieth et al. (2015a, b) as well as findings from Gerbaulet et al. (2012a) and Reitz, Felix, Clemens Gerbaulet, von Hirschhausen, Claudia Kemfert, Casimir Lorenz, and Pao-Yu Oei. 2014. "Verminderte Kohleverstromung könnte zeitnah einen relevanten Beitrag zum deutschen Klimaschutzziel leisten." 47. Wochenbericht. Berlin, Germany: DIW.

[24]50Hertz, Amprion, TenneT, and TransnetBW. 2018. "Szenariorahmen für den Netzentwicklungsplan Strom 2030 (Version 2019)—Entwurf der Übertragungsnetzbetrreiber."

power installations. This is in line with the provisions in the Ministry for Economic Affairs and Energy's Green Paper and White Paper on the future development of the German electricity market (BMWi 2015a, b). This underscores that not "every last kilowatt hour of power generated" should be transmitted but rather that economic motivations should play a more central role in grid planning.

8.7   Conclusions

The electricity transmission infrastructure has to play a certain role in the energiewende process. The transmission infrastructure (as well as the distribution infrastructure) is one element in a renewables-based energy system, as it provides flexibility between different producing technologies, e.g., intermittent renewables and dispatchable gas plants, and allows for geographical smoothing between different regions (e.g., East and West, North and South). Infrastructure is important in any development context, and it is always better to be slightly oversupplied—in particular in periods of system transformation, such as the first years of the energiewende. Taking these concerns further, some (particularly zealous) TSOs have even expressed concerns that electricity transmission is a potential Achilles' heel of the energiewende. The focus on transmission planning, especially the early years of the energiewende, was natural, since the previous system had been highly intransparent and not open to public policy debates. As Ignacio Perez-Ariaga has put it, transmission investment should be as "unexciting" ("boring" or "uneventful") as possible, in order to avoid indecision by investors and high capital costs (Olmos and Pérez-Arriaga 2009, 5286).

This chapter shows that these concerns are less of a concern than stated, and that—on the contrary—transmission expansion demand can be met due to incentives provided to the TSOs. Model-based analysis of the German electricity grid, as well as case study experience from close to a decade of almost daily work on the topic, helps to allay fears about the lack of network investment becoming a major barrier to the energiewende. In fact, transmission development has proceeded smoothly (even though somewhat delayed, yielding high costs for redispatch and infeed management) over the last years, leading to a system, with the highest quality indicators in Europe.

Looking back over the last decade, electricity transmission in Germany has developed steadily. Important connections have been realized, in particular between the former East and West German grids. Network congestion, once feared to become a critical issue, has remained within tolerable levels. Thus, transmission infrastructure does not threaten to create substantial long term bottlenecks that could impede or stall the energiewende. The state of discussion also suggests that splitting up the German electricity network into several zones, debated at the European level, is currently not a relevant issue for the energiewende.

This chapter also provided at least two explanations for why transmission expansion has not been an obstacle so far: One is the continued institutional setting of a mostly congestion-free "copper plate" network, where all electricity fed in according to the merit-order principle has to be integrated independently of the location. The other driver is similarly strong, i.e., a rate of return on equity of 9.05% (and 6.91% as of 2019), far beyond what can be gained by investing in a similar level of risks. This incentive pushed the TSOs towards high levels of grid expansion.

One finding discussed in this chapter that may offer potential for conceptual innovation is that of an explicit link between transmission planning and climate targets. Previously, the energy mix was taken as given in the transmission planning procedure. Since 2015, the German regulator has introduced an explicit CO_2 target into the scenarios as the basis for network planning. By tightening this constraint over time, the German government should be able to facilitate the achievement of CO_2 emission targets for the energy sector, and at the same time, plan the network effectively to achieve the energiewende objectives.

References

BMWi. 2015a. *An Electricity Market for Germany's Energy Transition – White Paper by the Federal Ministry for Economic Affairs and Energy.* Berlin.
————. 2015b. *The Energy of the Future, Fourth 'Energy Transition' Monitoring Report – Summary.* Berlin.
Boll, Georg. 1969. *Geschichte des Verbundbetriebs: Entstehung und Entwicklung des Verbundbetriebs in der deutschen Elektrizitätswirtschaft bis zum europäischen Verbund.* Frankfurt/Main: Verlags- u. Wirtschaftsgesellschaft der Elektrizitätswerke.
Breuer, Christopher, and Albert Moser. 2014. *Optimized Bidding Area Delimitations and Their Impact on Electricity Markets and Congestion Management.* In IEEE proceedings. Kraków: IEEE.
Burstedde, Barbara. 2012. *From Nodal to Zonal Pricing: A Bottom-up Approach to the Second-Best.* In EEM proceedings, Florence.
Consentec, and Frontier Economics. 2011. *Relevance of Established National Bidding Areas for European Power Market Integration – an Approach to Welfare Oriented Evaluation.* Study commissioned by the Federal Network Agency (BNetzA), Cologne and Bonn.
dena. 2005. *dena-Netzstudie I (Endbericht) – Energiewirtschaftliche Planung für die Netzintegration von Windenergie in Deutschland an Land und Offshore bis zum Jahr 2020.* Cologne: Deutsche Energie-Agentur GmbH.
————. 2010. *dena-Netzstudie II (Endbericht) – Integration erneuerbarer Energien in die deutsche Stromversorgung im Zeitraum 2015-2020 mit Ausblick auf 2025.* Berlin: Deutsche Energie-Agentur GmbH.
Egerer, Jonas, Lucas Bückers, Gregor Drondorf, Clemens Gerbaulet, Paul Hörnicke, Rüdiger Säurich, Claudia Schmidt, et al. 2009. *Sustainable Energy Networks for Europe – The Integration of Large-Scale Renewable Energy Sources until 2050.* Electricity Market working papers WP-EM-35, Dresden.
Egerer, Jonas, Clemens Gerbaulet, Richard Ihlenburg, Friedrich Kunz, Benjamin Reinhard, Christian von Hirschhausen, Alexander Weber, and Jens Weibezahn. 2014. *Electricity Sector Data for Policy-Relevant Modeling: Data Documentation and Applications to the German and European Electricity Markets.* DIW data documentation 72, Berlin.
Egerer, Jonas, Jens Weibezahn, and Hauke Hermann. 2016. Two price zones for the German electricity market – market implications and distributional effects. *Energy Economics* 59 (September): 365–381.

ENTSO-E. 2010. *Ten-Year Network Development Plan 2010 – 2020*. Final report. Brussels: European Network of Transmission System Operators for Electricity.

———. 2018. *First Edition of the Bidding Zone Review*. Final report, Brussels.

EnWG. 2011. *Gesetz über die Elektrizitäts- und Gasversorgung (Energiewirtschaftsgesetz - EnWG) Novelle 2011*.

Eyre, Sebastian, and Michael G. Pollitt. 2016. *Competition and Regulation in Electricity Markets*, The International Library of Critical Writings in Economics 315. Cheltenham: Edward Elgar Publishing.

Gerbaulet, Clemens. 2017. *Electricity Sector Decarbonization in Germany and Europe - a Model-Based Analysis of Operation and Infrastructure Investments*. Berlin: Technische Universität Berlin.

Gerbaulet, Clemens, Jonas Egerer, Pao-Yu Oei, Judith Paeper, and Christian von Hirschhausen. 2012. *Die Zukunft der Braunkohle in Deutschland im Rahmen der Energiewende*. DIW Berlin, Politikberatung kompakt 69. Berlin: Deutsches Institut für Wirtschaftsforschung (DIW).

Jarass, Lorenz, and Anna Jarass. 2016. *Integration von erneuerbarem Strom: Stromüberschüsse und Stromdefizite – mit Netzentwicklungsplan 2025*, MV-Wissenschaft. 1st ed. Münster: Verl.-Haus Monsenstein und Vannerdat.

Jarass, Lorenz, and Gustav M. Obermair. 2012. *Welchen Netzumbau erfordert die Energiewende? – Unter Berücksichtigung des Netzentwicklungsplans 2012*. Münster: Verl.-Haus Monsenstein und Vannerdat.

Kemfert, Claudia, Friedrich Kunz, and Juan Rosellón. 2016. A welfare analysis of electricity transmission planning in Germany. *Energy Policy* 94: 446–452.

Leuthold, Florian, Hannes Weigt, and Christian von Hirschhausen. 2012. A large-scale spatial optimization model of the European electricity market. *Networks and Spatial Economics* 12 (1): 75–107.

Mieth, Robert, Clemens Gerbaulet, Christian von Hirschhausen, Claudia Kemfert, Friedrich Kunz, and Richard Weinhold. 2015a. *Perspektiven für eine sichere, preiswerte und umweltverträgliche Energieversorgung in Bayern*. DIW Berlin, Politikberatung kompakt 97. Berlin: Deutsches Institut für Wirtschaftsforschung (DIW).

Mieth, Robert, Richard Weinhold, Clemens Gerbaulet, Christian von Hirschhausen, and Claudia Kemfert. 2015b. *Electricity Grids and Climate Targets: New Approaches to Grid Planning*. DIW Economic Bulletin 5/2015, Berlin.

Olmos, Luis, and Ignacio J. Pérez-Arriaga. 2009. A comprehensive approach for computation and implementation of efficient electricity transmission network charges. *Energy Policy* 37 (12): 5285–5295.

Schröder, Andreas, Pao-Yu Oei, Aram Sander, Lisa Hankel, and Lilian Laurisch. 2013. The integration of renewable energies into the German transmission grid - a scenario comparison. *Energy Policy* 61: 140–150.

The Brattle Group. 2007. *International Review of Transmission Planning Arrangements*. A report for the Australian Energy Market Commission. Brussels: The Brattle Group.

Trepper, Katrin, Michael Bucksteeg, and Christoph Weber. 2013. *An Integrated Approach to Model Redispatch and to Assess Potential Benefits from Market Splitting in Germany*. EWL working paper 19/2013. Essen: Chair for Management Science and Energy Economics University of Duisburg-Essen.

———. 2015. Market splitting in Germany – new evidence from a three-stage numerical model of Europe. *Energy Policy* 87: 199–215.

Weber, Alexander, Thorsten Beckers, Patrick Behr, Nils Bieschke, Stella Fehner, and Christian von Hirschhausen. 2013. *Long-Term Power System Planning in the Context of Changing Policy Objectives – Conceptual Issues and Selected Evidence from Europe*. Study commissioned by Smart Energy for Europe Platform (SEFEP).

Chapter 9
Sector Coupling for an Integrated Low-Carbon Energy Transformation: A Techno-Economic Introduction and Application to Germany

Jens Weibezahn

> *"We will [...] advance the integration of the heat, mobility and electricity sectors in conjunction with storage technologies."*
> Coalition Agreement for the 19th Legislative Period of the German Bundestag, March 14, 2018 (CDU, CSU, and SPD. 2018. "Ein neuer Aufbruch für Europa—Eine neue Dynamik für Deutschland—Ein neuer Zusammenhalt für unser Land. Koalitionsvertrag zwischen CDU, CSU und SPD. 19. Legislaturperiode." Authors' translation, p. 72)

9.1 Introduction

The previous chapters have shown that the first phase of the energiewende, focusing on the electricity sector, was largely successful. In fact, it was relatively easy to increase the share of renewables in electricity, now almost 40%, and to close down nuclear power plants, albeit at the cost of temporarily high CO_2 emissions. Yet, in order to reach the climate goal of a 55% reduction in greenhouse gases by 2030 and

This work was carried out as part of the project "Long-term planning and short-term optimization of the German electricity system within the European framework: further development of methods and models to analyze the electricity system including the heat and gas sector", funded through grant "LKD-EU", FKZ 03ET4028A, German Federal Ministry for Economic Affairs and Energy. The author would like to thank Clemens Gerbaulet, Leonard Göke, Martin Kittel, Casimir Lorenz, Pao-Yu Oei, and Christian von Hirschhausen for comments. The usual disclaimer applies.

J. Weibezahn (✉)
TU Berlin, Berlin, Germany
e-mail: jew@wip.tu-berlin.de

© Springer Nature Switzerland AG 2018
C. von Hirschhausen et al. (eds.), *Energiewende "Made in Germany"*,
https://doi.org/10.1007/978-3-319-95126-3_9

an 80–95% reduction in the coming decades until 2050 (base year 1990, BMWi and BMUB (2010)), the second phase needs to focus on all energy usage, especially heat, transportation, and usage as a raw material in the chemical industry. In that context, intensified "sector coupling" will be required, accompanied by a further shift from fossil fuels to renewable ones.

This chapter provides an overview of the upcoming challenges in the next phase of the energiewende, by focusing on the technical and economic challenges of coupling electricity, heat, and transportation, in an attempt to advance the low-carbon transformation. We apply the concepts to the ongoing energiewende in Germany. By intensifying the links between the sectors, one can harvest "low-hanging fruits" in terms of flexibility and fuel switching from fossil to renewable energies. This is a precondition to attain the ambitious targets of the energiewende with respect to CO_2 emission reductions. While this chapter focusses on Germany, the technical and economic arguments are valid at a broader scale, and apply to other transformation processes as well.

The chapter is structured in the following way: The next section describes the basic idea of "sector coupling", until recently a widely unknown concept, including a schematic stylized scheme. In Section 9.3 we describe how sector coupling might evolve in the transportation and heating sectors, and that far-reaching electrification is at the core of the process. Section 9.4 provides some concrete quantitative scenarios for sector coupling for the case of Germany to 2030 and to 2050, based on a rapidly growing body of recent literature. While there is consensus on the feasibility of reaching ambitious decarbonization targets, different models suggest different pathways of reaching them. The role of synthetic fuels (domestic and/or imported) is controversially discussed. Section 9.5 concludes.

9.2   The Basic Idea of "Sector Coupling"

In 2016, Germany had a total primary energy demand of more than 3700 TWh. About 93% of this primary energy was consumed by the energy sector. Usage as a raw material, mainly in the petrochemical industry, accounted for 7%. 34% of primary energy came from oil, 23.6% from coal (12.3% hard coal and 11.3% lignite), 22.6% from fossil gas, 6.9% from nuclear, and 12.6% from renewable sources.[1] The largest source of CO_2 emissions was coal (lignite and hard coal), accumulating to a share of 41% in 2016, followed by mineral oil with 34%, and fossil gas with 22%, based on total emissions of 751.7 Mt.[2] Due to conversion and

---

[1]AGEB. 2017. "Auswertungstabellen." Arbeitsgemeinschaft Energiebilanzen e. V.

[2]BMWi. 2018. "Gesamtausgabe der Energiedaten—Datensammlung des BMWi." Berlin: Bundesministerium für Wirtschaft und Energie.

other losses, only 68% of the primary energy was used as final energy. Although precise differentiation between sectors is difficult, it is estimated that about half the energy was used for heat, one third for fuels, and only one fifth for electricity (Agora Energiewende 2018).

The first conclusion from this statistic is that increasing energy efficiency and halving primary energy usage until 2050 (compared to 2008, Bundesregierung (2010)) will be one of the critical success factors of the low-carbon energy trans formation. The second conclusion is that due to the limited potentials for solar thermal and geothermal energy, biomass, and biofuels, the increased use of renew able power from wind and photovoltaics is the predominant strategy to further decrease greenhouse gas emissions in all energy sectors. However, this strategy requires an increased coupling of energy sectors and is the corner stone for an integrated energy transformation.

The basic idea of sector coupling is to facilitate a more sustainable use of different types of energy across sector boundaries, that is, electricity, heat, and transportation. In addition, the objective of sector coupling is to substitute fossil fuels by renew ables, both electricity and fuels. Thus, sector coupling targets a more rational use of energy, in the techno-economic sense, and lower greenhouse gas emissions. In addition, sector coupling can activate additional degrees of freedom in the energy system, and therefore introduce more flexibility into the system—facilitating the further integration of intermittent renewable energy sources like wind or solar (Wietschel et al. 2018).

As such, the coupling of sectors is nothing new and has been practiced for a long time, for example, by means of combined heat and power (CHP) plants or electricity used in rail transport. Advanced coupling can be achieved by different technology options, with the most efficient one being the direct usage of electricity in battery electric vehicles (BEV), rail transportation, trolley trucks and buses in the transpor tation sector, and power to heat (PtH) and heat pumps in the heat sector. The indirect (and therefore less efficient) usage of electricity is via a conversion into synthetic fuels (power to gas (PtG) and power to liquid (PtL)). Also other synthetic fuels produced from biomass are conceivable, yet not mature for commercial applications. Figure 9.1 shows a schematic overview of a future coupled energy system, primarily based on electricity from wind and solar PV. Consequently, the distinct energy sectors coalesce and have to be assessed in an integrated way.

One of the benefits of a decarbonized and integrated energy sector are new business models for energy utility companies, service providers, and new market players. Additional economic value will be added within Europe and Germany, decreasing commodity dependence from other parts of the world.

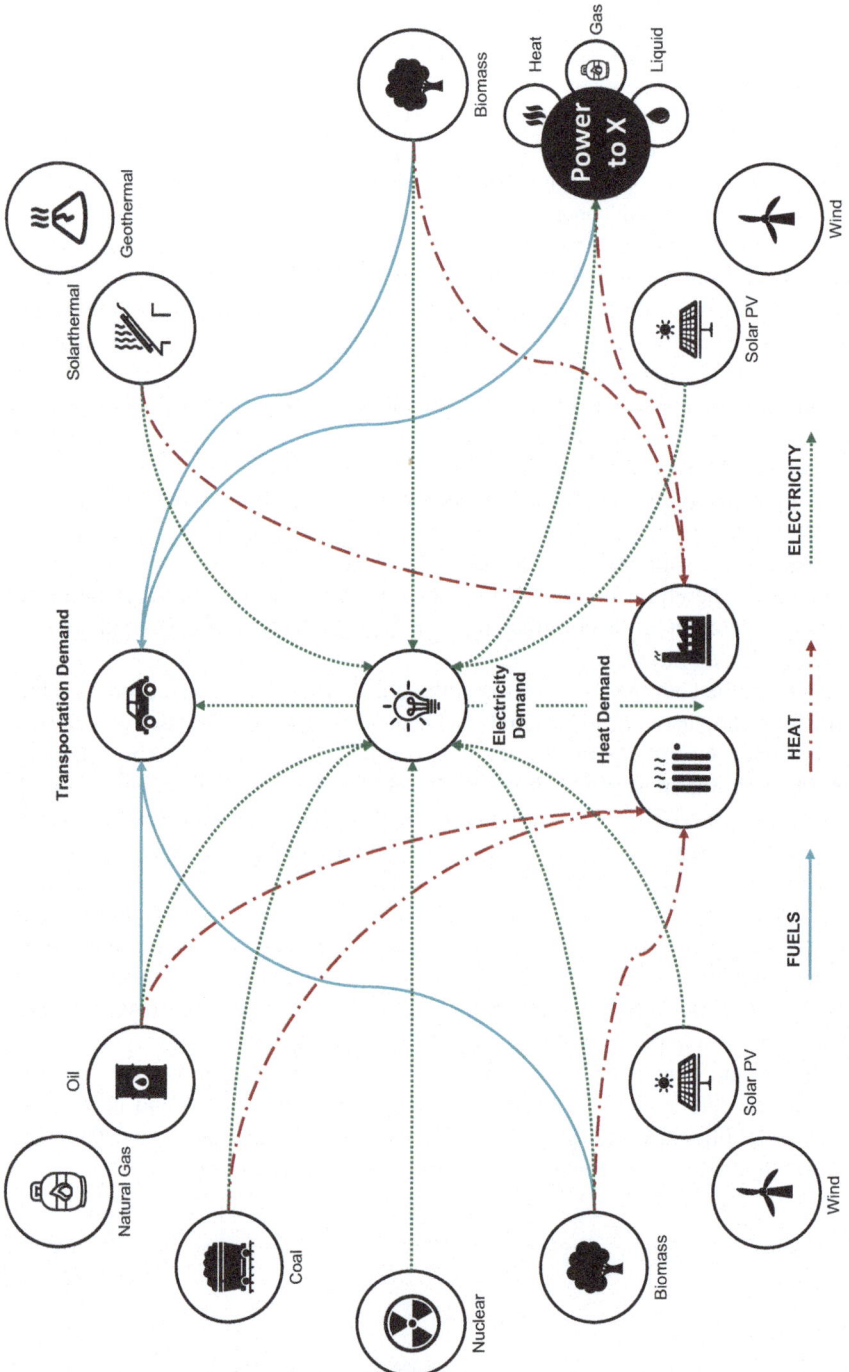

Fig. 9.1 Schematic overview of the shift towards a decarbonized energy sector. Source: Own depiction based on SRU (2017)

9.3   Sectors

The different sectors in sector coupling can be delimited in different ways, yet most of the literature agrees on the definition of three sectors: electricity, heating and cooling, and transportation. Within the sectors a further distinction can be made, mostly into industrial, commercial and service, and household consumers. The following subsections provide a more detailed view on the transportation and heating/cooling sectors, their current energy consumption (see Fig. 9.2) and the technology options for direct or indirect electrification. It concludes with the intersectoral interdependencies with the electricity sector.

9.3.1   Transportation

The German transportation sector accounts for a final energy consumption of about 750 TWh/year. Currently, 95% is based on mineral oil while only 1% is based on electricity (mostly rail transportation, not necessarily from renewable sources) and 4% on renewable energy, mainly biofuels as addition to gasoline; fossil gas has a negligibly small share (Fig. 9.2). While the German government foresees a reduction of consumption by 10% in the year 2020 and 40% in 2050 compared to 2005 levels (BMWi and BMUB 2010), the actual energy demand and consequently also greenhouse gas emissions in the transportation sector are steadily growing. This is mostly due to the fact that the transportation demand for goods and passengers is increasing year by year. At the same time, CO_2 emissions increased to 165 Mt in 2016 despite

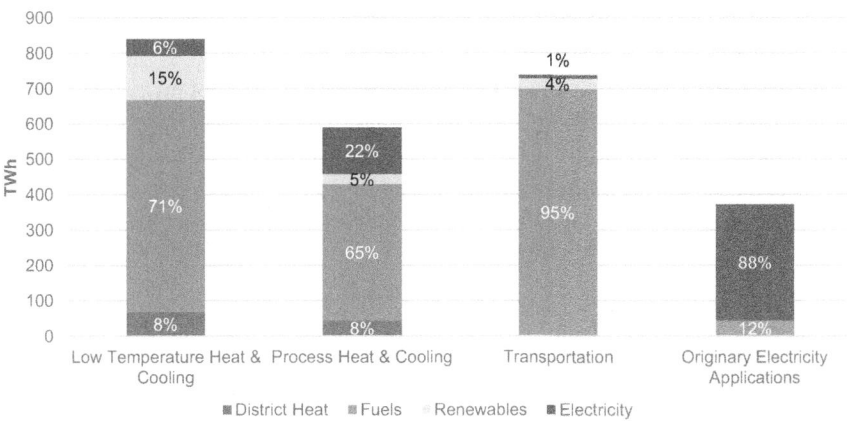

Fig. 9.2 Final energy usage by application and energy carrier. Source: Own depiction using data from AGEB (AGEB. 2018. "Anwendungsbilanzen für die Endenergiesektoren 2013–2016." Arbeitsgemeinschaft Energiebilanzen e. V.) based on acatech et al. (2017)

emission standards for vehicles being tightened and the first driving bans in place or planned in Hamburg and Stuttgart in the light of the emissions scandal.

The adverse trend in the transportation sector requires a definite low-carbon transformation strategy, resting on at least two pillars: (1) the transformation of mobility behavior, leading to a shift in the modal split and (2) the transformation of the fuel mix towards more renewables. While a future decrease in passenger kilometers of transportation demand seems unrealistic, the current trend towards urbanization and digitalization could be used to increase the share of public transportation (local and long-distance) and bikes in the modal split. Significant investments in infrastructure from bike lanes to high-speed rail lines in combination with digital solutions would be necessary to incentivize this shift. Additionally, nudges like subsidies for public transportation season passes and congestion charges for cars and parking could be supporting measures. A side effect of this strategy is an increase in the quality of life in cities due to less air pollution from fine particles, nitric oxide, and airborne gases as well as noise pollution. As aforementioned, "efficiency first" has been declared to be the leading principle by the German government (BMWi 2016, see also Chap. 7).

Cars

To decarbonize the transportation sector, different technological options are available or are currently being developed. The most prominent are probably battery electric vehicles (BEVs), directly electrifying passenger transportation and thereby increasing efficiency compared to conventional combustion engines. Assuming an efficiency factor of about 30% for internal combustion engines compared to about 80% for electric engines, the final energy usage for passenger road transportation could be reduced from about 400 TWh to 150 TWh (acatech et al. 2017). In 2016, the share of electric vehicles of all new registrations in Germany was less than 0.4%. In absolute values they account for less than 12,000 out of 3.5 million newly registered vehicles[3] and less than 35,000 in total stock.[4] Other countries like Norway and China have higher shares in registrations and stock, mostly thanks to generous subsidies or regulations. The German subsidies of up to 4000 €, shared by the government and as discounts by the manufacturers, appear to not provide sufficient incentives, which is why the reserved public funds of 600 million € for this program have not been exhausted (less than 120 million € distributed between May 16, 2016 and April 30, 2018 for roughly 66,000 vehicles[5]).

A technology that has caught on in Germany on a larger scale than BEVs are (plug-in) hybrid electric vehicles ((P)HEVs). By January 1, 2017, there were already

---

[3]KBA. 2017. "Neuzulassungen von Pkw im Jahr 2016 nach ausgewählten Kraftstoffarten." Kraftfahrt-Bundesamt.

[4]KBA. 2017. "Bestand an Pkw am 1. Januar 2017 nach ausgewählten Kraftstoffarten." Kraftfahrt-Bundesamt.

[5]BAFA. 2018. "Elektromobilität (Umweltbonus): Zwischenbilanz zum Antragstand vom 30. Juni 2018." Bundesamt für Wirtschaft und Ausfuhrkontrolle.

more than 165,000 hybrid vehicles in stock[6] with almost 50,000 new registrations in 2016 alone.[7] HEVs have a combustion engine combined with an electric engine with battery so their range is considerably extended compared to BEVs, thus they are to date still fueled completely by fossil fuels. PHEVs on the other hand have the additional option to be charged directly with electricity. Some BEVs are also equipped with so-called range extenders, an additional combustion engine that is activated when the battery power is exhausted.

Fuel cell (electric) vehicles (FC(E)V) use electric engines powered by hydrogen fuel cells. Their advantage is the faster refueling process and a longer range due to the higher energy concentration of a hydrogen tank. Although their development is ongoing for many years already, the technology is still not available for mass production, making them more expensive. Since hydrogen needs to be produced from electricity by electrolysis and then converted back into electricity, the whole process suffers from significant conversion losses in the order of 50%.

Natural gas vehicles (NGV) use fossil gas as a fuel (emitting CO_2) or can be powered by bio methane from biogas or synthetic methane. Both alternatives have a lower energy content. The advantage of NGVs is the already existing (yet sparse) infrastructure of gas stations across Germany.

The last option is the replacement of conventional diesel or petrol by biofuels or synthetic fuels with similar properties that could be distributed via the existing, well permeated network of gas stations and can be burned in only slightly retrofitted internal combustion engines. The major drawback of this process is the additional conversion step for synthetic fuels. While a BEV has an approximate overall energy efficiency (from electricity generated by renewable sources to wheel) of 69%, FCEVs only reach about 26%. Yet, this is still a higher efficiency rate compared to the 13% of power to liquid processes (acatech et al. 2017). Translated into a km per kWh scale (comparable to the "miles per gallon" concept in the US) a conventional internal combustion engine can reach about 1.5 km/kWh from mineral oil, while a fully electric car will yield 5 km/kWh. Power to liquid and power to gas concepts with a combustion engine or with an electric engine achieve 1 km/kWh and 2 km/kWh, respectively. It is essential to use the most efficient technology options available since additional electricity demand from the transportation sector alone would amount to more than 1000 TWh per year if fossil fuels were mostly substituted by synthetic fuels. Neglecting the rivalry with food production, a rough estimate shows that the current energy demand from the transportation sector could also not be supplied from biofuels produced only on agricultural sites within Germany, even if the entire available agricultural area in Germany was used for fuel production only. (Quaschning 2016)

---

[6]KBA. 2017. "Bestand an Pkw am 1. Januar 2017 nach ausgewählten Kraftstoffarten." Kraftfahrt-Bundesamt.

[7]KBA. 2017. "Neuzulassungen von Pkw im Jahr 2016 nach ausgewählten Kraftstoffarten." Kraftfahrt-Bundesamt.

One of the key success factors of BEVs and PHEVs, aside from the currently prohibitively expensive price, is the availability of a sufficient charging infrastructure with an adequate level of standardization and interoperability so that vehicles are able to use a high number of charging stations. However, current infrastructure does not yet suffice to provide for a large number of potential users, predominantly for those living in apartment buildings with no access to a charger connection in their own garage.

While BEVs are an option for short-range transportation, mostly in urban areas where they are being used already, heavy-duty and long-range transportation reverts to different technology options. This is due to the undue weight of batteries and high time consumption of charging processes needed for these high capacities. One option that is been tested in different pilot projects are trolley trucks, using a contact wire along their route, which could be used along major transportation corridors. To avoid the need to transship for the first and last mile, those vehicles would need to be equipped with additional short-range batteries or hybrid solutions or fuel cell engines. Assuming a subsidized introduction phase for the infrastructure on German highways, studies show that about 80% of heavy-duty trucks could be converted to trolley trucks in an economic viable way, only requiring about 30% of the German national highways to be equipped with contact wires.[8] A shift towards more freight traffic on electrified rail corridors can further alleviate the problem.

Aviation and Maritime Transportation

A special case is air and maritime transportation. Fully battery electric airplanes are not very likely to achieve market maturity within the next decades since the specific energy content of currently available batteries is too low and their weight is too high. Also, planes depend on short turn-around times at the airports since they are very capital intense assets and only earn money while airborne, which would be prevented by long recharging cycles. Hydrogen is, due to its comparatively low energy content, also not a probable option. Therefore, liquid fuels with a high specific energy content will still be needed. Instead of using fossil fuels, they could be synthetic or of organic origin like algae (Adeniyi et al. 2018). Likewise, maritime transportation can at least partially been switched to biofuels.

In all cases, the degree of decarbonization ultimately depends on the electricity mix present in the system. In order to achieve a reduction in greenhouse gas emissions, renewable energy capacities need to be tremendously expanded. Otherwise, coal or fossil gas capacities will be used to power vehicles, only lowering local emissions and improving the quality of life of the local population, but adversely affecting the climate.

---

[8]Wietschel, Martin, Till Gnann, André Kühn, Patrick Plötz, Cornelius Moll, Daniel Speth, Jan Buch, et al. 2017. "Machbarkeitsstudie zur Ermittlung der Potentiale des Hybrid-Oberleitungs-Lkw." Studie im Rahmen der Wissenschaftlichen Beratung des BMVI zur Mobilitäts- und Kraftstoffstrategie. Karlsruhe: Fraunhofer ISI, Fraunhofer IML, PTV Transport Consult, TU Hamburg-Harburg, M-Five.

9.3.2   Heating and Cooling

In the heating and cooling sector, two major issues can be distinguished: on the one hand there is space heating and cooling and the provision of warm water, all at comparably low temperatures, on the other hand there is process heating and cooling for industrial and commercial purposes at extremely high or low temperatures. In 2016, the German heating and cooling sector used about 1430 TWh of final energy (Fig. 9.2).[9] Figure 9.3 shows the technical options of providing heat using renewable energies.

Space Heating, Cooling, and Warm Water
Space heating, cooling, and warm water accounts for 33% or about 840 TWh of the final energy consumption (Fig. 9.2). According to political objectives, the energy usage of buildings is supposed to be reduced by 20% by 2020 compared to 2008, while until 2050 all buildings are set to be "climate neutral" (BMWi and BMUB 2010). One way to achieve this and limit the energy usage and carbon emissions of buildings is to enhance insulation. Since there are technical and economic limitations in this field, carbon emissions of the used energy need to be lowered as well. Modern condensing boilers have reached a yield level for the calorific value of burned fuels that cannot be increased any further through innovation (acatech et al. 2017), which is why only combined heat and power units could increase the efficiency. Consequently, a fuel switch towards either organic or synthetic fuels or renewable energies is necessary.

Using rooftop solar thermal panels for heat generation is one option to achieve this switch. Those panels however can only contribute a limited share of the required heat (mostly for warm water generation) since there is a seasonal offset between high supply in the summer and high heat demand in the winter. Geothermal heat generation is another option, yet there is a very limited potential in Germany (acatech et al. 2017).

Replacing the natural gas used in gas boilers by biogas or gas from power to gas processes can serve as a bridging technology for houses that have not yet been refurbished with other technologies. Electric heat pumps are more efficient in generating heat, though so far this technology is not very prevalent and mostly used in newly built or renovated single-family homes. One of the reasons are the still very high investment costs compared to a gas boiler and, compared to natural gas or heating oil, high consumer prices for electricity. Heat pumps have a higher efficiency with lower final temperatures which means that underfloor heating systems using low temperature levels are most efficient. However, warm water in apartment buildings, which—for sanitary reasons—needs to be at a minimum temperature of 60 °C, is more difficult to supply. Therefore, also hybrid systems or biogas fired heat

---

[9]AGEB. 2018. "Anwendungsbilanzen für die Endenergiesektoren 2013–2016." Arbeitsgemeinschaft Energiebilanzen e. V.

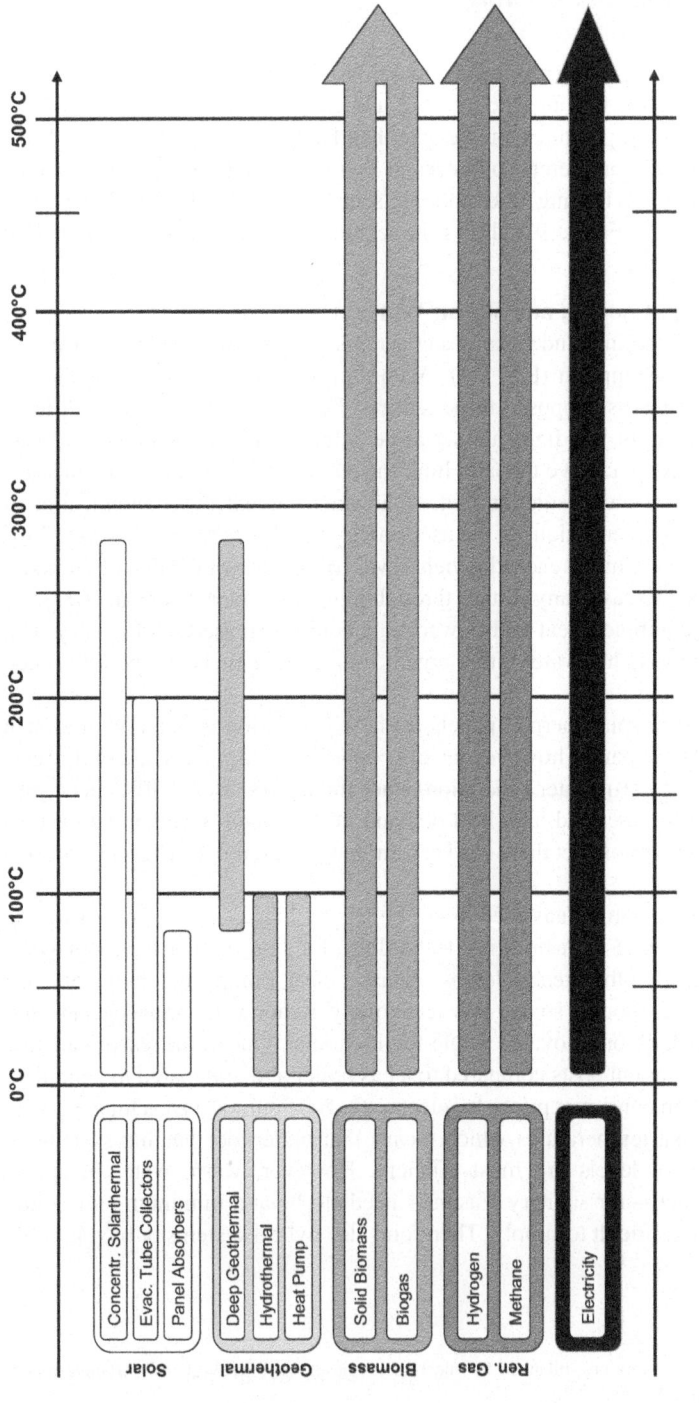

Fig. 9.3 Obtainable heat levels by renewable sources. Source: Own depiction based on Naegler et al. (2015, 2016)

pumps or communal heating and power stations will be required for applications where heat pumps are not technically or economically viable options.

Currently, the modernization rate of German buildings amounts to about 1% annually, while most studies suggest a necessary rate of at least 2% in order to reach climate targets. Otherwise, the refurbishment of heating systems will not advance fast enough. At the same time, about 3% of all home renovations each year lack any energy improvements. Thus, there is a potential for an increased rate of energy related modernization, but it needs to be promoted by suitable political measures. (SRU 2017)

District heating systems will still play a certain role in the future, provided sufficiently dense demand. A centralized provision of heat, distributed via a low-temperature heat grid in densely populated areas, has tremendous efficiency advantages over decentralized heating systems. Power to heat technologies (Bloess et al. 2018), that is, generation of heat from excess electricity, for instance in times of high renewable production, can be used in those facilities. At the same time, waste heat from industrial processes can be used in the residential sector. In combination with large-scale (and long-term) heat storages based on water or salt, this system would provide a lot of flexibility to the overall energy system.

Space cooling currently has a neglectable share, mostly already being generated from electricity. Due to rising temperatures in the wake of climatic change, and to more extreme summers to be expected, the demand for air conditioning (AC) is likely to rise significantly.

Process Heating and Cooling

Process heating and cooling for industrial and commercial purposes account for 23% or about 590 TWh of the final energy consumption in Germany (Fig. 9.2). More than 90% thereof is from process heat, mostly generated from fossil natural gas and coal. Only industrial demand is of relevance here, since the commercial demand and the demand from households is already mostly generated directly from electricity (e.g. cooking). The industrial demand for process heat can be split into the high-temperature range above 500 °C, mid-range temperatures and the low-temperature range below 100 °C. The low-temperature range accounts for only about 25% of the heat demand, while the high-temperature range has the largest share of more than 57% (Naegler et al. 2015).

Whereas the low-temperature range could mostly be replaced by efficient heat pumps, temperatures above 200 °C cannot be achieved by this technology. In these processes fossil fuels need to be replaced by biomass or synthetic fuels or they should be directly electrified wherever possible (see Fig. 9.3). Certain processes require very high temperatures above 1500 °C that are hard to reach using electricity as energy carrier. Moreover, currently used energy carriers might have additional purposes within the process. Coke in blast furnaces for example provides the necessary stability of the materials in the furnace (acatech et al. 2017).

Again, a rise of process efficiency is crucial to achieve the energy transformation in the industry sector. However, the energy used for many processes in the basic substance industries is thermodynamically required for physical phase transitions or chemical conversions and constitute an elementary component of the final product (e.g. glass, ceramics, and plastics), which cannot be replaced. In addition, the electrification of some processes might be a lot less efficient than the current methods. Hence, process optimization potentials are limited in a twofold manner: minimizing heat losses and waste heat being used for space heating in the companies' buildings or redirected into neighboring district heating networks.

Increasing the quota of recycled materials in the German economy would yield a further decrease in energy demand from industrial processes since the recycling of raw materials like glass, paper, plastic, aluminum, or steel is usually less energy intensive compared to new production. Yet, many of today's recycling technologies lead to a so-called "down cycling", reusing the material in a lower quality form. Plastic water bottles for example are down cycled into fibers for clothing production or park benches. With those proceedings, the need for new high-quality plastic is not being reduced. These emissions can only be abated by switching to a different production process.[10]

Another important aspect is the formation of CO_2 as a byproduct. For example, major emissions come from burning in the cement production. These emissions can only be abated by switching to a different production process. Alternatively, the CO_2 can be separated and deposited with carbon capture, transport, and storage (CCTS) technologies or used as a base material in other processes, for example, carbon capture and utilization (CCU). Neither of the two is sufficiently developed to yield promising results, and it is likely that the energiewende will have to do without them.

In conclusion, the decarbonization of the industrial sector including heating and cooling is a challenge compared to the other sectors. Only a minor part is already electrified, a major increase in electricity demand can be expected and for some processes, there is currently no alternative to the usage of synthetic fuels.

9.3.3 The Electricity Sector in the Core of Interdependencies

The lower-carbon sector coupling is likely to evolve around the electricity sector. In fact, the fuel switch from fossil fuels to electricity of the described sectors transportation, heat, and industry has wide-ranging consequences for the electricity sector. These sectors are highly dependent on efficiency gains, but also on the flexibility options those sectors provide for the system and the assumed scenarios and pathways for sector coupling. The overall goal of decarbonization leads to a high demand for renewable energy from competing sectors and applications. In general, a large number of options for sector coupling are available and conceivable. Figure 9.4 provides a detailed overview of the possibilities to couple transportation, heat, and industry via the electricity sector.

---

[10]One example of a new binder with significantly reduced energy usage and CO_2 emissions is Celitement, developed at Karlsruhe Institute of Technology (KIT): www.celitement.de.

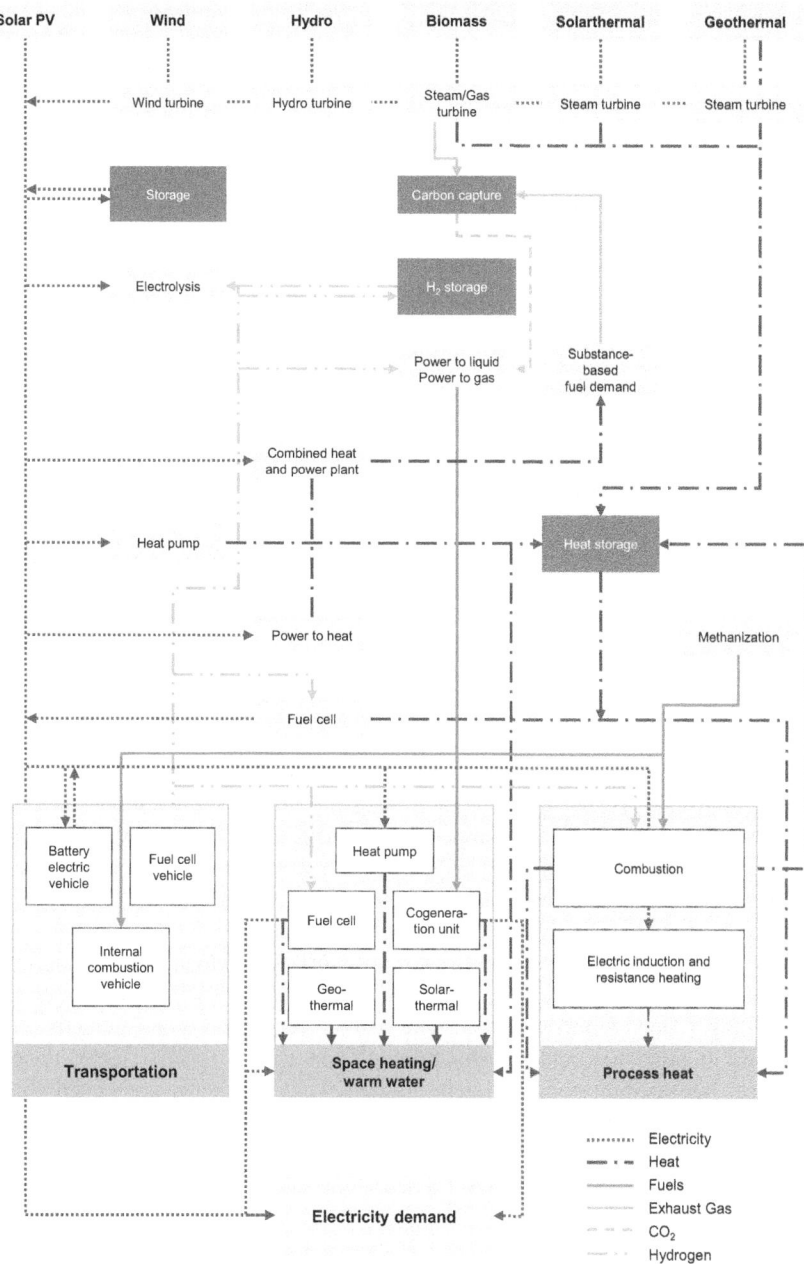

Fig. 9.4 Options for sector coupling in a decarbonized energy system. Source: Own depiction based on SRU (2017)

When shifting towards a higher degree of electrification, the decarbonization of the electricity system is the key success factor. Otherwise, the current CO_2 emissions from other sectors would only be shifted to an electricity generation from fossil fuels, implying only a locally emission free energy use. Fluctuating and intermittent renewable energy sources can be employed for power production, using photovoltaics, onshore and offshore wind generation, biomass, or geothermal technologies, in combination with storage. Via the electricity transportation and distribution grid, the electricity can be directly used in all sectors or be stored in long-term storages like pumped hydro storages or short-term storages like batteries. An advance in sector coupling will further increase the needs for flexibility in the system.

For applications where no direct electrification is possible, indirect electrification is an option, using synthetic fuels produced with the help of electricity. This path can also be used for a long-term chemical storage of energy. As mentioned above, synthetic fuels are a viable replacement for some cases in the transportation sector but also for the substance-based use of primary energy.

Heat can be directly produced from electricity, via so-called power to heat (PtH) applications (Bloess et al. 2018). Those can be small-scale or large-scale electric boilers or heat pumps. The generated heat can then be used directly for heating buildings, warm water generation or in industrial applications. It can also be generated centrally and transported via district heating networks. Heat storages in homes or at a larger scale can decouple supply from demand.

The advantages of synthetic fuels are low costs of refurbishing (cutover costs) of the existing technology. Most applications like gasoline and diesel cars could be easily adapted to synthetic fuels. Yet, synthesizing fuels using electricity is an additional conversion step in the value chain, thus lowering the overall efficiency of the used energy and therefore increasing the amount of additional electricity required. This exacerbates the competition between sectors for renewably generated electricity even further. In the long term, the costs for new technologies and infrastructure necessary for a direct electrification might therefore outweigh the increased costs of electricity due to the higher demand.

The dimensions and cost of the energy system are also highly dependent on the flexibility present in the system, that is, of electricity generation, of electricity and heat storages, and of electricity load. The cost of energy provision is directly related to gains in efficiency and the flexibility of the whole system. Sector coupling increases the flexibility in the system in many ways but is also associated with an increased need for flexibility due to the higher electricity demand.

A more flexible demand for electricity lowers the amount of required generation capacities and storage technologies by peak-load shaving and load shifting options. This can be achieved by flexible heat pumps for space heating and warm water production or a regulated charging of electric vehicles. The achieved savings can compensate the higher costs associated with these demand side flexibility technologies. Inflexible demand on the other hand would lead to a larger necessary dimensioning of generation and storage capacities to absorb the associated high load peaks. This would add to the costs of electricity generation, also via the need to generate the additional renewable energy at less favorable and therefore more expensive locations.

A key factor is flexibility from industrial processes in energy intensive sectors, especially when they are electrified. To date most production processes are optimized to run on a steady basis without any flexibility. A so-called "flex-efficiency" production (Agora Energiewende 2018), increasing efficiency and adding flexibility to the system, is necessary for a successful integration. Incentives for energy optimized production processes will be a prerequisite for businesses to adapt them.

Power to X technologies provide further flexibility and storage options for electricity in other forms. Power to heat allows the production of heat from electricity via heat pumps or boilers. Power to gas and power to liquid can be used for the generation of synthetic gas and fuels from electricity, utilizable in the above-mentioned fields and most notably as long-term and seasonal chemical energy storages (see Buttler and Spliethoff (2018) for an overview). The production of those synthetic gas and fuels could also happen in North Africa or the Middle East, providing oil- and gas-exporting countries with new non-fossil-based business models (Agora Energiewende and Agora Verkehrswende 2018).

Newly developed inexpensive storage technologies like Carnot batteries can help to store large amounts of electricity over a longer period. Those batteries use high temperature heat pumps to generate heat that is then stored in water or salt storage tanks. Electricity can be regenerated from the heat via thermal engines. Alternatively, the heat can be directly dispensed for heating and cooling. Researchers predict a cycle efficiency of 75% for this technology.[11]

The choice of technologies in one sector has therefore implications on the flexibility needs and selection of energy carriers in the other coupled sectors. The possibility to shift loads between the sectors in conjunction with high flexibility lowers the need for demand side flexibility. In many cases there is a trade-off between flexibility and efficiency, for example between direct and indirect electrification. Coupling of the sectors increases the degrees of freedom of the overall system, shifting the attention to the efficiency of the system components.

9.4 Some Model-based Evidence

Although the discussions about far-reaching sector coupling are only emerging, some detailed studies already provide some evidence of the potential effects: SRU (2017) and Ausfelder et al. (2017) provide an overview of the most prevalent analyses for Germany. Brown et al. (2018) extend the literature for the European case. Although these studies vary in the set boundaries of the energy system, they

---

[11]"DLR arbeitet an Gigabatterie." VDI nachrichten, May 3, 2018. https://www.vdi-nachrichten.com/Technik/DLR-arbeitet-an-Gigabatterie.

concur that a far reaching decarbonization (80–95%) of all the regarded sectors until 2050 is technically and economically feasible via a comprehensive electrification.

9.4.1 Electrification is Key

Most studies assume an increase in energy efficiency and additional electricity demand from the transportation, heat, and industry sectors. By 2050, the final electricity demand will grow to about between 780 TWh and 1450 TWh, that is, an up to twofold increase compared to today's values. Some studies even reach about 3000 TWh, assuming no efficiency gains or a demand fully supplied by domestic generation, see Quaschning (2016). The calculated yearly peak demands do not differ much from today's: 60 GW to 80 GW. Only one study reaches 110 GW. The storage demand varies between 8 GW and 15 GW with an outlier at 75 GW. This flexibility demand is mostly met by batteries or the storage technology is not further specified.

The different growth rates in electricity demand also yield different installed renewable capacities depending on the transformation path (Fig. 9.5). For 2030 the studies assume a photovoltaic capacity between 68 GW and 109 GW, onshore wind capacities between 51 GW and 97 GW, and offshore capacities between 11 GW and 22 GW, while for 2050 a photovoltaic capacity between 75 GW and 290 GW, onshore wind capacities between 64 GW and 204 GW, and offshore capacities between 15 GW and 70 GW are being calculated. Electricity imports and exports do not exceed 50 TWh per year, limiting the possibilities to shift emissions to neighboring countries.

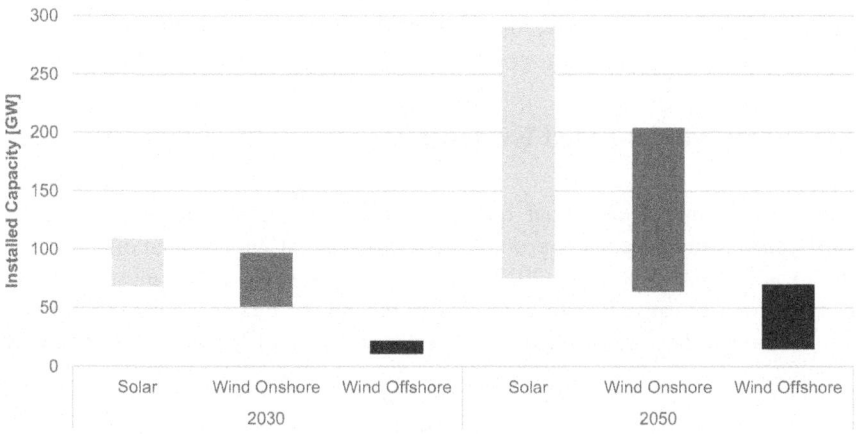

Fig. 9.5 Range of installed capacities in Germany from study scenarios. Source: Own depiction based on SRU (2017)

While electricity could be generated to a large extent in Germany, synthetic fuels might be supplied in part from abroad, between 20 TWh and 1200 TWh. Necessary generation capacities and the area required is shifted outside the country's borders.

9.4.2   "Efficiency First"

Thus, the consumption of electricity will rise significantly with an increasing level of sector coupling and electrification of loads. A multifold increase in capacity expansion of renewable generation technologies is key for decarbonization. Otherwise the sectors would be electrified but for example electric vehicles—perceived to be clean means of transportation—would only shift their greenhouse gas emissions to an electricity production from gas and coal (Schill and Gerbaulet 2015). Since renewable energy sources are in fact limited domestically or come with increasing marginal costs of capacity additions due to the need to draw on less favorable production sites, even renewably generated electricity has to be used as efficiently as possible. The principle "efficiency first" (Agora Energiewende 2018) with all conceivable process improvements applies to all sectors—every kilowatt-hour not consumed is a kilowatt-hour saved.

The energy concept of the federal government therefore also states ambitious efficiency goals for electricity: 10% less final energy use in 2020 and 25% less in 2050 compared to the year 2005 (BMWi and BMUB 2010). Efficiency gains in the electricity sector will be eaten up entirely by the mentioned increase in demand from heat and transportation. At the same time the increased need for flexibility that comes with a higher share of and higher generation amounts from renewables have to be considered when deciding for specific decarbonization pathways for the sectors in question. The concept of flex-efficiency (Agora Energiewende 2018) therefore needs to be implemented: energy savings in times of low renewable generation is especially valuable. Efficiency is extended by a temporal component.

9.4.3   Role of Synthetic Fuels Uncertain

While the verdict is clear on the role of electrification, the role of synthetic fuels is discussed more controversially. Synthetic fuels may become essential for a deep decarbonization of the energy sectors (Agora Energiewende and Agora Verkehrswende 2018). While space heating and cooling can be supplied directly from electricity using heat pumps, high-temperature applications in industrial processes are not flexible enough for direct electrification. In the interplay with the need for synthetic fuels in industry for substance-based usage and applications that cannot be directly electrified due to chemical or physical reasons, the deployment of synthetic or organic fuels might contribute to the flexibility needs. Furthermore, the chemical energy storage capabilities of synthetic fuels could help to bridge phases of so-called "dark doldrums", when there is not enough electricity generation from photovoltaics and wind power over a longer period of time (i.e. a couple of days up to two weeks) and where shortfalls in electricity and heat supply might coincide. The higher energy costs of those fuels, coming from the lower efficiency of additional conversion steps, could be outweighed by the lower need for generation capacities and storages to cushion those periods that do not occur often but would need to be anticipated. Efficient combined heat and power plants could be fired with those synthetic fuels and gases to produce electricity and heat. The same level of security of supply therefore can be reached with less installed capacities.

While synthetic fuels may play a certain role in industry and the electricity sector, their relevance is controversial for the transportation sector. Only applications where a very high specific energy content is essential should rely on indirect electrification. Those are namely air transportation and, in part, maritime transportation. Still, the transportation sector can contribute significantly to the flexibility of the system by offering demand side flexibility from controlled charging processes of electric vehicles.

9.4.4 Digitalization and Smart Infrastructure

Aside from the big picture of necessary electrification and investments in renewables and flexibility, some other aspects will play a role in a successful sector coupling. The energy sector needs to increase the level of its digitalization, moving towards a smart, efficient use of infrastructure. An increasingly decentralized structure of "prosumers", customers who are producers of energy or electricity at the same time, relies on an interconnected system using new technologies (Agora Energiewende 2018). *Smart markets* with real-time smart metering of electricity will give incentives to customers to offer flexibility for the system. Regional price mechanisms can enhance an efficient utilization of distribution and transmission grids. Taxes and levies may not overlay those price effects.

Smart homes will be able to optimize the energy usage in buildings in combination with rooftop PV generation, battery storages, flexible heat pumps, and electric vehicle charging. *Smart mobility* will offer new concepts of transportation, avoiding unnecessary trips and increasing the efficiency of the transportation infrastructure. Following the trends for example in the ICT industry, more and more products will be offered "as-a-service", such as mobility-as-a-service in shared mobility concepts, calling for a change in consumer mentality. New business models will arise around sector coupling and the digital energy industry—enhancing a more flexible and intelligent use of energy in general.

Electricity transmission networks will be expanded to a certain degree. Yet, an efficient utilization should limit the amount of required capacity expansions. Due to the progressing decentralization and higher energy demands in homes from heat pumps and electric vehicles, load profiles and flows in the distribution grids will change drastically, possibly leading to the need for further investments.

District and local heat networks will still play a role especially in densely popu-lated areas. Due to the improvements in building insulations, heat demand will decrease. Grids should gradually be retrofitted to decentralized low-temperature networks, accommodating waste heat from industrial processes and biogas plants, geothermal generation, heat pumps, and other power to heat applications.

9.4.5 Other Issues: Fossil Gas, Transportation, and Market Design

The demand for natural fossil gas will gradually decrease with the switch to renewable generation technologies. While gas fired power plants can act as very flexible back-up capacity with comparably low CO_2 emissions, in a fully decarbonized energy world there is no CO_2 budget available for converting fossil gas into electricity anymore. The natural gas grid infrastructure will therefore shrink, with a possible withdrawal from sparsely populated areas. Still, the gas grids can be retrofitted and used for transportation and distribution of synthetic green gases like hydrogen and methane that will be used in the industry sector. It can also support the energy sector with additional flexibility.[12] An alternative to this are biofuels.

In the transportation sector the electrification and expansion of rail networks should be expedited. A sufficient charging infrastructure for fast-charging electric vehicles along major transportation corridors and in densely populated cities needs to be established to facilitate the switch to electric mobility, especially for longer trips and for people not living in single-family homes. Heavy-duty freight trans-portation could be taken over by hybrid trolley trucks, requiring a major infra-structure implementation of overhead contact systems along major highways.

An adapted market design, coherent in pricing and taxation for all sectors and fuels, will be the foundation of a coupled energy sector, accompanied by investment incentives like public technology funding and regulatory frameworks. Consistent and sufficiently high CO_2 prices will facilitate the shift towards renewables in all sectors. The EU Emissions Trading System (ETS) could be advanced accordingly or a tax on CO_2 emissions could be introduced to speed up the development on a national level. Pricing CO_2 has the advantage of being a technology neutral policy measure, promoting the most cost efficient abatement options, avoiding lock-in effects, and anticipating not yet known new technologies (acatech et al. 2017). Other emissions have to be considered as well.

---

[12]Frontier Economics, IAEW, 4 Management, and EMCEL. 2017. "Der Wert der Gasinfrastruktur für die Energiewende in Deutschland." Studie im Auftrag der Vereinigung der Fernleitungsnetzbetreiber (FNB Gas e. V.). Köln.

9.5 Conclusion

The low-carbon transformation of the German energy system (but also of others) has so far focused on the electricity sector. As described in the previous chapters, it was relatively easy to attain, and even to surpass, the goals on renewables, and taking nuclear plants from the grid, while the reduction of greenhouse gas emissions and the phasing out of coal will take somewhat more time than expected. However, as the energiewende enters the next phase, even more efforts are required to work towards the large-scale introduction of renewables, which is required in all sectors in order to attain the decarbonization targets.

In this chapter we have provided a broad survey of "sector coupling", that is, the combination of technical and economic interdependencies between electricity, transportation, and heat, accompanied by a larger share of renewables. Both elements are necessary (though not sufficient) to succeed the energiewende: without technical interdependencies, transportation and heating are likely to remain largely fossil, whereas introducing renewables into electricity alone is insufficient, too. We observe and describe a rapidly growing literature on sector coupling: while ambitious targets are agreed upon, they can be reached, concretely for the German case, by deepening sector coupling.

Further research is required to translate sector coupling into more concrete policy instruments, to accompany and steer the process. It is clear that a stronger carbon price helps the general trend, but more specific instruments are needed to electrify transportation and heating, and to internalize the adverse environmental effects of fossil fuels in all three sectors, which are the very reason for this exercise. SRU (2017) and Ausfelder et al. (2017) include early suggestions of targeted policy instruments, but they need to be deepened to translate the rather abstract idea of sector coupling into real life.

References

acatech, Leopoldina, and Akademienunion. 2017. *Sektorkopplung – Optionen für die nächste Phase der Energiewende*. München.

Adeniyi, Oladapo Martins, Ulugbek Azimov, and Alexey Burluka. 2018. Algae biofuel: current status and future applications. *Renewable and Sustainable Energy Reviews* 90 (July): 316–335.

Agora Energiewende. 2018. *Energiewende 2030: The Big Picture – Megatrends, Targets, Strategies and a 10-Point Agenda for the Second Phase of Germany's Energy Transition*. Impulse. Berlin.

Agora Energiewende, and Agora Verkehrswende. 2018. *The Future Cost of Electricity-Based Synthetic Fuels*. Study. Berlin.

Ausfelder, Florian, Frank-Detlef Drake, Berit Erlach, Manfred Fischedick, Hans-Martin Henning, Christoph Kost, Wolfram Münch, et al. 2017. *Sektorkopplung – Untersuchungen und Überlegungen zur Entwicklung eines integrierten Energiesystems*. Schriftenreihe Energiesysteme der Zukunft. München.

Bloess, Andreas, Wolf-Peter Schill, and Alexander Zerrahn. 2018. Power-to-heat for renewable energy integration: a review of technologies, modeling approaches, and flexibility potentials. *Applied Energy* 212 (February): 1611–1626.

BMWi. 2016. *Grünbuch Energieeffizienz – Diskussionspapier des Bundesministerium für Wirtschaft und Energie.* Berlin.

BMWi, and BMUB. 2010. *Energiekonzept für eine umweltschonende, zuverlässige und bezahlbare Energieversorgung.* Berlin.

Brown, T., D. Schlachtberger, A. Kies, S. Schramm, and M. Greiner. 2018. Synergies of sector coupling and transmission reinforcement in a cost-optimised, highly renewable european energy system. *Energy* 160 (October): 720–739.

Buttler, Alexander, and Hartmut Spliethoff. 2018. Current status of water electrolysis for energy storage, grid balancing and sector coupling via power-to-gas and power-to-liquids: a review. *Renewable and Sustainable Energy Reviews* 82 (February): 2440–2454.

Naegler, Tobias, Sonja Simon, Martin Klein, and Hans Christian Gils. 2015. Quantification of the European industrial heat demand by branch and temperature level. *International Journal of Energy Research* 39 (15): 2019–2030.

Naegler, Tobias, Sonja Simon, Hans Christian Gils, and Martin Klein. 2016. Potenziale für erneuerbare Energien in der industriellen Wärmeerzeugung. *BWK – Das Energie-Fachmagazin* 68 (6/2016): 20–24.

Quaschning, Volker. 2016. *Sektorkopplung durch die Energiewende – Anforderungen an den Ausbau erneuerbarer Energien zum Erreichen der Pariser Klimaschutzziele unter Berücksichtigung der Sektorkopplung.* Berlin.

Schill, Wolf-Peter, and Clemens Gerbaulet. 2015. Power system impacts of electric vehicles in Germany: charging with coal or renewables? *Applied Energy* 156 (October): 185–196.

SRU. 2017. *Umsteuern erforderlich: Klimaschutz im Verkehrssektor.* Sondergutachten. Berlin.

Wietschel, Martin, Patrick Plötz, Benjamin Pfluger, Marian Klobasa, Anke Eßer, Michael Haendel, Joachim Müller-Kirchenbauer, et al. 2018. *Sektorkopplung – Definition, Chancen und Herausforderungen.* S 01/2018. Working Paper Sustainability and Innovation. Karlsruhe: Fraunhofer ISI.

Part III
The German Energiewende in the Context of the European Low-Carbon Transformation

Chapter 10
The Electricity Mix in the European Low-Carbon Transformation: Coal, Nuclear, and Renewables

Roman Mendelevitch, Claudia Kemfert, Pao-Yu Oei, and Christian von Hirschhausen

> *"Renewables rise substantially, ...*
> *Carbon capture and storage has to play a pivotal role in*
> *system transformation, ...*
> *Nuclear energy provides an important contribution."*
> European Commission (EC 2011d): Energy Roadmap.

This chapter is based on modeling analyses and analytical work on the European energy and electricity sector and builds on several research projects, see references. We thank Jonas Egerer, Clemens Gerbaulet, Christian Hauenstein, Casimir Lorenz, Karsten Neuhoff, Wolf-Peter Schill, and Thure Traber for comments and discussions; the usual disclaimer applies.

R. Mendelevitch (✉)
Resource Economics Group, Humboldt-Universität zu Berlin, Berlin, Germany

German Institute for Economics Research (DIW Berlin), Berlin, Germany
e-mail: roman.mendelevitch@hu-berlin.de

C. Kemfert
German Institute for Economics Research (DIW Berlin), Berlin, Germany

Hertie School of Governance, Berlin, Germany

German Advisory Council on the Environment (SRU), Berlin, Germany

P.-Y. Oei
Junior Research Group "CoalExit", Berlin, Germany

TU Berlin, Berlin, Germany

German Institute for Economics Research (DIW Berlin), Berlin, Germany

C. von Hirschhausen
DIW Berlin, Berlin, Germany

TU Berlin, Berlin, Germany

10.1   Introduction

The European Union, too, has embarked on the transformation of its energy and electricity system to low-carbon energy sources, just like Germany and many other countries. Germany makes up a large part of the European energy system, despite being just one of the EU's 28 Member States. But the EU has more options and rules to pursue its energy and climate policies than even a single Member State like Germany does. Thus, while there are strong interdependencies between the energiewende in Germany and the low-carbon energy transformation in Europe, the two still have to be analyzed as two distinct processes. From a legal perspective, any reform in Germany is subject to European laws such as the Single Market Act, state aid regulations, competition rules, the European Emissions Trading System (EU ETS), and other legal instruments. European institutions also exert influence on Member States' energy policies both directly and indirectly through their roadmapping exercises and forecasts, legislative and jurisdictive authority, and through political negotiations in the broader context of European energy and climate policies. However, the choice of a country's energy mix is first and foremost a national prerogative, as stipulated by Article 194 of the Amsterdam Treaty on the Functioning of the European Union, which assigns decisions on the energy mix to the Member States. This sovereignty will be useful in explaining the differences between the energy mix of the German energiewende and that of the European energy system.

This chapter analyzes the European strategy for low-carbon transformation in relation to specific aspects and features of the German energiewende. Due to the different preferences, objectives, and institutional settings of decision-making processes in Germany and Europe, lessons from the German context are not directly applicable to the European context and vice versa. While some lessons apply to both—such as the German experience with ambitious CO_2 reduction targets—others do not, such as the potential role of coal and nuclear energy in the longer-term energy mix. The most striking difference is that Europe appears to be maintaining an energy mix that combines shares of coal, nuclear, and renewables, whereas the German energiewende is focusing on renewables and phasing out both coal and nuclear. The chapter focuses on issues of generation, while the next Chap. 11 deals specifically with the energy transmission infrastructure.

The history of the European Union has been a history of energy policy since the very beginning. This chapter therefore begins with a brief survey of European energy (and later climate) policies going back to 1951, with the decisions to establish the European Community for Steel and Coal (ECSC) and subsequently Euratom in 1957. Section 10.2 of the chapter covers the creation of the European internal market in the 1990s and its application to the energy sectors (mainly electricity and natural gas): The unbundling rules set out in 1996 (electricity) and 1998 (natural gas) have in fact laid the foundations of the low-carbon transformation by allowing natural gas and later renewables to enter a market dominated by the incumbent coal and nuclear monopolists. Section 10.2 also covers more recent discussions, such as the energy and climate package to 2020, the 2030 targets, and the parallel discussion about longer-term orientations up to 2050. We describe the triad of objectives of the European low-carbon transformation, which strongly resemble those of the German energiewende: (1) the 20-20-20 targets (20% reduction of GHG emissions, 20% increase in renewables, and 20% increase in energy efficiency), (2) GHG emission reduction targets (-40% up to 2030, and -80 to 95% up to 2050), and (3) targets for energy efficiency (40% increase) and renewables (27% share of total primary energy consumption by 2030). Like Germany, Europe is on track to reach both the GHG target and the renewable target, while energy efficiency is lagging behind somewhat.

Sections 10.3, 10.4, and 10.5 analyze the three pillars of European transformation towards a low-carbon energy system: coal with CO_2 sequestration, nuclear power, and renewables, respectively. In this context, we discuss a major difference between the European transformation to a low-carbon economy and the German energiewende: The two energy sources that Germany has banned from its energy mix, coal and nuclear, are still high on the European agenda. We therefore seek to identify economic and/or technical arguments that might explain this choice. Our results show that carbon capture, transport, and storage (CCTS) has made no progress thus far and that all European CCTS pilot projects have failed. Likewise, nuclear power is not an economic option for a low-carbon energy transformation due to its high costs and inherent risks; EU scenarios featuring high shares of nuclear power up to 2050 are not economically but politically driven. Meanwhile, the potential of renewables has been systematically underestimated in European scenario documents, due mainly to an overestimation of costs and an underestimation of the technical potential. We identify key factors that have shaped a European energy mix that appears to stand at odds with the probable unavailability of CCTS technologies in the foreseeable future and the high costs of nuclear energy. Here we find that many of the differences in the different Member States' assessments are due to differing interpretations of the costs and benefits of these technologies and specific national factors. Section 10.6 then compares two alternative scenarios for a low-carbon transformation in Europe: one is the EU Reference Scenario, which is based on the traditional triad of coal (with CCTS), nuclear, and renewables. In the other scenario, based on our own modelling work, neither CCTS nor nuclear are available at a reasonable cost and renewables carry the major burden of decarbonisation. Section 10.7 concludes.

10.2   European Energy (and Climate) Policies Since 1951

10.2.1   The Past: 1951 to Today

10.2.1.1   1951–1992

Energy issues were at the core of European discussions after World War II. The focus was initially on coal, later also on nuclear, and still later it extended to energy infrastructure and issues of energy efficiency. Since the 1990s, climate policy became a major issue. In fact, the two founding organizations of the EU each had a specific energy carrier at its core:

- In 1951, France, West Germany, Italy, and the three Benelux countries, Belgium, Luxembourg, and the Netherlands, founded the European Coal and Steel Community (ECSC).[1] The ECSC was an attempt both to harmonize and to protect heavy industry, which at that time was expected to become an engine of European reconstruction and development. Although a great deal of coal production was subsidized, there was a broad consensus among the six founding members to maintain coal as the dominant energy carrier.
- With the signing of the EURATOM Treaty in Rome (1957), Europe attempted to harvest the benefits of nuclear energy, a technology pursued intensively by the USA after the end of the Second World War. At the time, the hope was that nuclear energy would become a cheap and ubiquitous source of energy supply ("too cheap to meter") and that it would boost development not only in Europe but also in other emerging partner countries.[2]

Thus, the European energy system was initially based on coal—just as it was in the "two Germanies" after World War II— and later on significant shares of nuclear capacities, which were added from the 1960s onwards. During the period of the Cold War, the national energy mixes on both sides of the Wall were determined primarily by the domestic availability of resources and resource-related industries such as coal in the UK, Germany, Poland, and the Czech Republic, and hydropower in Austria, France, and Switzerland. Natural gas was discovered in and extracted from the North Sea region, mainly by the UK, the Netherlands, as well as non-EU Member State Norway. Some East European countries also had to rely on the Soviet Union for their supply of fossil fuels such as coal and later natural gas, as well as for nuclear technologies.

Figure 10.1 shows the electricity generation mix of Western Europe between 1950 and 2015. Until 1990, it is heavily dominated by coal and, to a lesser extent, other fossil fuels like natural gas and oil. Nuclear energy becomes significant in the 1980s and expands further in the 1990s, with a market share of over 25%. Also note the relatively constant contribution of hydroelectricity over time.[3]

---

[1]The European Community for Steel and Coal (ECSC) (1951) was subsequently integrated into the European treaty (EEC Treaty 1957) and became part of the European Union; the ECSC ended in 2002.

[2]The signatories of the EURATOM Treaty (1957) even wrote in the preamble that this had been concluded ". . .recognising that nuclear energy represents an essential resource for the development and invigoration of industry and will permit the advancement of the cause of peace..."and"...desiring to associate other countries with their work and to cooperate with international organizations concerned with the peaceful development of atomic energy. . .".

[3]Hydroelectricity is of course a renewable resource, although it will play a minor role in subsequent discussions due to technical constraints on its further expansion in Europe.

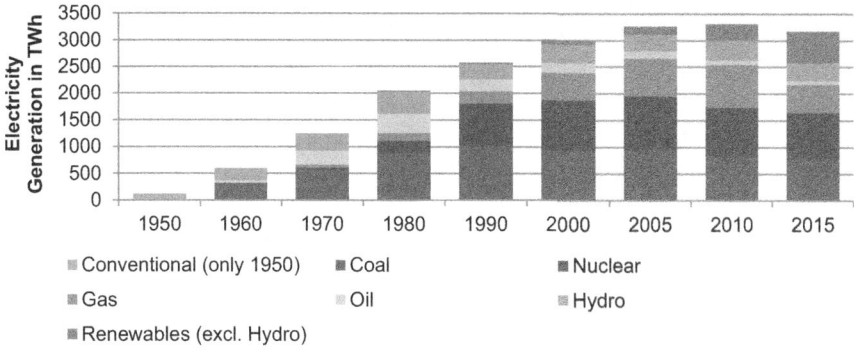

Fig. 10.1 Electricity generation in Europe 1950–2015 in TWh. European OECD countries until 1980, then countries of EU-28. Sources: Data from IEA (2009), and Raack et al. (1957), EC. 2018. "Energy Datasheets: EU28 Countries." Energy Statistics. European Commission

10.2.1.2   1992–2020

The low-carbon transformation in Europe and the energiewende in Germany would never have been possible without two key developments in the early 1990s:

- The 1992 Sustainability Summit in Rio de Janeiro allowed the European Union to establish itself at the forefront of a global *climate policy* movement. Five years later, it resulted in the Kyoto Protocol, which included provisions on the burden-sharing of greenhouse gas emission (GHG) reductions among EU Member States. Europe was also responsible for introducing three flexible instruments into the international climate agreements: emission trading, joint implementation (JI), and carbon offsets (Grubb et al. 2014);
- The European Union pushed for the creation of an *internal energy market* based on the Anglo-American concepts of vertical unbundling and creating markets in sectors previously dominated by monopolies. In that respect, the founding of the European Single Market in 1992 was a landmark. The single market initiative reached the energy sectors with a slight delay as well, and both the electricity and natural gas sectors were included in the single-market legislation, with the groundbreaking Directives 96/92/EC (electricity) and 98/30/EC (natural gas) laying the foundation for two decades of work towards an integrated European energy market.

Looking back, the emergence of climate policy and the development of legislation on competition and the internal market were clearly intertwined: In fact, the latter was a necessary (though not sufficient) condition for breaking with the old system of conventional-based incumbent and state-owned monopolists. They would probably have sought to maintain the old generation structures focusing on coal and nuclear energy and resisting changes such as the market entry of natural gas, not to mention renewables.[4] Vertical unbundling of generation and transmission, as prescribed in the European Directives, and the initiation of competition can be considered as preconditions for both the German energiewende and the European low-carbon transformation. Without vertical unbundling, the promotion of competition, the guarantee of market access for newcomers, and the emergence of cross-border policy considerations, the incumbent utilities would have maintained the old generation structures. Thus, even though the European internal market has not been fully achieved, the first Electricity Market Directive (1996) can still be considered a broad success two decades later, having pushed through sector-wide energy reforms against opposition from the incumbent energy trusts.

Climate policies and the deepening of the internal market continued through the 2000s. In fact, the two main energy policy trends in the 2000s were the development of the internal market with two "acceleration directives" for the electricity and natural gas sectors (in 2003 and 2009, respectively) and the intensified efforts to develop and implement climate policies. Gradual implementation of the decisions of the Kyoto Protocol and the first Renewables Directive (1991) led to widespread awareness of the urgency of climate policy and progress in this areal, culminating in the 2007–2009 integrated energy and climate package. In this period, the European institutions set their targets for 2020 and also developed a longer-term strategy of broad decarbonization to be achieved by 2050.

While the integrated energy and climate package did not set legally binding targets up to 2050, it did define precise measures for 2020: the famous "20-20-20" goals. Based on a Commission communication from 2007 and a Council decision from 2008, the European institutions adopted a package of directives for energy and climate conservation, in 2009.[5] Their objectives were:

- A 20% reduction in greenhouse gas (GHG) emissions by 2020 (compared to 1990 figures). To achieve these targets, the intention was to reform emissions trading as a key instrument for reducing greenhouse gases.
- A 20% improvement in energy efficiency over current forecasts (defined as either an increase in energy productivity, as economic output per unit of energy used, or

---

[4]Unbundling in Germany was pushed forward by the Ministry for Environment (led at that time by SPD Minister Sigmar Gabriel) in an effort to jump-start implementation of the European Directives by creating a more sustainable energy mix among the unbundled energy companies (see BMUB. 2007. "Gabriel Welcomes European Commission's Legislative Package for the EU Electricity and Gas Markets." Press release 251/07. Berlin, Germany; and Theobald, Christian, and Christiane Theobald. 2013. Grundzüge Des Energiewirtschafts-rechts. 3rd ed. Munich, Germany: Verlag C.H. Beck.)

[5]European Climate and Energy Package.2009. This includes Directives 2009/28/EC on the promotion of the use of energy from renewable source, 2009/29/EC on emissions trading, 2009/31/EC on carbon storage, and Decision 406/2009/EC on effort sharing.

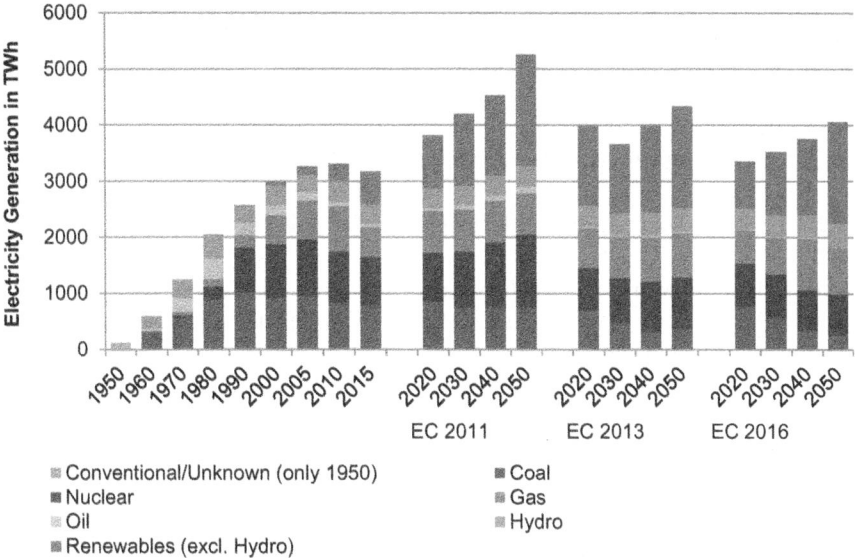

Fig. 10.2 The historical electricity mix since 1950 in Europe and Reference Scenario results until 2050 (in TWh). European OECD countries until 1980, then countries of EU-28. Scenarios for 2020–2050: according to the Reference Scenarios in the Energy Roadmap 2050 (EC 2011d), the EU Reference Scenario 2013 (EC 2013), and the EU Reference Scenario 2016 (EC 2016). Sources: Data from EC (2011a, b, 2013, 2016); IEA (2009) and Raack et al. (1957), EC. 2018. "Energy Datasheets: EU28 Countries." Energy Statistics. European Commission

a decrease in energy intensity). The energy efficiency targets were later set out in the Energy Efficiency Directive[6] and the Energy Efficiency Plan.[7]
• An increase in the proportion of renewables in overall energy consumption (gross final energy consumption for electricity, heat and transport) to 20%.

These 20% targets were all to be reached by 2020, and were given the catchy and appealing name "20–20–20 goals." The GHG emission reduction goal is on track. It has even benefited from the economic crisis, which reduced emissions considerably. The EU is also on track to reach its renewables target, but its efficiency targets are still far from being achieved. Figure 10.2 shows the evolution of electricity generation in Europe since 1950 and its projection until 2050. Clearly, renewables as a third pillar of electricity supply emerged by 2020 (estimate from EU Reference Scenario), but the two others, coal and nuclear, still maintained their respective

---

[6]See EC. 2012. Directive 2012/27/EU of the European Parliament and of the Council of 25 October 2012 on energy efficiency, amending Directives 2009/125/EC and 2010/30/EU and repealing Directives 2004/8/EC and 2006/32/EC, Brussels, Belgium: European Commission.

[7]See EC. 2011. Energy Efficiency Plan 2011. COM (2011) 109 final, Communication from the Commission to the European Parliament, the Council, the European Economic and Social Committee and the Committee of the Regions, Brussels, Belgium: European Commission.

output levels. This was also a "golden age" of natural gas, with a market share of about 20%.

10.2.2 Objectives Moving Forward: 2030 and 2050

10.2.2.1 2030: More Individual Targets, Less Coherence

In 2014, the European Commission followed suit on its 2030 targets and proposed a framework for energy and climate policies to bridge the 2020 targets with a (non-binding) vision for 2050, following a detailed impact assessment (see Box 10.1). This included a Europe-wide GHG emissions reduction target of 40% by 2030 compared to 1990 as well as an EU-wide renewable energy target of at least 27% of final energy consumption[8] and an efficiency target of 40%.[9]

However, while these targets can be considered ambitious, the decision-making process has lost its previous coherence. In fact, progress towards targets in the three areas of climate, renewables, and efficiency differs significantly due to the respective political processes: the most progress has been achieved toward the 40% reduction of GHG emissions because there is now corresponding legislation in place.[10] As has been the practice to date, it is left up to Member States to set the targets for the sectors not included in the EU-ETS. On the other hand, there is no roadmap for how the European Union or the Member States plan to meet the 27% renewables target.[11]

---

[8]See EC. 2014. A policy framework for climate and energy in the period from 2020 up to 2030. Commission Staff Working Document Impact Assessment COM(2014) 015 final, Communication from the Commission to the European Parliament, the Council, the European Economic and Social Committee and the Committee of the Regions, Brussels, Belgium: European Commission.

[9]To put these goals into context, the European Commission has defined the following "dimensions" for the Energy Union strategy of 2015: (1) a fully integrated European energy market; (2) energy security, solidarity and trust; (3) energy efficiency contributing to moderation of demand; (4) decarbonising the economy; and (5) research, innovation and competitiveness. See EC. 2015. A Framework Strategy for a Resilient Energy Union with a Forward-Looking Climate Change Policy. COM(2015) 080 final, Communication from the Commission to the European Parliament, the Council, the European Economic and Social Committee and the Committee of the Regions, Brussels, Belgium: European Commission.

[10]In order to keep the target in sight despite a high surplus of allowances, the annual reduction factor for the ETS sector was increased from 1.74% to 2.2%. In addition, a mechanism was put in place to stabilize the price of emission trading through a market stability reserve for the European Emissions Trading System, see EC. 2014. Proposal for a decision of the European Parliament and of the Council concerning the establishment and operation of a market stability reserve for the Union greenhouse gas emission trading scheme and amending Directive 2003/87/EC. COM(2014) 20/2, Brussels, Belgium: European Commission.

[11]One might consider the renewables target ambitious, considering that the target for renewable energy in primary energy of 27% by 2030 translates into an equivalent of approximately 50% of the electricity sector. This is fairly similar to the German target.

Box 10.1   Economic Scenario Analysis of the 2030 Targets: The European Commission's Impact Assessment of 2014

The economic and energy-specific consequences of different policy scenarios are derived from model-based analysis using an "impact assessment" (IA). Here we discuss the IA for the 2014 package focusing on the 2030 targets. The EC's 2014 legislative package was based on a comprehensive IA, including a set of scenarios and different targets that European policies could potentially work towards (EC 2014). Alongside key energy indicators, the IA also evaluates the development of macroeconomic variables based on the reference scenario "EU Energy, Transport and GHG Emissions Trends to 2050" (EC 2013).[12]

Kemfert et al. (2014) provide a detailed overview of the Commission's 2014 IA. The "default" reference scenario assumes that by 2030 there will be a 32.4% reduction in GHG emissions compared to 1990 levels, noting that in this baseline, the reduction in GHG emissions to be achieved by 2050 is only 44%, a 24.4% share of final energy consumption generated by renewable energy sources, and also energy savings of 21% compared to the 2007 reference development forecast.

The IA does not draw any clear conclusions with regard to the advantages of particular policies. Energy system costs, for example, are very similar across all scenarios: the additional average annual energy system costs are only €34 billion (1.6%) higher in the most ambitious scenario than in the reference scenario. Annual investments are €93 bn higher than in the reference scenario (€816 bn) and furthermore, €27 bn can be saved over the reference scenario through reduced fuel imports. More ambitious targets in the fields of climate change, renewable energy, and energy efficiency may result in a positive net impact on the overall economy, for instance through increased investment activity or decreased imports of fossil fuels from abroad.

Depending on the model used and the assumed use of CO_2 revenue, the EC's IA shows either slightly positive or negative net impact on GDP and employment. The most ambitious scenario results in the most significant growth in income and employment (44% emissions reduction, 35% share of renewable energy sources, and increased efficiency measures). According to this scenario, compared to the reference scenario, a positive employment effect of 1.25 mn people can be expected across Europe by 2030. The Commission's

(continued)

---

[12]The scenario builds on statistical data from 2010 and assumes a continuation of current economic trends and future demographic developments. Further, policy proposals that were agreed on or implemented prior to spring 2012 were also taken into consideration.

Box 10.1 (continued)
IAs illustrate that the scenarios with the most ambitious targets for emissions reductions and expansion of renewable energy would probably be only slightly more expensive than other scenarios and might even entail macroeconomic advantages.
    Source: Kemfert et al. (2014).

10.2.2.2 Reform Proposals for the Emission Trading System (ETS): Too Late and Too Slow

The reduction of GHG emissions is a major objective at the European level and one on which a certain longer-term consensus exists at the European level as well.

Figure 10.3 shows the EU 28 GHG emissions between 1990 and 2016 and the targets for 2020 (−20%), 2030 (−40%), and 2050 (−80 to 95%). The European Commission's reference scenario assumes that current policy measures alone will reduce emissions by 24% by 2020 and 32% by 2030 compared to 1990 levels.[13] However, given the implausible assumptions the scenario is based on, there is a real danger that the EU will face major difficulties in achieving its 2030 emissions target and will completely miss the 2050 target of reducing GHG emissions by 80 to 95%. Although the 2020 objective of a 20% reduction will be met thanks to the economic crisis, the EU's energy, transport, and heating sectors currently lack instruments for achieving the 2050 targets. Given the energy sector's durable capital stock and the danger of a carbon lock-ins, the question arises why a large share of the energy sector's reduction has to be deferred to 2030–2050 in order to meet the long-term emissions reduction target of 80 to 95%.

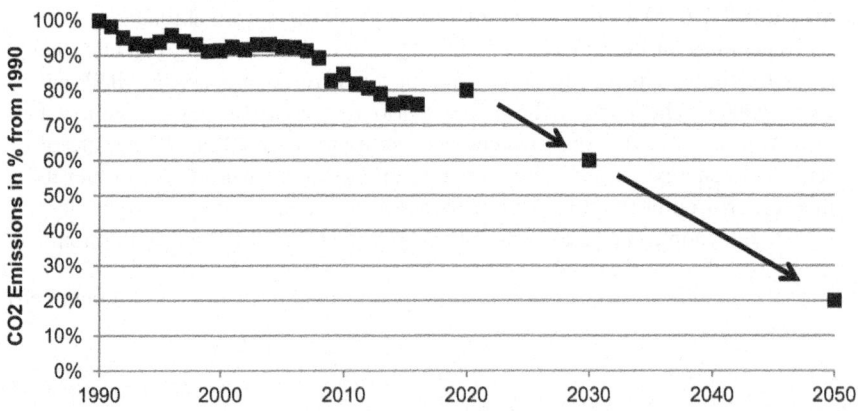

Fig. 10.3 EU-28 greenhouse gas emissions and targets for 2020, 2030, and 2050. Source: Own depiction, based on data from EEA 2018. "EEA Greenhouse Gas – Data Viewer." European Environment Agency

---

[13]See EC (2014).

Between 1990 and 2016, GHG emissions in the current EU-28 countries dropped by 24% compared to 1990. The lion's share of the reduction occurred during the crisis years between 1990 and 1993 (the transformation crisis in former East Germany and Central and Eastern Europe) and 2008 and 2012 (financial and economic crisis), whereas GHG emissions levels remained largely unchanged. Achieving emissions reductions in the sectors covered by the EU ETS is more cost-effective than in those not part of the scheme, such as transport or the private household sector. This made it relatively easy for the emissions trading sector to meet the emissions reduction target of 1.74% per year, with a large number of surplus permits still remaining in circulation. According to the European Commission, the emissions reduction target of 40% compared to 1990 levels should be met solely by implementing internal EU measures. This figure does not take into account GHG emissions produced abroad, however. Along with a 43% reduction in industries covered by the EU ETS, achieving the target is also contingent on a 30% reduction in other sectors (against 2005 levels)[14].

Achieving emissions reductions in the sectors covered by the EU ETS is more cost-effective than in those not part of the scheme, such as transport or the private household sector. This made it relatively easy for the emissions trading sector to meet the emissions reduction target of 1.74% per year, with a large number of surplus permits still remaining in circulation. According to the European Commission, the emissions reduction target of 40% compared to 1990 levels should be met solely by implementing internal EU measures. This figure does not take into account GHG emissions produced abroad, however. Along with a 43% reduction in industries covered by the EU ETS, achieving the target is also contingent on a 30% reduction in other sectors (against 2005 levels)[15].

The achievement of the ambitious climate targets hinges on at least two elements: a more stringent target for the emissions trading sector and a mechanism to coordinate Member States' targets for the non-ETS sector. In particular, the EU ETS will have to undergo far-reaching structural reforms, at least if it is to maintain its position as an international model (see also Chap. 3). Carbon emissions trading was first introduced in the EU in the early 2000s because the EU-wide carbon tax project would have required a unanimous decision by all Member States, which would have been impossible to achieve. The hope was that the system would facilitate an effective and efficient reduction in emissions. Following a pilot phase (2005–2007) and trading period with largely free allocation among the Member States (2008–2012), the system is now in its third trading period (up to 2020). It

---

[14] A problematic area is effort-sharing among the Member States of the EU with regard to sectors not covered by the EU-ETS. Currently, this is based *inter alia* on per capita GDP to reduce impact on the poorer countries. In view of these distribution issues, protracted negotiations can be expected in the future (Kemfert et al. 2014).

[15] A problematic area is effort-sharing among the Member States of the EU with regard to sectors not covered by the EU-ETS. Currently, this is based *inter alia* on per capita GDP to reduce impact on the poorer countries. In view of these distribution issues, protracted negotiations can be expected in the future (Kemfert et al. 2014).

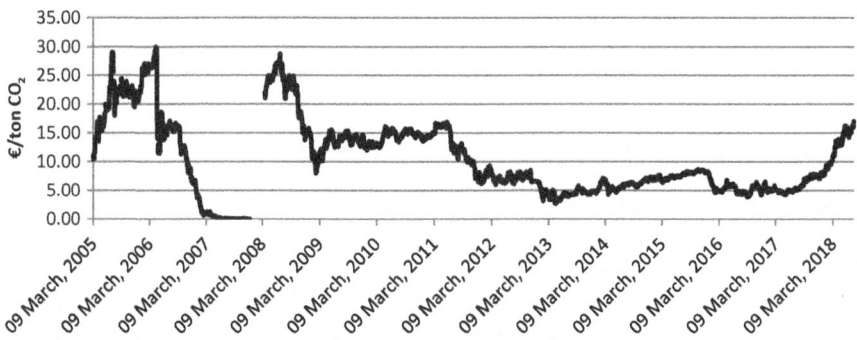

Fig. 10.4 Development of certificate prices in the EU ETS. Source: EEX

involves auctioning a significant share of emissions allowances throughout Europe and harmonizing the rules for free allocation.

If the positive impact of the EU ETS in promoting investment has been only moderate since its introduction in 2005, this impact has now been almost completely lost as a result of the economic crisis and the large number of credits from outside the EU. Figure 10.4 shows the price development of CO_2 certificates on the spot market. Apart from the high level of volatility, the collapse of the CO_2 price as a result of the 2008 economic crisis is also striking: Due to the slump in demand for certificates, which was not accompanied by a corresponding adjustment of supply, prices plummeted. It was only the trading participants' speculation and hedging strategies that prevented the price from hitting zero euros per ton of CO_2 (Neuhoff and Schopp 2013). Since then, the accumulated surplus of unused permits has hit around two billion tons (about 40% of annual emissions). Many market observers as well as the European Commission itself assume that a significant surplus will remain for some years, probably until the late 2020s (see Fig. 10.5).

The "market stability reserve" (MSR), which was added to the system in 2015, is unlikely to have a significant effect on prices (see also Sect. 4.4) In fact, although it is being considered as the institutional frame for the future of the EU ETS, the MSR seems ill-suited to making the ETS an effective instrument of climate policy. The MSR will start operating in January 2019 and is expected to increase or reduce the supply of carbon credits depending on the state of the market.[16] In the first 8 months of 2019 starting as of 1 January, 16% of the total number of allowances in circulation

---

[16]The plan is to announce the level of surplus certificates on mid-May each year (permits issued + credits from abroad—verified emissions—permits in the market stability reserve = permits in circulation). Based on this calculation, permits have accumulated since 2008, i.e., since the beginning of the second trading period. If the cumulative surplus exceeds 833 mn permits, up to 12% of the surplus certificates to be auctioned in that particular year (i.e., at least 100 mn) will be transferred to the reserve. The maximum certificates which can be transferred is temporarily doubled from 12% to 24%. Conversely, if the number of permits in circulation dips below 400 mn, the Commission will release 100 mn permits from the reserve back into the market the following year. The remaining long-term surplus should correspond to the hedging demand of the power sector. It is assumed that this occurs because in many cases power producers have sold their power production up to three years in advance and issue it with certificates at the point of sale.

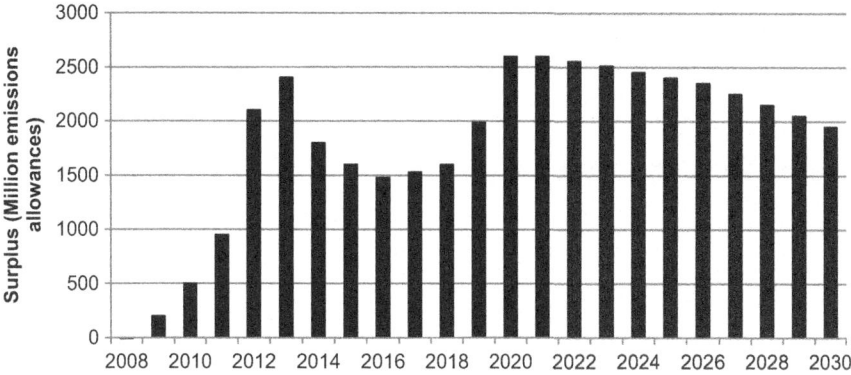

Fig. 10.5 Expected surplus permits in the EU ETS. Source: Kemfert et al. (2014, 23)

will therefore be placed in the reserve.[17] However, since the effects of the MRS will only result in a gradual reduction of the surplus thereafter the mechanism is unlikely to have lasting effects.

10.2.2.3 Renewables Targets: Modest Ambitions and No Roadmap

Although some progress has been made since the beginning of the century, the future of renewables is in peril in Europe. Between 2004 and 2011, the renewable share of gross final energy consumption in EU countries rose from 8% to only 13%,[18] but there are still concerns about whether the overall target for 2020 (20%) will be achieved. To achieve this goal, renewables in Europe would have to grow by an average of more than 6% per annum.

After some struggles, the European Council set the renewables target for 2030 at a 27% share of gross final energy consumption across Europe. This can be interpreted as ambitious given the current low levels of renewable energy shares and the fact that it implies a 50% share of renewables in the electricity sector. Given the huge potential of renewables, however, it can also be interpreted as a lack of ambition.

A major problem of the renewables policy is the lack of a comprehensive institutional framework prescribing how the goals should be attained. The 2030 package does not contain any specific objectives for the individual EU Member States. Consequently, they are not directly bound to particular targets. The Commission's proposal of an EU-wide binding target is still vague. It is particularly unclear how the target should be met: There is no sharing of the objective across the Member States, no coherent approach to implementation, and there are no sanctions for non-compliance. Although the Commission stated in its Communication that there will be a new governance system based on national energy plans to ensure the

---

[17]See EC (2018).

[18]Eurostat, Europe 2020 indicators (2013).

target is achieved, this governance structure, with its iterative voting process between the Commission and the Member States, still remains unclear.[19] In this respect, there is no evidence of the framework having a binding effect.

10.2.2.4 Energy Efficiency: An Important Goal, But Not a Current Priority

Energy efficiency is defined as the ratio of output of goods or services to input of energy[20]. Improvements in energy efficiency are indicated by a rise in energy productivity (economic output per unit of energy used) or a fall in energy intensity (energy use per unit of economic output)[21]. Improving energy efficiency makes it easier for countries to meet their relative targets for increasing the share of renewables in overall consumption. Although the EU has made bold proposals to increase energy efficiency—for instance, by prohibiting the sale of inefficient light bulbs—the overall effects in terms of efficiency are deceiving.

In 2008, the Commission agreed to a 20% reduction of primary energy consumption by 2020 compared to the reference scenario.[22] The plan was to meet this target primarily through efficiency improvements in the building, services, transport, and energy sectors (especially through increased use of cogeneration). An Energy Efficiency Directive went into effect at the end of 2012[23] stipulating that, in 2020, the EU-28's primary energy consumption should not exceed 1483 Mtoe (mn tons of oil equivalent) (see Fig. 10.6) and final energy consumption should be no more than 1086 Mtoe. Member States were obliged to implement this Directive in their national legislation in 2014 and to submit "National Energy Efficiency Action Plans" describing measures implemented to meet these targets.

In terms of the efficiency target, the consensus here is that increased efforts are necessary to achieve the 20% target. The progress made in the field of energy

---

[19]A new governance system has been proposed, based on national plans, with the aim of facilitating a competitive, secure, and sustainable energy supply. Improvements are needed in competitiveness, transparency, security of investment, and EU-wide coordination. These plans are to be implemented in an iterative process between the Commission and the Member States to facilitate compliance with legal requirements and provide long-term prospects." European Commission. 2030. Climate and energy goals – Press Release, Brussels, Belgium: European Commission.

[20]See EC. 2012. Directive 2012/27/EU of the European Parliament and of the Council of 25 October 2012 on Energy Efficiency, Amending Directives 2009/125/EC and 2010/30/EU and Repealing Directives 2004/8/EC and 2006/32/EC.

[21]Diekmann, Jochen, Wolfgang Eichhammer, Anja Neubert, Heilwig Rieke, Barbara Schlomann, and Hans-Joachim Ziesing. 1999. Energie-Effizienz-Indikatoren. Statistische Grundlagen, theoretische Fundierung und Orientierungsbasis für die politische Praxis. Heidelberg, Germany: Physica-Verlag Heidelberg.

[22]See EC. 2008. Communication from the Commission – Energy efficiency: delivering the 20% target. COM/2008/0772 final, Brussels, Belgium: European Commission.

[23]EC. 2012. Directive 2012/27/EU of the European Parliament and of the Council of 25 October 2012 on Energy Efficiency, Amending Directives 2009/125/EC and 2010/30/EU and Repealing Directives 2004/8/EC and 2006/32/EC.

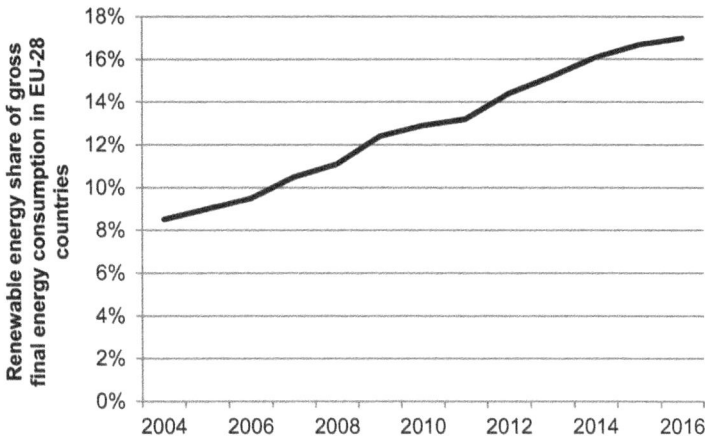

Fig. 10.6 Renewable energy share of gross final energy consumption in EU-28 countries. Source: Eurostat

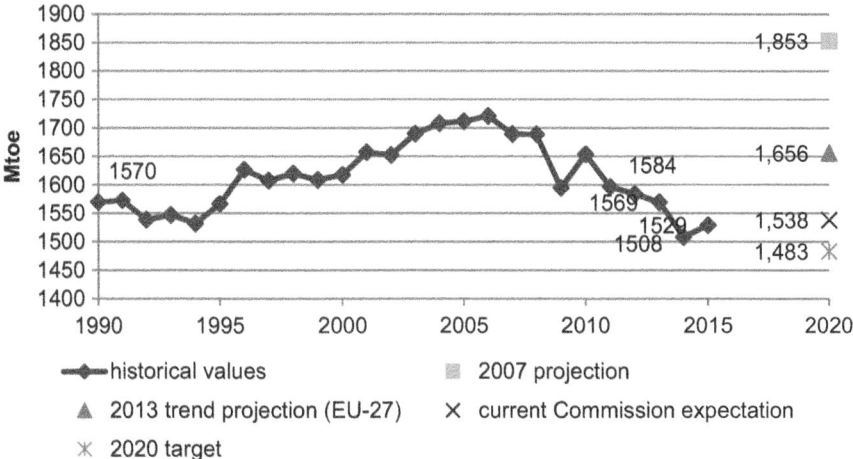

Fig. 10.7 Primary energy consumption in EU-28 and projections to 2020. Source: Kemfert et al. (2014, 24) and EEA. 2017. "Trends and Projections in Europe 2017 – Tracking Progress towards Europe's Climate and Energy Targets." 17/2017. Luxembourg: European Energy Agency

efficiency to date is noteworthy but does not go far enough (see Fig. 10.7) and varies across the Member States: In Italy, France, and Spain, relative improvements since 2001 are below the EU average, whereas Germany and the UK recorded above-average improvements. Poland's energy productivity is relatively low but has increased significantly since 2001 (see also Chap. 6).

If the EU does not increase its efforts, it will fail to meet the target of a 20% reduction in primary energy consumption by 2020 compared to the reference scenario (see Fig. 10.8). According to a 2013 trend projection, unless the Energy

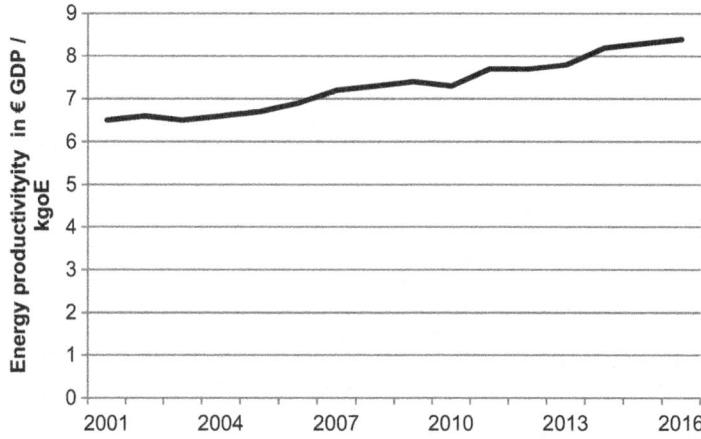

Fig. 10.8 Development of energy productivity in EU-28 countries. Source: Eurostat

Efficiency Directive is successfully implemented, a maximum reduction of only 10% would be achieved. Currently, the Commission anticipates that the targets for energy efficiency could only be reached, if the decreasing trend from 2005 would continue.[24] However the recent increase in primary energy consumption between 2014 and 2016 makes it unprobable that the 2020 target will be reached. Energy savings achieved up to 2014 were not only a result of energy efficiency improvements but partly also due to the economic crisis. Primary energy consumption in 2012 was at around the same level as in 1990.

Going forward, the European Council has agreed on a 40% efficiency target for 2030. After a revision of the first Energy Efficiency Directive in 2014, the decision was made to raise the objective. Thus, in the run-up to 2030, a concerted effort is needed to make significant progress on energy efficiency targets and work toward achieving a sustainable energy system.

---

[24]EC. 2017. "2017 Assessment of the Progress Made by Member States towards the National Energy Efficiency Targets for 2020 and towards the Implementation of the Energy Efficiency Directive as Required by Article 24(3) of the Energy Efficiency Directive 2012/27/EU." COM (2017) 687 final. Brussels, Belgium: European Commission. 2017 assessment of the progress made by Member States towards the national energy efficiency targets for 2020 and towards the implementation of the Energy Efficiency Directive as required by Article 24(3) of the Energy Efficiency Directive 2012/27/EU.

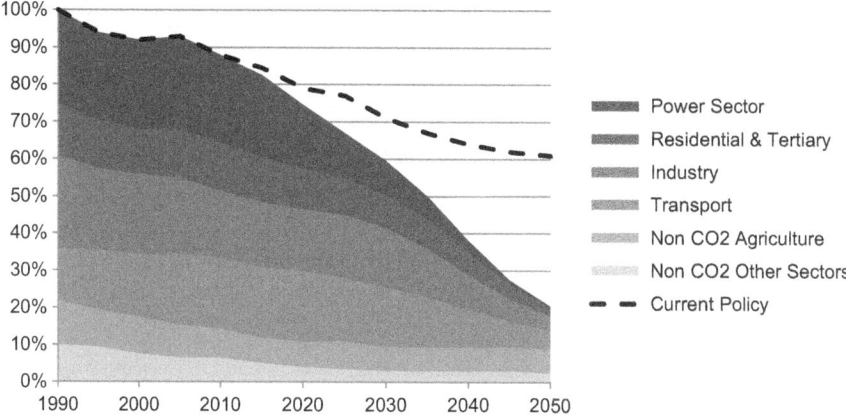

Fig. 10.9 Decarbonization of the European energy system, according to the 2050 Roadmap. Source: EC (2011c)

10.2.2.5 The Longer-Term Perspective to 2050: Far-Reaching Decarbonization

The European Commission defines the long-term perspective for the decarbonization of its energy system in its Energy Roadmap 2050. Its vision can be summarized as follows: it aims at the decarbonization of the European energy system through a GHG emissions reduction of 80 to 95% compared to 1990 (EC 2011d). Figure 10.9 presents what has become the most recognizable graph in EU energy and climate policy based on reference scenario calculations. It shows the GHG emission reductions under a "current policy" scenario of existing energy policies in the range of only 40%. However, if one takes the greenhouse gas targets for 2050 seriously, at least 80% if not 95% would have to be attained. In this scenario, as Fig. 10.9 clearly shows, the power sector needs to achieve full decarbonization because it has the lowest abatement costs. Transport, industry, and agriculture, on the other hand, maintain a certain GHG budget through 2050. The following three sections therefore focus on the three options available for the decarbonization of the European electricity sector: coal with CO_2 capture, nuclear, and renewables.

Following on the integrated energy and climate package of 2009, the European Commission undertook further efforts to reach firm consensus on a very long-term perspective: 2050. After the proposal of longer-term roadmaps by Member Countries such as the UK, the passage of legislation on climate change, and the German Energy Concept (BMWi and BMU 2010), the European Commission developed a series of roadmaps for 2050.[25] The three roadmaps, released between 2010 and

---

[25]Meeus et al. (2012) provide a comparison of the energy roadmap in relation with other policy documents on the same topic.

2011, formulate climate objectives in different areas: a Climate Roadmap[26], an Energy Roadmap[27], and a Transport Roadmap[28].

10.3   Coal and the Pervasive Search for Carbon Capture, Transport, and Storage (CCTS)

10.3.1   Coal-Producing Member States Seek Coal-Based "Decarbonization"

Not only German but also a great deal of European wealth has been produced through coal-powered manufacturing. The British industrial revolution would not have happened without coal, and the same holds for industrialization in France, Belgium, Germany, Poland, and other countries, in somewhat later periods. Central and Eastern Europe also has extensive coal resources. Countries such as Romania, Hungary, former Czechoslovakia, and Estonia (oil shale) built up a strategic supply during socialist times. Clearly, coal (and other fossil fuels) still constitute an important part of the European energy system.

It therefore comes as no surprise that many EU Member States considered the ambitious decarbonization policies of the European Union as a direct threat in the 2000s; this was particularly the case for the new EU Member States from Central and Eastern Europe. While the UK, France, Germany, and others could stretch the decline of their respective coal industry out over more than half a century, Central and Eastern European countries could not. They were faced with a dual challenge after the end of the Cold War and following the integration of the European energy markets: Their low-value, labor-intensive coal resources were no longer competitive, and climate objectives had become much tighter. This led them to resist ambitious climate targets.

The solution to this problem not only in Eastern Europe but also across the entire EU appeared in the form of a new technology, carbon capture, transport, and storage (CCTS), with the apparent potential to achieve the two different goals of meeting climate targets while maintaining a healthy coal industry. Debates about CCTS emerged parallel to EU enlargement (in 2007) with the IPCC Special Report on CCTS (IPCC 2005). During the second half of the last decade, there were high hopes that CCTS would enable coal use to continue. This hope also explains the important

---

[26]EC. 2011. A Roadmap for Moving to a Competitive Low Carbon Economy in 2050. COM(2011) 112. Brussels, Belgium: European Commission.

[27]EC. 2011. "Energy Roadmap 2050." Communication from the Commission to the European Parliament, the Council, the European Economic and Social Committee and the Committee of the Regions. Brussels, Belgium: European Commission.

[28]See EC. 2011. Roadmap to a Single European Transport Area – Towards a competitive and resource efficient transport system. White Paper COM(2011) 144, Brussels, Belgium: European Commission.

role of coal and natural gas in the Energy Roadmap 2050, which sought a far-reaching decarbonization of the energy system and an almost complete decarbonization of the power sector (EC 2011d). The Reference Scenario following the 2009 Climate and Energy Package, which formed the basis of the Energy Roadmap 2050, still reflected the enthusiasm: Expected CCTS power plant capacity increased from zero GW to more than 100 GW by 2050 in the base scenario, while in other scenarios, the corresponding figure was as high as 193 GW ("diversified supply technology scenario") (EC 2011b, 24). Even in a scenario where the availability of the technology was delayed, the capacity of CCTS power plants was still expected to be 148 GW by 2050 (EC 2011b, 24).[29]

10.3.2 No CCTS Demonstration Projects to Date

After sharing some of the initial enthusiasm, the authors of this chapter identified a widening gap between the high hopes in CCTS and the meager results. We developed several hypotheses about what we have called a "lost decade" for CCTS (von Hirschhausen et al. 2012a). Among the possible explanations were incumbent resistance to structural change, wrong technology choices, over-optimistic cost estimates, a premature focus on energy projects instead of industry, and the underestimation of transport and storage issues. The following two subsections summarize these trends. We also explain why CCTS is no longer an option for decarbonization in Europe, and argue that alternatives must be sought.

Since the emergence of "clean coal" technologies using CCTS in the early 2000s, the high hopes in these technologies have been dashed. In fact, the optimistic development scenarios described above run contrary to current developments: there are still no production chains on a demonstration scale anywhere in the world where carbon is captured in power plants, transported downstream, and then stored permanently underground.[30] Despite efforts in some countries to develop pilot projects over the last decade, there have been no significant successes. In continental Europe, all demonstration projects so far have been canceled or postponed indefinitely. In Germany, both industry and policymakers have dropped their plans for the large-scale industrial implementation of CCTS technology as part of the energiewende (von Hirschhausen et al. 2012a, b).

Table 10.1 provides a list of (failed) CCTS-projects in Europe, indicating the large discrepancy between the initial hopes and realities. The only two operating

---

[29]The updated 2013 EU Reference Scenario was more conservative about CCTS but still forecasted 38 GW of installed CCTS capacity in 2050 (EC 2013). This number was even further reduced in the 2016 EU Reference Scenario with 17 GW of installed CCTS capacity in 2050 (EC 2016, 66). In this EU Reference Scenario the introduction of CCTS in the EU is supposed to be based on three demonstration plants (White Rose, UK; Peterhead, UK; ROAD, NL), that were assumed to be running by 2020/25. However, all three named projects were canceled by the year 2017, see below.

[30]At the Boundary Dam Integrated Carbon Capture and Storage Project (CA) captured CO_2 is used for EOR. Only surplus CO_2 not needed for EOR is stored in the research storage project Aquistore.

Table 10.1 CCTS projects in Europe. Sources: Adapted from Oei et al. (2014), including information from GCCSI (2014) and GCCSI (2017)[a]

Project	Jänschwalde	Porto-Tolle	ROAD	Belchatow	Compostilla	Don Valley	Killingholm (C-GEN)	Longannet Project	Getica	ULCOS	Green Hydrogen
Country	DE	IT	NL	PL	ES	UK	UK	UK	RO	FR	NL
Technology	Oxyfuel	Post	Post	Post	Oxyfuel	Pre	Pre	Post	Post	Post	Pre
Storage	Aquifer	Aquifer	Oil-/gasfield	Aquifer	Aquifer	EOR	Aquifer	EOR	Aquifer	Aquifer	EGR
Capacity [MW]	250	250	250	260	320	650	450	330	250	Steel	H_2
Plan in 2011	2015	2015	2015	2015	2015	2015	2015	2015	2015	2016	2016
Status in 2018	Canceled 2011	Canceled 2014	Canceled 2017	Canceled 2013	Canceled 2013	Canceled 2015	Canceled 2015	Canceled 2011	Canceled 2014	Canceled 2012	Canceled 2012

	White Rose (UK Oxy)	Peel Energy	Peterhead	Teesside (Eston)[b]	Eemshaven	Pegasus	Maritsa	Mongstad	Caledonia Clean Energy[c]	Norway Full Chain CCS
Country	UK	UK	UK	UK	NL	NL	BG	NO	UK	NO
Technology	Oxyfuel	Post	Post	Various	Post	Oxyfuel	Post	Post	Post	Various
Storage	Aquifer	Oil-/gasfield	Oil-/gasfield	Aquifer	EOR	Oil-/gasfield	Aquifer	Aquifer	Aquifer/EOR	Aquifer
Capacity [MW]	430	400	400	0.8 Mtpa	250	340	120	630	3 Mtpa	1.3 Mtpa
Plan in 2011	2016	2016	2016	2016	2017	2017	2020	2020	–	–
Status in 2018	Canceled 2016	Canceled 2012	Canceled 2015	Mid 2020s	Canceled 2013	Canceled 2013	Canceled 2013	Canceled 2013	2024	2022

[a]Further sources about individual projects available online
[b]Power plant with CCTS canceled in 2014, now industrial park collective
[c]Formerly Captain Clean Energy

large scale CCTS projects in Europe are at the natural gas production facilities Sleipner and Snøhvit in Norway (GCCSI 2017). At both sites, the produced natural gas has a high CO_2 content which is reduced in the processing facilities and the separated CO_2 is captured and injected into offshore geological formations. It is unlikely that other project ideas will change this trend until the 2020s: The Norway Full Chain CCTS demonstration project is supposed to capture CO_2 at a cement plant and possibly at a waste-to-energy facility and store the CO_2 in an off-shore aquifer, but an investment decision is not expected before 2020. In the UK Caledonia Clean Energy considers a new natural gas power plant (up to one GW) with CO_2 capture (ca. 3 Mtpa), with the CO_2 to be stored in offshore geological formations; Teesside Collective (UK) consists of a number of energy intensive companies which are investigating the possibilities for capturing CO_2 in their industrial zone. The CO_2, approximately 0.8 Mtpa, is supposed to be stored in offshore geological formations (GCCSI 2017).

The discouraging prospects of CCTS technology in Europe were confirmed in a Commission Communication on the future of carbon capture and storage in Europe.[31] The Commission noted that all efforts to date, despite having been offered lucrative financial support, have not led to the construction of a single demonstration plant. They assigned the blame for this to both the energy industry itself and the relatively unambitious policies of Member States. The Communication also noted that of all the planned demonstration projects to date, not one has taken the planned development path. Furthermore, there was little chance of a demonstration power plant being built any time soon.

The difficulties of CCTS are not limited to Europe but span the entire globe (Reiner 2016). Neither in emerging economies such as India or China nor in industrialized countries such as the USA, Canada, or Australia has any significant progress been achieved. There are only two power plants in the world that have succeeded in demonstrating carbon capture at scale, although with no or little (experimental) storage: one is the Boundary Dam Integrated Carbon Capture and Storage Project of SaskPower (Canada), whose operation started in October 2014;[32] the other one is Petra Nova Carbon Capture (USA) that started operating in January 2017.[33]

---

[31]See EC. 2013. The Future of Carbon Capture and Storage in Europe. COM/2013/0180 final, Communication from the Commission to the European Parliament, the Council, the European Economic and Social Committee of the Regions on the Future of Carbon Capture and Storage in Europe, Brussels, Belgium: European Commission.

[32]The major share of captured CO_2 is used for EOR, while a minor share is stored at an experimental geological storage site http://sequestration.mit.edu/tools/projects/boundary_dam.html. In the eyes of the operator, this constitutes sufficient reason to continue burning coal: "Through the development of the world's first and largest commercial-scale CCS project of its kind, SaskPower is making a viable technical, environmental and economic case for the continued use of coal." (http://saskpowerccs. com/ccs-projects/boundary-dam-carbon-capture-project/, downloaded April 1, 2016).

[33]At Petra Nova the captured CO_2 is used for EOR: https://www.nrg.com/case-studies/petra-nova. html, last accessed May 07, 2018.

10.3.2.1 Large Growth of CCTS in EU Scenarios Not Plausible

In a first analysis of European decarbonization scenarios, von Hirschhausen et al. (2013) were among the first to criticize the exponential development of CCTS in the EU Reference Scenarios and other documents. The findings showed that the significant increase of CCTS in the EU Reference Scenario in the Energy Roadmap 2050 (over 100 GW by 2050) was not the result of an optimization model but rather that the model was deliberately calibrated to produce such results. The main levers to achieve this were (1) the low estimates of capital costs, and (2) the assumption of positive and significant learning rates.

In fact, given that there has been no successful demonstration of large scale CCTS technology at any power plant with downstream carbon transport and storage, all cost estimates are speculative, and long-term cost forecasts in particular need to be made with serious caution. The capital cost of a CCTS power plant were generally estimated at €2100 to €3500/kW (EC 2011d). Irrespective of the selected carbon capture technology (post-combustion, pre-combustion, or oxyfuel), efficiency decreases by 21 to 33% compared to the reference power plant without carbon capture due to the additional energy demand. Overall, the carbon capture stage alone leads to an increase in power generation costs of 50%.[34] The cost reduction potential of this part of the technology chain is estimated to be very low. In addition to carbon capture costs, there are also costs of transport and storage. For a large-scale deployment of CCTS technology as envisaged in the scenarios in the Energy Roadmap 2050, a carbon transport network of many thousands of kilometers of pipeline would be required due to the distances between the emission sources and potential carbon storage sites.

In terms of future cost developments, it is unclear whether CCTS technology would have positive or negative learning rates. Analogous developments in other technologies would suggest positive learning rates, that is, a gradual decrease in the average cost of power generation. However, negative learning rates are also plausible, as is the case with nuclear energy, which would lead to cost increases. Already in 2010, researchers at Stanford University highlighted the risk that the positive learning effects expected for CCTS could in fact fail to materialize (Rai et al. 2010).

The rather inflexible mode of operation of CCTS power plants is likely to drive costs further upward. Given the increasing demand for flexibility of fossil fuel power plants in the context of the increasing share of supply from fluctuating renewable energies sources such as solar and wind power, even adjusted cost estimates may be too low because current calculations for the sensitive thermodynamic and chemical processes of carbon capture are designed for continuous base load operation.

---

[34]EASAC. 2013. "Carbon Capture and Storage in Europe." EASAC policy report 20. Halle (Saale), Germany: German National Academy of Sciences. The Crown Estate, Carbon Capture & Storage Association, and DECC. 2013. "CCS Cost Reduction Taskforce – Final Report." London, UK: UK Carbon Capture and Storage Cost Reduction Task Force.

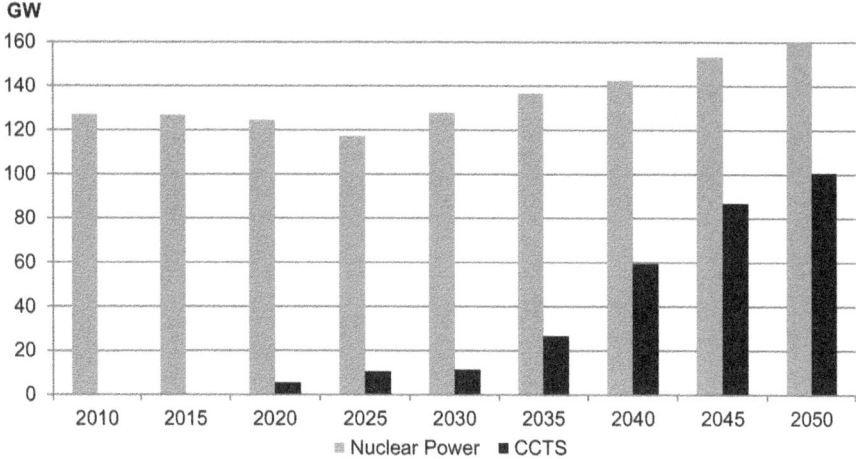

Fig. 10.10 Installed capacity of nuclear power and CCTS according to the 2011 Energy Roadmap (in GW). Source: EC (2011a)

Increasing the flexibility of CCTS power plants can only be achieved with significant cost increases (Rubin and Zhai 2012) (Fig. 10.10).

In addition to the underestimation of capital costs and failure to consider the transport and storage costs, another outdated assumption was made in the model of the European Reference Scenario: a very high figure of 5.4 GW of generating capacity for CCTS was hypothesized for 2020. This figure assumed the successful implementation of all CCTS projects that have applied for funding under the European Economic Program for Recovery.[35] The Energy Roadmap also assumed significant learning rates beyond the year 2020. Given the presumed high growth rates, specific investment costs were expected to fall to €2064/kW by the year 2020, which might have generated additional capacity at CCTS power stations; this additional capacity would further reduce investment costs, so at the end of the observation period in 2050, the price per kilowatt would be down to €1899 and installed CCTS power plant capacity would exceed 100 GW (Fig. 10.11).[36] The updated Reference Scenario 2016 still calculates with 17 GW of installed CCTS capacity the year 2050 (EC 2016, 66). The introduction of CCTS in the EU was

---

[35]See EC. 2009. List of 15 energy projects for European economic recovery. MEMO/09/542, Brussels, Belgium: European Commission.

[36]The estimates for capital costs and the timing of the CCTS roll-out were slightly modified in the subsequent Reference Scenario 2013, but this had little impact on the overall trend in the scenario, see Kemfert et al. (2014).

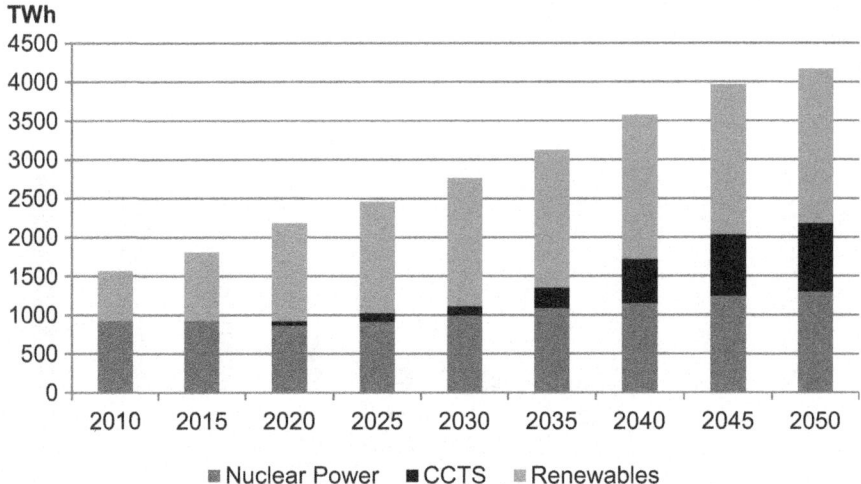

Fig. 10.11 Gross electricity production from nuclear power, CCTS, and renewable energy sources according to the 2011 Energy Roadmap (in TWh). Source: EC (2011a)

supposed to be based on three demonstration plants (White Rose, UK; Peterhead, UK; ROAD, NL), that were assumed to be running by 2020/25.

Reality looks very different at the end of the decade, as laid out above. The new EU initiatives are likely to end up similarly effectless as those of the early 2010s. Although the possibility of one day having CCTS should not be ruled out entirely, it is very unlikely that this technology will contribute significantly to the low-carbon transformation in Europe.

10.4   Nuclear Power in the European Electricity Mix

10.4.1   Development of Nuclear Power in Europe

As noted above, the European Community was born with the signing of a treaty on nuclear power: the EURATOM treaty of 1957. At that time, the race for technological supremacy had begun between the USA and the Soviet Union, as well as a few other countries, over the military and civil use of nuclear power. In fact, the commonly held idea that nuclear power is cheap—one even shared by some experts—is not based on empirical evidence. Rather, it was propagated to serve the political objectives of the USA and the prospective European nuclear powers of the 1950s whose aim was to monitor civilian and military use of nuclear power worldwide. In his historic "Atoms for Peace" Speech to the United Nations General Assembly on December 8, 1953, then President of the United States of America Dwight D. Eisenhower proposed the idea of collective management of radioactive material under the supervision of an international authority. The International Atomic Energy Agency (IAEA) in Vienna was subsequently founded to prevent the misuse of fissionable material to build atomic bombs. The now widely accepted notion of cost-effective nuclear power was advocated at that time as a basis for fruitful cooperation.[37] However, Atoms for Peace soon showed visible signs of failure, since neither the Soviet Union nor the emerging countries (such as India and Pakistan) had any intention of complying with the proposed division of labor. Along with the UK and France, which forged ahead with military and civilian use parallel to the USA, the Soviet Union launched its own nuclear program and worked steadily on it throughout the Cold War. In other countries, too, the military and civilian use of nuclear power has been introduced, for instance in India, Pakistan, and China.

In Europe, too, the concept of cost-effective nuclear power was associated with the objectives of political cooperation and economic development. The Treaty establishing the European Atomic Energy Community (Euratom) signed in Rome in 1957 was therefore intended to promote international cooperation on atomic energy as a basis for modernization and industrialization.[38] As shown in Fig. 10.1 above, nuclear power started to emerge in Europe the 1960s, and increased significantly in the 1970/1980s. After attempts to develop their own national technology, most countries reverted to reactors licensed by the USA or Soviet Union: in Western Europe, most countries adopted the General Electric boiling water reactor (BWR) or the Westinghouse pressurized water reactor (PWR); only the UK continued to build a model of its own design, a gas-cooled, graphite-moderated reactor. In Eastern Europe, reactors were built using Soviet-style technology (See Wealer et al. (2018) for an analysis of the technology developments and the worldwide diffusion patterns of nuclear power from 1951–2017).

In the early years of nuclear power, little thought was given to the back-end of the plant cycle, that is, the phase of dismantling the facilities and storing the nuclear waste. Although operators and/or states made some provisions for future costs, no political effort was undertaken to adapt organizational models to the specifics of nuclear power production. Added to this was a general failure to account for long-term economic costs, which were simply ignored or denied. In fact, in the traditional economic analysis of nuclear power, the costs of decommissioning plants and disposing of nuclear waste are generally discounted and matter very little in the investment decision; a typical example of this kind of analysis is (D'haeseleer

---

[37]See Lévêque (2014, 172): "Who can doubt, if the entire body of the world's scientists and engineers had adequate amounts of fissionable material with which to test and develop their ideas, that this capability would rapidly be transformed into universal, efficient, and economic usage."

[38]The signatories even wrote in the preamble of the contract that this had been concluded "...recognising that nuclear energy represents an essential resource for the development and invigoration of industry and will permit the advancement of the cause of peace..." and "...desiring to associate other countries with their work and to cooperate with international organizations concerned with the peaceful development of atomic energy...." (EURATOM Treaty 1957).

2013).[39] This explains why today, almost 70 years later, no European country has a storage site available for its high-level radioactive waste (Brunnengräber et al. 2015).

10.4.2 Costs of Nuclear Energy Prohibitively High

10.4.2.1 Private Costs High and Generally Rising

There has been intensive debate over the economics of nuclear power and whether nuclear has the potential to one day become an economically viable option. When evaluating the economic viability of nuclear power, distinctions must be drawn between private and social operational costs, with the latter also including environmental effects and technical risks, and between short-term and long-term costs, the latter including investment, insurance, and so on. Construction of a nuclear power plant may be worthwhile from a microeconomic perspective—for instance, from an investor's point of view—as long as the government or the energy customers bear a large share of the private and the social costs. Operation of an existing nuclear power station can be profitable, provided that the government takes responsibility for the safety risks—the cost of which cannot be calculated—as well as for dismantling, final disposal of nuclear waste, and investment in R&D.

However, a look at the history of nuclear power indicates that even when reducing the analysis to private costs, private investment has never been forthcoming (Davis 2012; von Hirschhausen 2017). When taking into account the operational risks and the immense costs of R&D, dismantling of power plants, and final disposal of radioactive waste, this form of energy has never been economical. Furthermore, to this day, over six decades since the first civilian use of nuclear power, the question of disposal of high-level radioactive waste remains unresolved.[40] In view of still unresolved technical and institutional problems such as safety and final disposal, as well as the primacy of political over economic considerations, it is not surprising that not a single nuclear power plant in the world to date has been fully financed and constructed by private investors under competitive market economy conditions (Wealer et al. 2018). The high requirements for R&D, capital investment, insurance against the risk of nuclear accidents, and final disposal of radioactive waste make nuclear power unprofitable.

---

[39]Two arguments can be lodged against this approach. First, discounting uncertain, and highly dangerous, events in the future is ethically not possible, so that for these events a discount rate of 0% should be used (Schulze et al. 1981). Second, these costs eventually arise, and without making careful provisions, they may pose an existential challenge to nuclear energy companies as currently observed in several European countries such as Germany, France, and Belgium.

[40]The cost of disposing of spent fuel elements is still largely unknown because even after six decades of nuclear energy use, there are no permanent disposal sites anywhere in the world that guarantee the safe storage of nuclear fuel rods for tens of thousands of years.

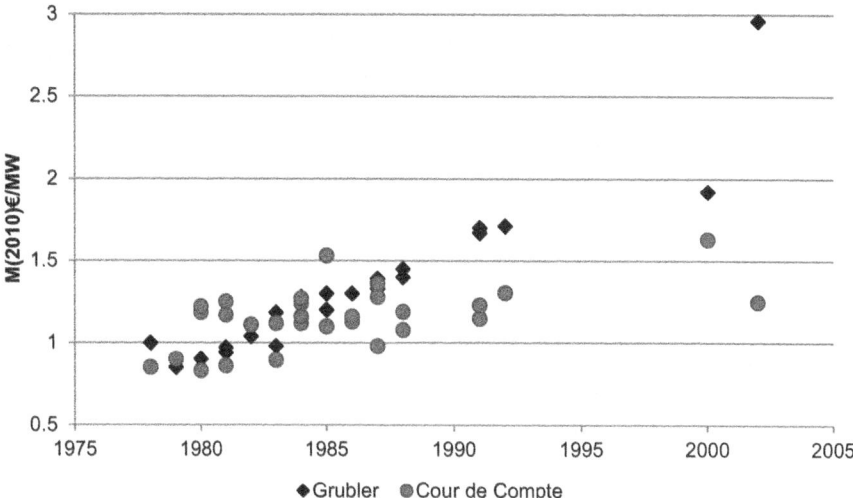

Fig. 10.12 Historically specific investment costs of French nuclear power plants (€mn/MW). Based on 2010 prices. Figures related to "second generation" nuclear power plants. Source: Grubler (2010), and Rangel and Lévêque (2015)

The hope for cheap nuclear energy ("too cheap to meter") turned out to be relatively short-lived. In the 1960s it had already become evident that the technical and economic risks of nuclear power were too high to be borne by the private sector (Radkau and Hahn 2013). Then, in the 1970s and 1980s, the hope for economies of scale and cost reduction vanished as well, with costs rising in all of the nuclear energy countries, in particular in the USA (Cantor and Hewlett 1988) and France (Grubler 2010, and Rangel and Lévêque 2015). In contrast to all other power generation technologies, nuclear has not become cheaper over the decades; its capital costs have increased many times over. For example, the output-specific investment per kilowatt in France in 1980 was approximately €1000/kW; in 1990 it was between €1300–1600/kW; and in 2000 it was between €1500–3000/kW (see Fig. 10.12).[41] In the USA, too, output-specific investment rose significantly from 1973 (approx. US $1000/kW) to 1990 (approx. US$8000/kW in year 2010 USD) (Davis 2012).[42] Past experience of rising capital costs appears to apply to the current stage of development in the "third generation" (European Pressurized Reactor, EPR) of nuclear power plants. The cost estimates for the two nuclear power plants currently under

---

[41]Based on 2010 prices. See Rangel and Lévêque (2015) and Grubler (2010), based on cost data from the French Court of Auditors (Cour de Compte).

[42]The reasons for this were, in particular, changing standards, a lack of continuity in the construction of nuclear power plants, and more stringent safety regulations.

construction in Olkiluoto (Finland) and Flammanville (France) are continually increasing.[43]

Thus, the development of nuclear power in Europe cannot be explained by economic rationality, but by what (Lévêque 2014, 212) identified as the main driving force behind nuclear technology since the 1930s: "nuclear technology was born of science and war." In fact, there simply is no economic case for nuclear and never has been (von Hirschhausen and Reitz 2014). The most plausible explanation for the divergent attitudes and policies on nuclear power within Europe goes back to 1945, when the UK and France, winners of the Second World War and allies of the first atomic power, the USA, decided to develop nuclear power for military purposes, implying the co-development of extensive civil applications as well. Their motives for developing nuclear technology were more than economic. The close ties between military and civil uses of nuclear power as well as an array of political factors in the policy making process—and not economic arguments—were behind the decisions to expand the use of nuclear power. It was a nuclear bomb that brought the Second World War to an end, and the potential civil use of nuclear power was the main argument for developing nuclear technology after the war. This remains a key motivation to pursue the nuclear chain to this day (Lovins and Lovins 1980; Cox et al. 2016; von Hirschhausen 2017).

Another important but often overlooked cost factor is insurance against potential major accidents. The costs of such major accidents at nuclear power plants can be extremely high and difficult to quantify.[44] Currently, due to the very low insurance requirements for nuclear power plant operators, these costs are borne primarily by society, and the government and/or uninsured private citizens bear the cost of risks. Irrespective of the most economically advantageous form of insurance (public, private, or a mix of the two), such costs must be included in the economic calculations.[45]

---

[43]In 2006, the original estimate was €1500/kW. Since then it has risen to €4500/kW (mid-2008) (Thomas 2010), has climbed to €5100/kW (December 2012, see EnergyMarketPrice. 2012. EdF Unveils a Sharp Rise in Costs for Flamanville Nuclear Reactor Construction), and was estimated somewhere in the range of €8000/kW in 2016. Specific reasons for this increase include planning errors, problems with the automatic control systems, and also revised safety requirements, see Reuters. 2012. Finland's Olkiluoto 3 Reactor Delayed Again." July 16, 2012.

[44]See for example Diekmann, Jochen. 2011. "Verstärkte Haftung und Deckungsvorsorge für Schäden nuklearer Unfälle – Notwendige Schritte zur Internalisierung externer Effekte." Zeitschrift für Umweltpolitik und Umweltrecht 34 (2).

[45]The economic viability of nuclear power is also diminished by a further tightening of safety regulations that are currently being developed at the European level. EU Energy Commissioner Günther Oettinger responded to the nuclear accident in Fukushima by recommending the mandatory stress testing of European nuclear power plants, which revealed an urgent need for some to be retrofitted. A draft regulation will form the basis for binding rules on liability and compulsory inspection routines to be introduced in all countries; see EC (2013). Draft proposal for a Directive amending Nuclear Safety Directive IP/13/532. Press Release, Brussels, Belgium: European Commission.

10.4.2.2   UK Hinkley Point Nuclear Power Plant Project as a Turning Point of the Public Debate

The failure of the nuclear power industry to make an economic case for nuclear is well known among specialists, as discussed by Davis (2012) among other authors, but this fact has not become part of the public debate until recently. A turning point in the diverging perceptions can be seen in the huge cost overruns of the 2010s, both from existing plants (Olkiluoto and Flamanville, as described above) and from a new project in the UK, Hinkley Point C. The latter, the first nuclear new build in the UK in the last 30 years, clearly illustrates the absence of economic rationality in the political decision-making on nuclear power. The project entails the construction of a twin-unit power plant using a French-designed EPR nuclear reactor, the first twin unit of this kind to be built on European soil, with a total output of 3200 MW.[46] The Hinkley Point project was developed by a consortium consisting of the French energy companies EdF and Areva and two Chinese state-owned corporations.[47] Initially, EdF estimated the cost of the project at 16 bn GB pounds[48], the equivalent of approximately 5000 GB pounds of specific investment per kW (around €6000–7000/kW) (von Hirschhausen 2017).

Negotiations were held between the government and the French state company EdF over the level of financial security to be received by the latter to build a new third-generation nuclear power plant. It was rapidly becoming apparent that the potential investor was not keen on making market-based investments, and was also calling for a very high price guarantee (Toke 2012). The proposed nuclear new-build project illustrates the enormous volume of overt and hidden subsidies required to construct a new nuclear power station today. The 2015 package negotiated between the British government and the consortium (2015) comprises various direct and hidden subsidies:

- A "strike price" of 92.50 GB pounds for every megawatt hour of power that the reactors generate over a 35-year period, adjusted to inflation[49] (the equivalent of around €120/MWh).[50] Should the power station's output have to be reduced for energy-system-related reasons, the operating company would receive financial compensation which *de facto* equates to a guaranteed minimum payment. In addition, the British government has offered the consortium a credit guarantee

---

[46]The project is an attempt to update the UK's outdated nuclear power stations; in the medium-term, the UK's nuclear program envisages new builds with a total output of 16 GW (HOC-ECCC 2013).

[47]The two state-owned companies are China General Nuclear Power Group (CGN) and China National Nuclear Corporation (CNNC).

[48]Clercq, Geert De, and Karolin Schaps. 2013. "UK Gives Unprecedented Support to £16 Billion Nuclear Deal." Reuters UK. October 21, 2013.

[49]DECC. 2013. "Initial Agreement Reached on New Nuclear Power Station at Hinkley." Press release. London, UK.

[50]By way of comparison, this roughly the same as the "strike price" for onshore wind turbines in the UK, but this has only been granted for 15 years.

to underwrite up to ten billion GB pounds of debt on the project at preferential terms. Consequently, investors do not need to rely as heavily on expensive bank loans, which are subject to the relevant risk premiums.[51]

- The British government will also protect the investor from changes in nuclear liability and insurance obligations at the European level.[52]
- Further, discussions are underway as to whether the completion risk will also be borne by the British government (HOC-ECCC 2013).
- Finally, the agreement between the British government and the investors also allows for possible increases of the strike price under certain, as yet unspecified, conditions.

When the agreement between the British government and operator consortium was announced, the European Commission launched a formal investigation into proposed state subsidies for the plant. Its initial statement on the project was highly critical: in it, the Commission contended that there was no proven need for the project from an energy economy perspective since the power stations were unlikely to be operational until the mid- to late 2020s at the earliest, by which time the imminent excess demand was likely to have diminished again.[53] However, bowing to pressure from the UK government, the Commission finally proceeded to a political vote, in which they decided these conditions would not constitute illegal state aid. The UK has declared it needed the technical competences, and perhaps the spent fuel for producing plutonium, but still has not made any final decision to go ahead with the Hinkley Point project. Whatever happens at Hinkley Point in the coming years, it is clear that nuclear power would come at very high costs to the government and consumers alike, and that a potential project at the site would not serve as either a positive example or a signal to any other country pondering the nuclear option, but rather quite the opposite.

10.4.3 Renaissance of Nuclear Power in EU Scenarios Not Plausible

Given the lack of an economic case for nuclear and the skyrocketing costs of recent new-build projects, how can the "renaissance" of nuclear power in almost all EU scenarios be explained? One must recognize, first, that there is no longer any serious discussion on the global level of a "renaissance" of nuclear power. Nuclear power is currently being used for energy production primarily in Western industrialized

---

[51]Gosden, Emily. 2014. "Nuclear Setback as EC Attacks Hinkley Point Subsidy Deal." The Telegraph, January 31, 2014.

[52]The European Commission is planning to make liability insurance mandatory after the Fukushima accident.

[53]See EC. 2013. State aid SA. 34947 (2013/C) (ex 2013/N) – United Kingdom Investment Contract (early Contract for Difference) for the Hinkley Point C New Nuclear Power Station, p. 18.

countries, in post-Soviet states, in Japan and Korea, as well as in the emerging countries China and India. The oldest nuclear power park in the world is located in North America. After two surges in growth following the oil crises of the 1970s, the American nuclear construction boom ended under the shadow of the Three Mile Island and the Chornobyl disaster of 1986; the last reactor to be built was Watts Bar 1 in Tennessee, which came online in 1996. Asian countries, however, have continued to regularly construct nuclear power plants. In the first decade of this century, there was a general expectation of a global "nuclear renaissance," both in the Western world and in the emerging countries (Joskow and Parsons 2012, 201). The nuclear accident in Japan in March 2011, as well as the economic disaster of nuclear power plants in recent years, dashed all hopes of such a renaissance.

The chances of a nuclear renaissance in Europe at this stage are very low. Across the continent, there is growing consensus that nuclear energy is not cost-effective, and a general understanding that both the open and hidden risks are high. In addition to Germany, Belgium and Switzerland have also opted for a nuclear phase-out, and Italy voted in a 2011 referendum against plans to revive its nuclear program. Lithuanians rejected a proposal to build a nuclear power plant in cooperation with the other Baltic States and Poland; the project is now on the brink of being abandoned. In 2012, the Bulgarian government halted construction of the Belene nuclear power plant with its two planned reactor units after only sporadic efforts at implementation since the 1980s. Similarly, in 2013, a Slovakian court rescinded the building permits for two nuclear reactors. Among the countries that still have plans for new builds, neither Poland (with new-build plans for up to three plants, each with a capacity of 1.6 GW by 2030,[54] nor Hungary, where the government is currently negotiating with the Russian Rosatom corporation over a new power station to replace the one in Paks, have the financial means or the technical skills to develop their own nuclear chain, and both countries would need to rely on (subsidized) technology from abroad.[55]

How, then, can the lack of economic competitiveness be reconciled with the fact the EU Reference Scenario still includes nuclear power in the future electricity mix and even forecasts a "renaissance" in the 2030/2040s? The explanation is quite simple: The EU does not calculate "economically" optimal generation mixes but calibrates the model to represent political preferences of the Member States, then adds these together to provide a European aggregate. In such a behavioral approach, cost estimates and other parameters are chosen to obtain the politically preferred results. The 2011, 2013, and 2016 EU Reference Scenarios, respectively, illustrate

---

[54]Polish Ministry Of Economy. 2014. "Polish Nuclear Program: Program Polskiej Energetyki Jądrowej." Warsaw, Poland: Ministerstwo Gospodark.

[55]Particularly in the wake of the nuclear disaster in Fukushima, the European Commission has been striving to improve safety standards and liability conditions, although, in accordance with the Euratom Treaty, the oversight of nuclear power plants remains the responsibility of the individual Member States. In an initial step, in 2011, all nuclear reactors were subject to a "stress test" and safety provisions were reviewed. As a result, virtually all nuclear power stations would have to be upgraded at a cost of approximately €25 bn for the 132 reactor units investigated.

this process. In fact, these scenario analyses conclude that nuclear power is cost-effective and depict it as a key pillar of power supply in the run-up to 2030 and 2050. This applies to both the Energy Roadmap (for moving to a low-carbon economy in 2050) and the December 2013 Reference Scenario for 2030/2050 (EC 2013). The Reference Scenario, which serves as the basis for the White Paper (policy framework for climate and energy in the period from 2020 to 2030), forecasts that nuclear power capacity for 2030 will be similar to today's levels, although from a current level of 125 GW, the capacity of existing nuclear power stations is forecasted to fall on a level of 97 GW by 2025 and to rise again to 142 GW by 2050.

The Reference Scenario envisages an "economic" situation of nuclear energy that is implausible with regard to the investment costs and resulting economies of scale and the assumed investment decisions, and that also fails to take insurance, dismantling, and final disposal costs adequately into account. Previous cost escalations show that the cost estimates in the Energy Roadmap 2050 as well as subsequent EU documents, which assume comparatively low costs and high competitiveness for the technology, are unrealistic. Third-generation nuclear power plants currently under construction in Finland and France require an investment of approximately €6000–10,000/kW, which includes construction, decommissioning, disposal, and completion risk costs. Based on past empirical evidence of increasing safety requirements, this generation of power plants is not likely to see falling costs (at best one can assume constant capital costs). In addition, there are variable operating and maintenance costs of about €20–25/MWh. Even these figures, which correspond to an average cost of €109/MWh, show that nuclear energy is comparatively expensive.[56] There are substantial risk costs that are largely borne by the general public. The cost calculations in the Energy Roadmap 2050 are significantly lower: First, the starting value for the year 2010 is only €4382/kW, and second, it assumes significant cost reductions for the coming decades—both of which are contradicted by the experiences described above. These facts explain the surprising and systematic calculation of additional nuclear power capacity in the Reference Scenario of the Energy Roadmap.

Despite slight revisions between 2011, 2013, and 2016 the assumptions made in the Reference Scenario regarding investment costs remained overly optimistic. These figures were revised slightly upwards in the Reference Scenario compared with the 2050 Roadmap (€4350/kW versus €3985/kW). In the 2016 reference scenario, the investment costs for the third generation reactors were increased by a third but no specific figures are mentioned; refurbishment costs were also increased but with no indication on the magnitude (EC 2016). The algorithm for power plant expansion used in the Commission's model also ignores the risk-induced investment costs of private investors and therefore significantly underestimates the actual financing costs, which are of prime importance for capital-intensive technologies

---

[56]This value is calculated assuming a lifespan of 40 years, an interest rate of 10%, and a capacity factor of 83.3%. If a capacity of 50% is assumed, which may be quite realistic in a future with increasing feed-in from renewable energy sources, this figure increases to €165/MWh.

such as nuclear power. The model also fails to factor in regulatory risks, which in the private sector are of considerable significance and de facto reduce interest in capital-intensive and risky investment in nuclear power plants. In Sect. 10.6 below, we compare scenario results using alternative cost assumptions.

10.5   Renewables: Potential Underestimated

10.5.1   Increase in Renewables Since the 1990s

The idea that variable renewables could contribute to electricity supply goes back to about a century ago. After the Second World War, the development of the nuclear industry occurred in the broader context of ideas about a "solar economy"; after all, solar energy is generated by nothing other than nuclear fusion on the surface of the sun. In the 1950s and 1960s, popular and influential publications appeared touting the potential of (large-scale) solar energy. This enthusiasm was not followed by consistent R&D efforts or by any demonstration projects.

The emerging climate and environmental movement of the 1970s and 1980s "rediscovered" the potential contribution of renewables. They were identified as the key element of a "soft path" to energy policy and as an alternative to fossil electricity (Lovins 1976). Some pilot projects were developed, mainly as by-products of space missions, with an experimental decentralized supply of solar electricity.

At the EU level, policies were formulated in the 1980s leading to the 1990s regulation on renewables. The objective was to encourage the inclusion of some forms of renewables—not only hydro but also sun and wind—in Member States' electricity portfolios. Germany, for example, increased its renewables target for electricity from below 5% (1980s) to 15% based on the EU Directive. Figure 10.13 shows the development of renewable electricity (excluding hydroelectricity) in Europe between 1990 and 2015 in absolute terms (left axis) and relative terms (right axis). It clearly shows the increasing impact of the 2008 Directive, which was part of the 20-20-20 energy and climate package, setting an overall target of 20% of renewables in energy consumption by 2020. As described above, this target was raised to 27% for 2030.

10.5.2   The Breakthrough of Renewables is Now Occurring on the Global Level. . .

The 2010s have seen renewables go from niche player to the most dynamic and—from a social welfare perspective—most economic source of electricity. This trend is undeniable when considering the social costs of alternative energy carriers like coal and nuclear, and the sharp increase in the competitiveness of renewables, in

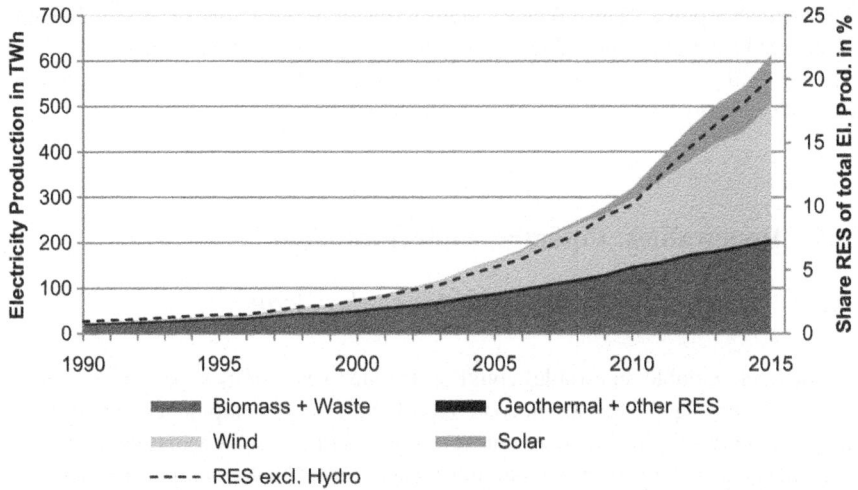

Fig. 10.13 Development of renewable energies in the EU-27 (1990–2015, without hydro). Source: Own depiction, based on data from EIA. 2018. "International Energy Statistics." U.S. Energy Information Administration. 2018

connection with available storage solutions, is also opening up private business models. International organizations and the IPCC have for a long time ignored this trend, as reported, amongst others, by Metayer et al. (2015), and Mohn (2018); they are now gradually turning around and recognize renewables as a key pillar of future energy supply, as expressed in the IEA World Energy Outlook 2015 (IEA 2015), or observed by Creutzig et al. (2017).

The case of photovoltaics is the most spectacular. The costs of this technology have declined significantly as efficiency has increased and plant costs have fallen. This has led to significantly lower average costs of photovoltaic power. Given some excess production capacity, the price pressure on photovoltaic modules, which make up the majority of total costs, has continued to rise.[57] Many studies point to annual cost reductions of 15% since 2008 (Wirth 2016; see also Grau et al. 2011). Unlike other technologies, learning rates in photovoltaics over the last few years have remained stable at 15 to 20%. This means that the specific costs have been falling by 15 to 20% when installed capacity is doubled. It can be generally assumed that this trend will continue for the foreseeable future. Thus, between 2013 (70 GW) and 2015 (150 GW), the installed capacity of solar photovoltaics worldwide has once again doubled. While numerous studies in the mid-2000s still assumed specific investments at around €3000/kW$_p$ (kW-peak), in 2015 the figure of €700/kW$_p$ for large-scale systems including installation costs was more realistic (Löffler et al.

---

[57]The cost of photovoltaics is made up of module costs, inverter costs, installation, maintenance, and area, also known as the "balance of system" (BOS). Module costs make up about 50% but are following a downward trend given the rapidly falling specific module prices.

2017). Depending on weather and climatic conditions, this translates into levelized cost of electricity generation (LCOE) below €5 c/kWh, thus undercutting many fossil generators even when external costs are ignored.[58]

As is the case with photovoltaics, the field of onshore wind turbines has also seen significant production increases and cost reductions in recent years. Most scenarios still assume possible cost reductions in the future. Different studies identify learning rates ranging from 5 to 15% (Pahle et al. 2012); however, these are likely to decline over time.[59] While investors had to raise more than €2000/kW in the early 2000s, specific investments have since fallen below half that. Furthermore, recent experience with different types of wind turbines has shown that it is possible to decrease the average production costs of wind power when using optimized turbine designs, even when specific investment costs remain constant. By adapting the design of the generator, rotor length, and mast height to locally prevailing wind conditions, significant gains in yield can be achieved. A lower specific capacity installation can lead to lower specific power generation costs. A smaller design also results in lower grid connection costs, since the required cable size decreases. Greater turbine utilization leads to a reduction in system costs (Molly 2012).

Another technological breakthrough can be observed at the level of storage technologies required to balance the intermittency of variable renewables such as solar or wind. Several storage solutions exist, and are technically proven, such as lead or lithium-ion batteries, or power-to-gas for longer-term storage (up to a few weeks); many of them are also becoming economically viable (Zerrahn and Schill 2015). There is a clear downward trend in the case of lithium-ion batteries (Burandt et al. 2018): Following first test applications in cars (Tesla), airplanes, and others, mass production was launched, leading to a cost decline for this technology that was still unexpected five years ago. Depending on the technical parameters and how they are managed, average costs of €3–5 c/kWh energy stored are expected in the 2020s. This puts renewables "baseload," where the intermittency is accommodated by storage, at about €6–8 c/kWh.

10.5.3 ... But Has Been Systematically Ignored in the EU Reference Scenarios

In contrast to other international organizations, the European Union has not yet adapted its scenarios to the changing nature and costs of renewables. This has led to systemic underestimation of renewables in the EU scenarios. In contrast to the optimistic forecasts for conventional energies (CCTS and nuclear), the future potential of renewables has been underestimated to date at the European level—mainly

---

[58]PV has won in auctions in Peru and Chile, in 2016, with bids of about 4 $cents/kWh, and further cost reductions are likely.

[59]Offshore wind farms will not be discussed here due to more uncertain cost estimates.

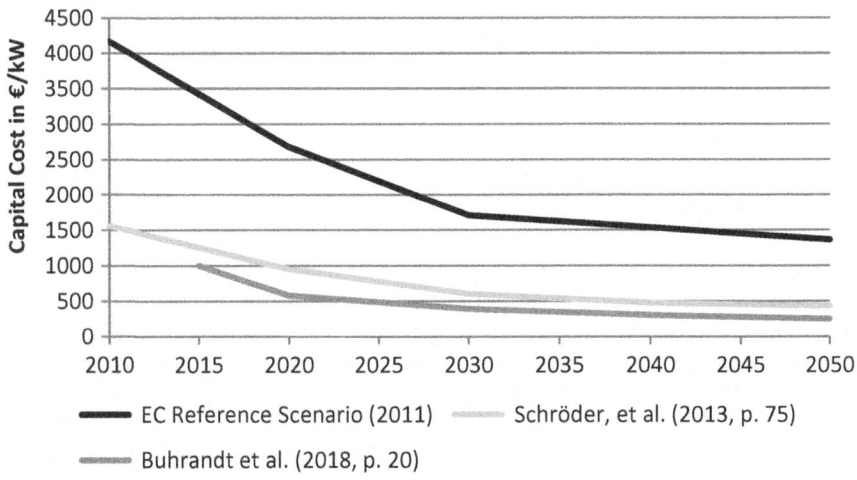

Fig. 10.14 Capital cost of solar PV. Source: Own depiction

due to high cost projections. Figure 10.14 shows an example of the overestimation of the costs of renewables in the Reference Scenario by comparing the estimate on specific investment costs contained in the Energy Roadmap 2011 with real-time estimates, in this case the documentation by Schröder et al. (2013) and update by Buhrandt et al. (2018). There are striking differences, e.g., for 2015: Whereas Schröder et al. (2013) forecasted €1200/W_p installed and the real figure was €700/W_p, the EU Reference Scenario assumed €3500/W_p.[60]

A similar discrepancy is also evident in the dynamic perspective. In line with the literature on learning rates, we assume that costs will fall by 20% between 2020 and 2030, by another 15% by 2030, and by 10% between 2040 and 2050. Here, the Energy Roadmap appears very conservative in its estimate of the cost reductions in photovoltaics beyond the year 2030: Although capital costs decrease linearly from 2010 (about €4000/kW) to 2030 (about €1660/kW), they will only drop slightly by 2050. Both the initial figures and the development of these cost estimates seem implausible: In 2015, the real capital costs were about half of the figure in the Reference Scenario projected for 2050.

Another discrepancy between the Energy Roadmap estimates and other analyses is seen in the investment costs for onshore wind: While most studies predict cost reductions, in the Energy Roadmap, the specific investment costs for onshore wind remain almost constant for the next four decades (€1106/kW in 2010 to €1074/kW in 2050). Last but not least, the breakthrough of storage technologies is ignored in the Reference Scenarios. Neither the 2011, or 2013, nor the 2016 versions include the

---

[60]See Meiß, Jan. 2013. "Prospective Energy Generation Costs – Topic 1: Solar." presented at the Workshop on Prospective Generation Costs until 2050, DIW Berlin, Berlin, Germany, March 8.

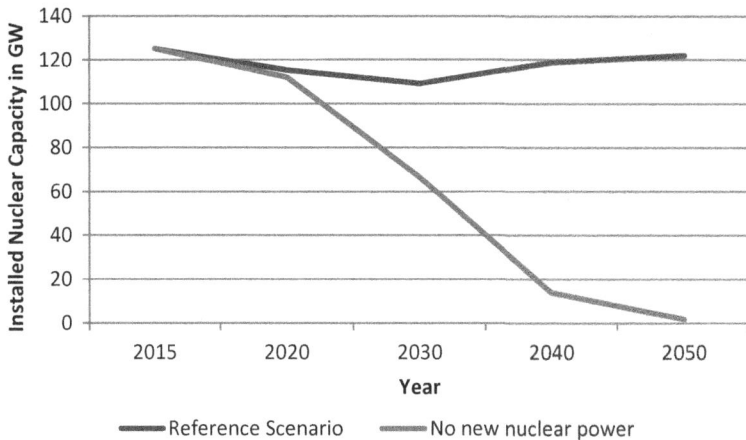

Fig. 10.15 Model assumptions on nuclear power. Source: Kemfert et al. (2015, 623)

newly available technologies that have led renewables from their previous niche role as "variable" sources to their central role as a baseload technology.

10.6   Comparing Alternative Low-Carbon Scenarios

Clearly there are different ways to reach the targets of the European low-carbon transformation. The objective of this final section is to contrast the EU Reference Scenario, based on the triad of coal (with CCTS), nuclear, and renewables, with a scenario in which neither CCTS nor nuclear are available at reasonable costs.

Kemfert et al. (2015) have compared two scenarios that differ from the EU Reference Scenario with respect to CCTS and nuclear power (Fig. 10.15):

- A "base scenario" takes the nuclear capacities from the reference scenario as given, and calculates an optimal power plant portfolio around that. In this scenario, numerous power plants are given extended service lives and new power plants are constructed across Europe, in particular after 2030. Relatively stable long-term capacity calls for new nuclear power plants with a capacity of around 120 GW by 2050, mainly in the period 2030 to 2040; approximately half the new installed capacity is located in France.
- A scenario "no new nuclear" assumes that nuclear power is too expensive, and that therefore, no new builds will take place. Existing nuclear power plants are decommissioned after 50 years.

Figure 10.16 compares the "base" scenario with the "no new nuclear" scenario. In both, a combination of renewables and some storage takes the lead position, and both also maintain a small stock of fossil capacity. Among the renewables, wind dominates solar power somewhat thanks to the higher running hours. In the "no new nuclear" scenario, storage takes over to assure continuous electricity supply. The

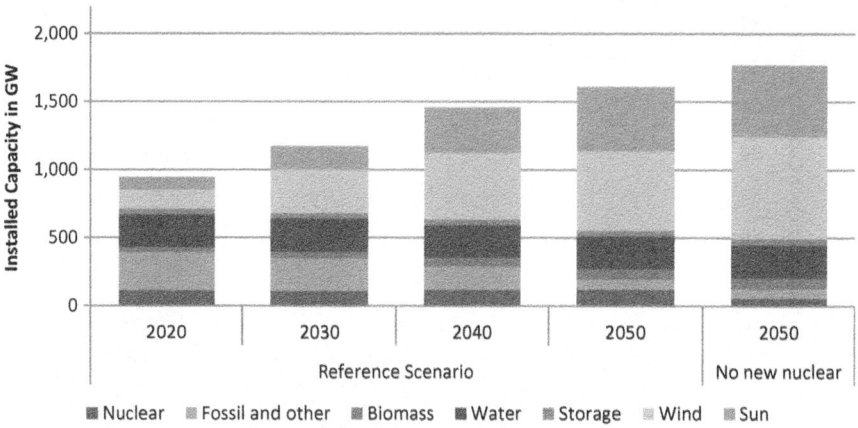

Fig. 10.16 Generation capacities in Europe to 2050 by scenario. Source: Kemfert et al. (2015, 624)

Reference Scenario has a significant share of nuclear electricity (924 TWh) and some CCTS (296 TWh). By contrast, the "no new nuclear" scenario relies to over 90% on renewables.

10.7   Conclusions

The European low-carbon transformation and the German energiewende have some points in common and also several key differences. Naturally, the German energiewende has to be considered in the European context, but must not be confounded with it. Therefore, this chapter has focused on the European approach to a low-carbon energy sector. The historical survey pointed to two interesting trends: first, Europe was "born" with two treaties on energy (coal and nuclear), which explains much of the continuity in these two energy sources through the decades. Second, even though European energy and climate initiatives have become stronger, the choice of the national energy mix has remained a national prerogative, thus limiting the potential scope of European policies.

Europe has set itself relatively ambitious targets for greenhouse gas emission reductions: −20% by 2020 (base year: 1990), −40% by 2030, and −80 to 95% by 2050. In contrast, the targets for renewables and energy are less stringent, currently targeting a 27% share of renewables and a 40% increase in energy efficiency by 2030. What is more worrying looking toward the immediate future is the loss of coherence in European energy and climate policies: After the coherent set of policies elaborated in 2007–2009, the economic and financial crisis; the return of the cold war with Russia, and the refugee crisis have cooled the enthusiasm for structured energy and climate policies to some degree.

The chapter also identified a major difference between the European and German pathways in terms of the energy mix. The European "low-carbon" transformation includes a significant number of coal plants with carbon capture and nuclear power, whereas the German energiewende is focusing on decarbonization through renewables. We have observed that the EU Member States' autonomy in determining their own domestic policy mix has given Europe a strong bias toward carbon capture, transport, and storage (CCTS), even though there has been no large scale power project with the full CCTS chain anywhere in the world so far and is unlikely to do so in the coming decades, and toward nuclear energy, the cost of which is systematically underestimated in the European Reference Scenarios. By contrast, the significant reduction in the costs of renewables, in particular solar photovoltaic, has been ignored and is still underestimated in European policy documents, thus leading to a structural underestimation of renewables. For the power sector, the Impact Assessment presents an outlook that is risky from a technology policy point of view and questionable from an economic perspective: Its climate targets can only be met by increasing the number of coal-fired plants with carbon capture and of nuclear power stations.

The German energiewende can therefore inform European energy and climate policies, but it should not be considered a potential model. There is a clear need to distinguish reform processes in other countries and at the European level from the German energiewende. While Denmark and the Nordic countries have set renewables targets far above the German 38% by 2020, many other EU Member States have opted for different approaches that are not compatible with the energiewende, as is the case with some Central and East European countries that are hanging on to coal electrification and countries that still consider nuclear energy a serious option, like France and the UK.

The modeling exercise in the last section clearly illustrated a point made in the preceding discussion: The decarbonization of the European electricity sector can theoretically be achieved using significant amounts of fossil fuels, mainly coal, in combination with carbon capture, and an equally significant amount of nuclear power. Some Member States have calibrated their respective energy policies in this way, and hence the EU Reference Scenarios (using a behavioral approach based on national preferences) contain substantial amounts of coal with carbon capture and nuclear electricity. On the other hand, economic optimization with our modeling approach (dynELMOD model) yields a different result, showing that the European low-carbon transformation targets can be met mainly with renewables. Clearly, different approaches to achieve low-carbon transformation co-exist within Europe.

References

BMWi, and BMU. 2010. Energy concept – for an environmentally sound, reliable and affordable energy supply. Berlin, Germany.

Brunnengräber, Achim, Maria Rosaria Di Nucci, Ana Maria Isidoro Losada, Lutz Mez, and Miranda A. Schreurs. 2015. *Nuclear waste governance: an international comparison*. Berlin, Germany: Springer.

Burandt, Thorsten, Konstantin Löffler, and Karlo Hainsch. 2018. GENeSYS-MOD v2.0 – enhancing the global energy system model. *DIW Data Documentation*, no. 94 (July).

Cantor, Robin, and James Hewlett. 1988. The economics of nuclear power: further evidence on learning, economies of scale, and regulatory effects. *Resources and Energy* 10 (4): 315–335.

Cox, Emily, Phil Johnstone, and Andy Stirling. 2016. Understanding the intensity of UK policy commitments to nuclear power. SPRU Working Paper Series 2016–16. Sussex, UK: University of Sussex.

Creutzig, Felix, Peter Agoston, Jan Christoph Goldschmidt, Gunnar Luderer, Gregory Nemet, and Robert C. Pietzcker. 2017. The underestimated potential of solar energy to mitigate climate change. *Nature Energy* 2 (9): 17140.

D'haeseleer, William D. 2013. Synthesis on the economics of nuclear energy – study for the European Commission, DG Energy. Final Report. Leuven, Belgium: KU Leuven.

Davis, Lucas W. 2012. Prospects for nuclear power. *Journal of Economic Perspectives* 26 (1): 49–66.

EC. 2011a. Energy roadmap 2050: impact assessment, Part 1/2. SEC(2011) 1565/2. Commission Staff Working Paper. Brussels, Belgium: European Commission.

———. 2011b. Energy roadmap 2050: impact assessment, Part 2/2. SEC(2011) 1565. Commission Staff Working Paper. Brussels, Belgium: European Commission.

———. 2011c. A roadmap for moving to a competitive low carbon economy in 2050. *COM* 2011: 112.

———. 2011d. Energy roadmap 2050. Communication from the Commission to the European Parliament, the Council, the European Economic and Social Committee and the Committee of the Regions. Brussels, Belgium: European Commission.

———. 2013. *EU Energy, Transport and GHG Emissions Trends to 2050: Reference Scenario 2013*. Brussels, Belgium: European Commission.

———. 2014. A policy framework for climate and energy in the period from 2020 up to 2030. Commission staff working document impact assessment COM(2014) 015 final. Communication from the Commission to the European Parliament, the Council, the European Economic and Social Committee and the Committee of the Regions. Brussels, Belgium: European Commission.

———. 2016. *EU Reference Scenario 2016: Energy, Transport and GHG Emissions – Trends to 2050*. Brussels, Belgium: European Commission.

———. 2018. Publication of the total number of allowances in circulation in 2017 for the purposes of the market stability reserve under the EU Emissions Trading System Established by Directive 2003/87/EC. C(2018) 2801 final. Brussels, Belgium: European Commission.

ECSC. 1951. Treaty establishing the European Coal and Steel Community, ECSC Treaty.

EEC Treaty. 1957. Treaty establishing the European Economic Community, EEC Treaty. United Nations Treaty Series (UNTS).

EURATOM Treaty. 1957. Treaty establishing the European Atomic Energy Community (EURATOM). United Nations Treaty Series (UNTS).

GCCSI. 2014. The global status of CCS: 2014. Canberra, Australia: Global CCS Institute.

———. 2017. *The Global Status of CCS: 2017*. Canberra, Australia: Global CCS Institute.

Grau, Thilo, Molin Huo, and Karsten Neuhoff. 2011. Survey of photovoltaic industry and policy in Germany and China. DIW Discussion Paper 1132. Berlin, Germany: DIW Berlin.

Grubb, Michael, Jean-Charles Hourcade, and Karsten Neuhoff. 2014. *Planetary Economics – Energy, Climate Change and the Three Domains of Sustainable Development*. London, UK: Routledge.

Grubler, Arnulf. 2010. The costs of the French nuclear scale-up: a case of negative learning by doing. *Energy Policy* 38 (9): 5174–5188.

HOC-ECCC. 2013. Building new nuclear: the challenges ahead – 6th report of session 2012–2013. House of Commons, Energy and Climate Change Committee. London, UK: House of Commons, Energy and Climate Change Committee.

IEA. 2009. Energy balances of OECD countries 2009. IEA statistics. Paris, France: OECD.

———. 2015. *World Energy Outlook Special Report 2015: Energy and Climate Change*. Paris, France: OECD.

IPCC. 2005. IPCC Special report on carbon dioxide capture and storage. Prepared by Working Group III of the Intergovernmental Panel on Climate Change. Cambridge, UK: Cambridge University Press.

Joskow, Paul L., and John E. Parsons. 2012. The future of nuclear power after Fukushima. *Economics of Energy & Environmental Policy* 1 (2): 99–113.

Kemfert, Claudia, Christian von Hirschhausen, and Casimir Lorenz. 2014. European energy and climate policy requires ambitious targets for 2030. *DIW Economic Bulletin* 4 (8): 17–26.

Kemfert, Claudia, Christian von Hirschhausen, Felix Reitz, Clemens Gerbaulet, and Casimir Lorenz. 2015. European climate targets achievable without nuclear power. *DIW Economic Bulletin* 5 (47): 619–625.

Lévêque, François. 2014. *The Economics and Uncertainties of Nuclear Power*. Cambridge, UK: Cambridge University Press.

Löffler, Konstantin, Karlo Hainsch, Thorsten Burandt, Pao-Yu Oei, Claudia Kemfert, and Christian von Hirschhausen. 2017. Designing a model for the global energy system—GENeSYS-MOD: an application of the Open-Source Energy Modeling System (OSeMOSYS). *Energies* 10 (10): 1468.

Lovins, Amory B. 1976. Energy strategy: the road not taken? *Foreign Affairs* 6 (20): 9–19.

Lovins, Amory B., and L. Hunter Lovins. 1980. *Energy/War: Breaking the Nuclear Link*. 1st ed. San Francisco, CA: Friends of the Earth.

Meeus, Leonardo, Isabel Azevedo, Claudio Marcantonini, Jean-Michel Glachant, and Manfred Hafner. 2012. EU 2050 low-carbon energy future: visions and strategies. *The Electricity Journal* 25 (5): 57–63.

Metayer, Matthieu, Christian Breyer, and Hans-Josef Fell. 2015. The projections for the future and quality in the past of the world energy outlook for solar PV and other renewable energy technologies. In *31st European PV Solar Energy Conference and Exhibition, September 14–18, 2015*. Hamburg, Germany.

Mohn, Klaus. 2018. The gavity of status quo: a review of IEA's world energy outlook. *Economics of Energy & Environmental Policy* 8: 2.

Molly, J.P. 2012. Design of wind turbines and storage: a question of system optimisation. *DEWI Magazin* 40: 23–29.

Neuhoff, Karsten, and Anne Schopp. 2013. Europäischer Emissionshandel: Durch Backloading Zeit für Strukturreform gewinnen. 11. DIW Wochenbericht. Berlin, Germany.

Oei, Pao-Yu, Claudia Kemfert, Felix Reitz, and Christian von Hirschhausen. 2014. Braunkohleausstieg – Gestaltungsoptionen im Rahmen der Energiewende. 84. Politikberatung kompakt. Berlin, Germany: DIW.

Pahle, Michael, Brigitte Knopf, Oliver Tietjen, and Eva Schmid. 2012. Kosten des Ausbaus erneuerbarer Energien: eine Metaanalyse von Szenarien. 23/2012. Umweltbundesamt. Dessau-Roßlau, Germany: Federal Environment Agency.

Raack, Wolfgang, Paul Schorn, and Emil Schrödter. 1957. *Jahrbuch des deutschen Bergbaus*. Essen: Glückauf GmbH.

Radkau, Joachim, and Lothar Hahn. 2013. *Aufstieg und Fall der deutschen Atomwirtschaft*. Munich, Germany: Oekom Verlag.

Rai, Varun, David G. Victor, and Mark C. Thurber. 2010. Carbon capture and storage at scale: Lessons from the growth of analogous energy technologies. *Energy Policy* 38 (8): 4089–4098.

Rangel, Escobar Lina, and Francois Leveque. 2015. Revisiting the cost escalation curse of nuclear power: new lessons from the French experience. *Economics of Energy & Environmental Policy* 4 (2): 103–126.

Reiner, David M. 2016. Learning through a portfolio of carbon capture and storage demonstration projects. *Nature Energy* 1 (1): 15011.

Rubin, Edward S., and Haibo Zhai. 2012. The cost of carbon capture and storage for natural gas combined cycle power plants. *Environmental Science & Technology* 46 (6): 3076–3084.

Schröder, Andreas, Friedrich Kunz, Jan Meiß, Roman Mendelevitch, and Christian von Hirschhausen. 2013. Current and prospective costs of electricity generation until 2050. DIW Data Documentation 68. Berlin.

Schulze, William D., David S. Brookshire, and Todd Sandler. 1981. The social rate of discount for nuclear waste storage: economics or ethics? *Natural Resources Journal* 21 (4): 811–832.

Thomas, S. 2010. *The EPR in Crisis*. London: University of Greenwich.

Toke, David. 2012. Nuclear power: how competitive is it under electricity market reform? Presentation given at the HEEDnet Seminar presented at the HEEDnet Seminar, London, UK, July 17.

von Hirschhausen, Christian. 2017. Nuclear power in the 21st Century – an assessment (Part I). DIW Discussion Paper 1700. Berlin, Germany.

von Hirschhausen, Christian, and Felix Reitz. 2014. Nuclear power: phase-out model yet to address final disposal issue. *DIW Economic Bulletin* 4: 27–35.

von Hirschhausen, Christian, Johannes Herold, and Pao-Yu Oei. 2012a. How a 'low carbon' innovation can fail – tales from a 'lost decade' for Carbon Capture, Transport, and Sequestration (CCTS). *Economics of Energy & Environmental Policy* 1 (2): 115–123.

von Hirschhausen, Christian, Johannes Herold, Pao-Yu Oei, and Clemens Haftendorn. 2012b. CCTS-Technologie ein Fehlschlag: Umdenken in der Energiewende notwendig. 6. Berlin, Germany: DIW Wochenbericht.

von Hirschhausen, Christian, Claudia Kemfert, Friedrich Kunz, and Roman Mendelevitch. 2013. European electricity generation post-2020: renewable energy not to be underestimated. *DIW Economic Bulletin* 3 (9): 16–28.

Wealer, Ben, Simon Bauer, Nicolas Landry, Hannah Seiß, and Christian von Hirschhausen. 2018. Nuclear power reactors worldwide – technology developments, diffusion patterns, and country-by-country analysis of implementation (1951–2017). DIW Berlin, Data Documentation 93. Berlin, Germany: DIW Berlin, TU Berlin.

Wirth, Harry. 2016. *Aktuelle Fakten zur Photovoltaik in Deutschland*. Freiburg, Germany: Fraunhofer ISE.

Zerrahn, Alexander, and Wolf-Peter Schill. 2015. A greenfield model to evaluate long-run power storage requirements for high shares of renewables. DIW Discussion Paper 1457. Berlin, Germany: DIW.

Chapter 11
Energy Infrastructures for the Low-Carbon Transformation in Europe

Franziska Holz, Jonas Egerer, Clemens Gerbaulet, Pao-Yu Oei,
Roman Mendelevitch, Anne Neumann, and Christian von Hirschhausen

*To place one's faith in purely permissive sequences and to
rely on the ability of SOC [social overhead capital, i.e.,
infrastructure] to call forth other economic activities, can,
under these circumstances, be just as irrational as the
so-called "Cargo Cult" that has been engaged in by some of
the New Guinea tribes after the lamented departure of the
Allied expeditionary force at the end of World War II: "Those
in coastal villages have built wharves out into the sea, ready
for the ships to tie up, and those in land villages have*

F. Holz
DIW Berlin, Berlin, Germany

Norwegian University of Science and Technology (NTNU), Trondheim, Norway

J. Egerer
Friedrich-Alexander-Universität (FAU), Erlangen, Nürnberg, Germany

C. Gerbaulet · C. von Hirschhausen
TU Berlin, Berlin, Germany

DIW Berlin, Berlin, Germany

P.-Y. Oei
Junior Research Group "CoalExit", Berlin, Germany

TU Berlin, Berlin, Germany

DIW Berlin, Berlin, Germany

R. Mendelevitch
HU Berlin, Berlin, Germany

DIW Berlin, Berlin, Germany

A. Neumann (✉)
University Potsdam, Potsdam, Germany

DIW Berlin, Berlin, Germany
e-mail: anne.neumann@uni-potsdam.de

© Springer Nature Switzerland AG 2018
C. von Hirschhausen et al. (eds.), *Energiewende "Made in Germany"*,
https://doi.org/10.1007/978-3-319-95126-3_11

283

*constructed airstrips out to the jungle for the planes to land.
And they have waited in expectancy for the Second Coming of
the Cargoes." Touching as it is, such a belief in the
propitiatory powers of social overhead capital should not be
the basis of development policy.
Albert O. Hirschman (1958, 10:94): The Strategy of
Economic Development.*

11.1 Introduction

Both in the German energiewende and in the European low-carbon energy system
transformation, infrastructure is generally considered as a *conditio sine qua non*: a
necessary though not sufficient condition for a low-carbon economy—and one
without which energy transformation may fail. Thus, as described in Chap. 8 on
German electricity transmission infrastructure, the speed of transformation has
sometimes been considered to hinge on the speed of infrastructure development.
At the European level, there is a general belief that "new, flexible infrastructure
development is a no-regrets option …" (EC 2011c, 15). It therefore came as no
surprise when a large number of continental-scale infrastructure development plans
were proposed, receiving substantial attention and interest from policymakers and
some from private investors as well. This led to the idea of "infrastructure
supergrids," not only for electricity but also for natural gas and a new sector, CO_2
pipeline infrastructure.

At second glance, there may be some doubt as to whether "big infrastructure" is
really the appropriate way to approach the low-carbon transformation. First, there is
no consensus in the literature on how to design infrastructure policies to support the
transformation process: in general, the impact of infrastructure on economic development cannot be easily assessed, as noted by Albert Hirschman (1958) in his book
"The Strategy of Economic Development." Second, the key issues in the
energiewende and the low-carbon transformation in Europe have turned out to be
low-carbon electricity *generation* rather than transport, and the costs of infrastructure expansion are very modest, when compared to the significant costs of
low-carbon generation (ECF 2010, 2011). Third, more infrastructure can even
harm sustainable development—for instance when infrastructure makes it possible
to integrate more coal-based electricity into the energy system, thus favoring a
carbon lock-in. And fourth, but no less important, the idea of European supergrids
looked appealing on paper but has not yet been implemented on the ground due to a
variety of obstacles, whereas regional or even local infrastructure development has
been more successful.

In this chapter, we analyze the role of physical infrastructure in the European
low-carbon transformation, with a special focus on large-scale transmission infrastructure for electricity, natural gas, and CO_2. Although these infrastructures can

play a certain role, they are not necessarily the critical factors in low-carbon transformation, and often low-cost measures such as improving regulation or tightening access rules are more effective than capital-intensive infrastructure expansion. In the next section, based on a review of the literature on infrastructure and development and on infrastructure in the low-carbon context, we find that although a majority of authors see infrastructure development as a no-regrets option, there are also arguments against an oversupply of infrastructure. Sections 11.3–11.5 provide model- and case study-based analyses of different infrastructure sectors. The models we discuss take a bottom-up approach with a high level of sector-specific granularity: Section 11.3 focuses on electricity transmission and compares the plans for pan-European electricity highways with other, more modest scenarios focusing on domestic upgrades and selected cross-country interconnectors. Section 11.4 is dedicated to natural gas infrastructure: Our results show no evidence of a substantial need for additional pipeline or LNG infrastructure, but rather a need for modest investment, given the diverse and global European supply of natural gas. Our analysis of infrastructure planning for carbon pipelines in Sect. 11.5 yields an even more striking result: After a first wave of enthusiasm and plans for CO_2 highways across Europe, dozens of thousands kilometers long, perhaps not a single cross-border pipeline may be required—except for perhaps a few in the North Sea—simply because the underlying technology, carbon capture, transport, and storage (CCTS), is unlikely to be used at the expected scale. Our conclusion in Sect. 11.6 is that the way forward is more likely to lie in regional and local cooperation in infrastructure.[1]

11.2 Infrastructure and Low-Carbon Development

11.2.1 Macroeconomic Perspective: Development by Infrastructure Oversupply or Shortage?

There is a long and detailed discussion about the potentially beneficial effects of infrastructure on economic development going back to Adam Smith's (1776)

[1]This chapter builds on previous studies and papers based on sectoral infrastructure models for electricity (Leuthold et al. 2012; Egerer et al. 2014; Egerer 2016), natural gas (Egging et al. 2008; Egging 2013; Holz et al. 2013), and CO_2 pipelines (Mendelevitch 2014; Oei et al. 2014; Oei and Mendelevitch 2016). We examined the European perspective in the infrastructure subgroup of the EMF 28 Model Comparison "Europe 2050: The Effects of Technology Choices on EU Climate Policy" (see Holz and von Hirschhausen 2013). The chapter also addresses issues raised in earlier publications on European infrastructure, including a presentation to the German "Verein fuer Socialpolitik" (Hirschhausen et al. 2013) and studies for the European Investment Bank, the Foundation "Notre Europe," and the MIT Center for Energy and Environmental Policy Research (CEEPR) (see von Hirschhausen 2010, 2011, 2012). We thank the numerous referees and conference discussants who provided useful comments on these papers, as well as Claudia Kemfert, Brigitte Knopf, Friedrich Kunz, Casimir Lorenz, Juan Rosellon, and Alexander Weber for in-depth discussions; the usual disclaimer applies.

"Wealth of Nations," where in Book V, he suggests different organizational models for how infrastructure provision can contribute to general welfare. Infrastructure indeed has many characteristics of a public good, or a "club collective good" (Buchanan 1965), implying that it has positive externalities and can be a precondition for economic development. Infrastructure planning and implementation is also a major focus of public policy given its high visibility and resonance with the public and voters in particular, and it lends itself easily to international projects and cross-border cooperation. Therefore, infrastructure investments are often considered "no regret" options guided by the maxim that "more is better" and "big is beautiful." There have been a number of macroeconomic analyses of the growth- and productivity enhancing effects of infrastructure investments including the Aschauer papers of the late 1980s (Aschauer 1989), but it is still difficult to identify a unilateral relationship between infrastructure and development, as Gramlich (1994) noted in a critical survey of this literature.

In the discussion on approaches to low-carbon transformation, an additional argument in favor of infrastructure investment has been put forward by the climate policy community itself: Infrastructure offers a "no-regrets option" for climate policy because carbon price revenues, infrastructure investments, and sustainable development are closely linked. In a keynote speech at the 2013 I.E. European Conference, Edenhofer argued that infrastructure can become an important means of reaping a "triple dividend": Countries could not only reduce CO_2 emissions and budget revenues but also achieve sustainable growth by earmarking revenues to infrastructure projects.[2] This growth model assumes a positive marginal contribution of the infrastructure investment toward sustainability. Rausch and Reilly (2015) put forward a similar argument linking CO_2 revenues to a reduction of budget deficits and thus a freeing up of resources for productive use, that is, infrastructure.

There are also arguments in the opposite direction. One of these maintains that too much infrastructure could harm the transformation process by using up large quantities of scarce resources. This argument originates from the literature on economic development and suggests that development can be spurred by infrastructure *shortage* rather than oversupply. The reasoning goes back to Hirschman's (1958) "Strategy of Economic Development": In a situation of great uncertainties, especially about the development of the economy and its "directly productive capital" (DPC), infrastructure investment can be risky because social overhead capital (his term for infrastructure) "is largely a matter of faith in the development potential of a country or a region." When the trajectory of economic activity and other investments is unknown, it might be both inefficient and dangerous to proceed prematurely with capital-intensive infrastructure investment. Hirschman argues that an uncontrolled strategy of infrastructure expansion might not be conducive to longer-term sustainable development and might also even be harmful, the reason being that it is less

---

[2]Edenhofer, Ottmar. 2013. "The Economics of Uranium, Fossil Fuels and Climate Change Stabilization – Trade-Offs, Synergies and Solutions. Keynote at the IAEE 2013 European Conference." Presented at the 13th European IAEE Conference, Düsseldorf, Germany.

risky to first develop the activities a society wants to engage in and "then let the ensuing pressures determine the appropriate outlays for SOC [social overhead capital, i.e., infrastructure] and its location." (Hirschman 1958, 10:93).[3] If one applies Hirschman's argument to the low-carbon energy transformation, there are good reasons to avoid excessive and premature infrastructure development before low-carbon electricity generation is used on a large scale.

11.2.2   Yet Another Perspective: "Cables for Carbon"?

Davis et al. (2010) have raised an additional argument for a cautious approach to infrastructure development, specific to the low-carbon transformation. Based on an analysis of the currently existing, very CO_2-intensive infrastructure systems, they argue that it is important not to add infrastructures that would contribute additional greenhouse gas emissions to the system. They warn that not only does infrastructure development fail to promote CO_2 abatement; it also supports the old, fossil-fuel based, CO_2-intensive generation mix. Physical energy infrastructures such as electricity transmission lines, natural gas pipelines, and CO_2 pipelines are capital-intensive and have long lifetimes ranging from several decades up to a century. As a result, the choice of infrastructure has a significant influence on the carbon intensity of the energy system overall. As Davis et al. (2010) report, an estimated 496 bn tons of CO_2 will be emitted over the next fifty years by existing energy and transport infrastructures alone. This means that one has to be particularly careful not to build the wrong infrastructures in the future, because "there is little scope for further fossil-fuel based infrastructures."[4]

The arguments presented by Davis et al. (2010) are supported by empirical observations that an increase in available infrastructure capacity often promotes rather than impedes the conventional, high- CO_2 system. In fact, it is not clear whether the transmission expansion projects proposed in the past decade would have been effective in increasing the integration of renewables, as they proposed to do, or whether—as Davis, et al. feared—they would have actually favored conventional fossil-fuel generation. A similar case was made by Members of the European Parliament in a note on transmission links between Southern Europe and North Africa, entitled "Cables for Carbon" (Turmes 2010). Their paper argued that the high-voltage transmission projects planned in the region were conducive not to renewables expansion but instead to the exchange of fossil-fuel-based electricity.

---

[3]Hirschman (1958, 10:95) continues by suggesting that "it would be illegitimate and wasteful to expand SOC facilities in anticipation of the kind of extremely rapid economic progress that does hit a city or area sometimes, but whose occurrence or continuation can never be predicted with confidence."

[4]Davis et al. (2010); see also the comment by Ottmar Edenhofer: http://wealthofthecommons.org/essay/atmosphere-global-commons, downloaded August 25, 2014.

Figure 11.1 provides some evidence on the suggestion that most of the transmission lines would transport fossil and not renewable electricity.

Brancucci (2013) reported evidence confirming that more infrastructure favors more CO_2-intensive electricity generation under certain circumstances. While this sounds counterintuitive, it corresponds to a very simple economic logic: Transmission lines do not favor the low-carbon transformation per se—they only serve the transport of electricity under the current energy mix. Thus, if the current energy mix is very CO_2-intensive, then the construction of more transmission lines automatically leads to a higher CO_2-output from electricity generation. Based on this finding, Brancucci presented a model of electricity network development at the European level for different scenarios of the electricity mix: In a scenario of low CO_2 prices, which is a plausible assumption for the year 2025, the construction of more transmission lines leads to a *higher* market share of low-cost, CO_2-intensive electricity generation, mainly coal: More transmission capacity makes it possible to utilize coal and lignite more fully at the cost of gas plants.[5]

The same effect is also detected in a general equilibrium analysis of the macroeconomic effects of transmission development: Abrell and Rausch (2015) report that European-wide transmission expansion in a system with a high share of coal generation and a low share of renewables leads to a higher share of coal and increasing CO_2 emissions. Thus, in the base scenario with increasing transmission capacities, CO_2 emissions increase by 1–3%. Even when the 20% renewables target for Europe (2020 goal) is reached, CO_2 emissions still increase with network expansion. The explanation provided by Abrell and Rausch is similar to the one offered by Brancucci (2013): Cross-border transmission expansion favors the production and export of CO_2-intensive electricity.[6]

In a study on the effects of network expansion in Germany carried out in the framework of the "Energy Strategy for Bavaria," Mieth et al. (2015) identified a similar CO_2-enhancing effect for the construction of a high-voltage direct current (HVDC) line from the lignite basins in East Germany to Bavaria, the "South-East Corridor" (Süd-Ost-Passage). If this HVDC infrastructure were built, it would lead to a significant increase in electricity produced from CO_2-intensive lignite

---

[5]"Higher cross-border transmission capacity throughout Europe has a negative environmental impact in this scenario: CO_2 emissions increase by 3.6%. The reason is that the marginal cost of coal and lignite plants is lower than the marginal cost of gas plants because the CO_2 price is not high enough to have a significant impact on the merit order of generation. More transmission capacity makes it possible to utilize coal and lignite more fully at the cost of gas plants." Brancucci (2013, 41).

[6]"For low and intermediate levels of renewables, CO_2 emissions increase irrespective of the magnitude of the transmission infrastructure expansion (TIP). The main driver of this result is that TIP increases economic incentives to export (and produce) cheap coal-fired electricity resulting in a decrease of gas-fired production. A second effect driving the emissions increase is the boost in overall economic activities brought about by the efficiency gains from cross-country electricity trade. Even for already ambitious year-2020 RE production targets, we thus find that the TYNDP [ten year network development plan] fails to yield reductions in CO_2 emissions at the European level." Abrell and Rausch (2015, 35).

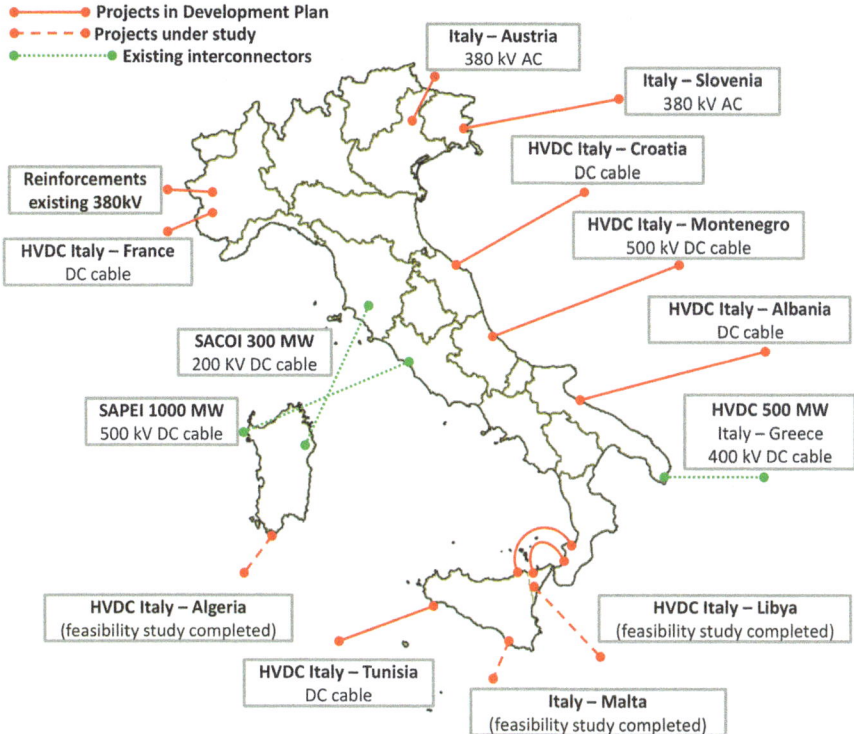

Fig. 11.1 "Cables for Carbon": Transmission expansion projects in the Meditarranean. Source: Own depiction, based on Turmes (2010)

amounting to around 2.5 TWh (or around 2.5 mn t of CO_2 annually). In this particular case, another effect could occur: The construction of large-capacity export transmission lines could induce investments in further lignite power plant capacity in the (previously export-constrained) region and lead to the opening of new lignite mines. Mieth et al. therefore argue that the construction of such a large transmission infrastructure would not increase supply security but instead serve future exports of CO_2-intensive electricity from lignite (Fig. 11.1).

11.2.3 Recent Empirical Evidence: Is "Big Really Beautiful"?

While the literature on infrastructure development is extensive, studies on the particular role of infrastructure in the low-carbon energy transformation are emerging only gradually. In particular, there is a substantial gap between the top-down models of climate and energy systems, which often have to make simplifying assumptions about the availability and costs of infrastructure, and the bottom-up realities of infrastructure development on-site in the real world. An example of this mismatch is the oversimplified role accorded to infrastructure in the renewables-based scenarios for Europe, which assume that almost all electricity consumption can be supplied by concentrated renewable sources spread across Europe, or even located in neighboring regions such as the Middle East and North Africa (MENA). Proponents of the "no regret" option have argued that transcontinental infrastructure is needed to reap the benefits of different low-carbon technologies such as solar resources in Southern Europe and wind on the North Sea. In this vision, huge infrastructure corridors are necessary to address the decarbonization challenge. An early modeling approach was the study by Czisch (2006) (a physicist by training), who argued for placing renewable capacities in "optimal" locations such as Saudi Arabia (for solar) and Iceland (for geothermal) and then connecting these locations through electricity infrastructure.

Many international organizations and think-tanks have picked up on this idea and developed ambitious projects such as the Desertec/Medgrid Initiatives (DLR 2006) or the "Grand Solar Plan" in the US, where the Southwest is connected with the rest of the country by 300,000 miles of high-voltage wires (Zweibel et al. 2008). A foundation created at the European level called "Friends of the Supergrid" (FOSG) has argued that supergrids are required to assure a low-carbon, renewables-based energy system, network stability, and European integration (FOSG 2015).[7] The natural gas industry has also developed ambitious expansion proposals, built on extensive gas consumption scenarios drawn up by Eurogas (EUROGAS 2010). A large-scale approach has also been deployed by the EU Joint Research Center (Morbee et al. 2010) for a European-wide pipeline infrastructure to transport CO_2 from carbon capture locations across Europe; the corresponding CO_2-pipeline network would stretch across 20,000 km and cover almost all of the countries of the EU with a high share of coal electrification.

While these supergrid infrastructure projects look impressive on paper and are generally attractive to policy makers, equipment suppliers, and the financial sector, they have encountered substantial problems getting off the ground and their appeal has diminished over time. The construction of transcontinental infrastructure corridors makes sense from a top-down perspective, but it overlooks the considerable transaction costs of financing, regulation, and project implementation. Looking at

---

[7]"FOSG encourages the efforts of the European Commission to create an integrated and strong liquid market in all timeframes and across all regions of Europe which will improve Europe's competitiveness and the secure supply of electricity. FOSG strongly supports an increased coordination of national policies that should ultimately lead to coordinated RES [renewable energy sources] support schemes and a common European approach to system adequacy assessment. These measures are crucial in achieving the European energy transition in a cost-effective way." (FOSG 2011).

this from an energy system perspective, Scheer (2010) suggested that the implementation of supergrids would be complicated and might undermine efforts at decentralized electricity distribution and thus support the incumbent electricity industry. Von Hirschhausen (2010) provided an early account of the obstacles that supergrid ideas were facing, and predicted that a much smaller percentage of projects would be realized than were being planned. In recent years, a growing number of papers and studies have reported evidence against large-scale, supergrid-type infrastructures. One study by a large TSO in the midwestern USA estimated the costs of transmission expansion and found that very concentrated generation structures would imply higher transmission costs (per unit of energy) than a combination of large- and small-scale electricity distribution (Midwest ISO 2010). A study by a German engineering company concluded that in terms of transmission costs, decentralized development of small-scale renewable installations would not be more expensive than highly centralized generation structures.[8] Another study published by the Association of German Electrical Engineers reported the same conclusion about a more decentralized approach to energy system development compared to centralized energy generation (VDE 2015).

There is another argument in favor of more modest infrastructure expansion: existing assets can often be used more efficiently by changing regulatory structures or increasing the technical efficiency of the existing system. As an example, shifting the regulation of natural gas pipelines from a regional "entry-exit" system to spatially differenciated prices (called "nodal pricing") can significantly enhance utilization of natural gas networks, thus making it unnecessary to add newbuild capacity (Ehrenmann and Smeers 2005) as argued by Makholm (2015) based on empirical evidence. Another low-cost means of increasing supply security is to introduce reverse flows on natural gas pipelines, which increase the flexibility of the system considerably, as demonstrated in cases of natural gas supply interruptions from Russia (Holz et al. 2014).

This evidence suggests that the expectations prevalent in the 2000s surrounding the idea of large supergrid infrastructures may have been unrealistic. Although some modeling work suggested that large-scale infrastructure projects might connect concentrated electricity generation across Europe, the reality on the ground has proven these predictions wrong. The following three sections trace developments in the three infrastructure sectors more concretely: electricity transmission (Sect. 11.3), natural gas (Sect. 11.4), and CO_2 pipelines (Sect. 11.5).

---

[8]Consentec, and Fraunhofer IWES. 2013. "Kostenoptimaler Ausbau Der Erneuerbaren Energien in Deutschland: Ein Vergleich Möglicher Strategien Für Den Ausbau von Wind- Und Solarenergie in Deutschland Bis 2033." Studie. Berlin, Aachen, Kassel: Agora Energiewende, Consentec, Fraunhofer IWES.

11.3   Electricity Transmission

11.3.1   Initial Conception: Supergrids

The initial conception of electricity transmission infrastructure for the European low-carbon transformation focused on "supergrids," i.e., European-wide, large-scale electricity transmission corridors, most often with high-voltage direct current technology (HVDC). The organization Friends of the Supergrid (FOSG) coined this term to describe a "pan-European transmission network facilitating the integration of large-scale renewable energy and the balancing and transportation of electricity, with the aim of improving the European market" (FOSG 2011). A vast system of electricity corridors would then connect generation and consumption sites across Europe.

Consequently, about a decade ago, supergrid projects emerged from a variety of sources. Among the first was a study by Gregor Czisch (2006) on transmission lines across Europe, and even beyond, to Russia and the MENA (Middle East North Africa) region. Figure 11.2 shows the extreme form of a pan-European supergrid: it includes the connection of North Africa, Russia, and Iceland to the rest of Europe for the large-scale transfer of electricity, whether from solar, wind, or natural gas. The study by Czisch became the model for very high-profile supergrid projects between Europe and neighboring regions, developed under various names including MedGrid, Desertec, and DII.[9] In a similar spirit, Egerer and Gerbaulet modeled supergrid corridors of 4–8 GW each to connect Algeria to Spain, Tunisia to Italy, and the Near East to Turkey (Egerer and Gerbaulet 2009; Egerer et al. 2009).

While Czisch chose a particularly technical approach and ignored costs altogether, other studies have derived similar network structures, for instance the ECF (2010, 2011), which used a technical-economic approach. These studies have all concluded that a pan-European, integrated DC network—of course linked with the already existing meshed AC network—is essential for the low-carbon transformation. In addition to the EU-MENA connection, two other popular supergrid projects have been the North Sea Grid, a meshed network of North Sea riparian countries, and

---

[9]See for MedGrid http://www.medgrid-psm.com/en/ (last download April 01, 2015), for Desertec http://www.desertec.org/de (last download April 01, 2015), and for DII http://www.dii-eumena.com/ (last download April 1, 2015). In a study for the "Union of the Mediterranean," the World Bank had concluded: "Besides the coordination of transmission expansion, there is still the need for substantial investments into generation facilities. The World Bank estimates investment needs of up to €23 billion per year until 2030," see Union for the Mediterranean (UfM) (2015). Fostering regional dialogue on energy: 3 UfM platforms on Gas, Regional Electricity Markets and Renewable Energy and Energy Efficiency are launched. Available online: http://ufmsecretariat.org/fostering-regional-dialogue-on-energy-launch-of-3-ufm-platforms-on-gas-regional-electricity-markets-and-renewable-energy-and-energy-efficiency/ Last accessed: May 27, 2015.

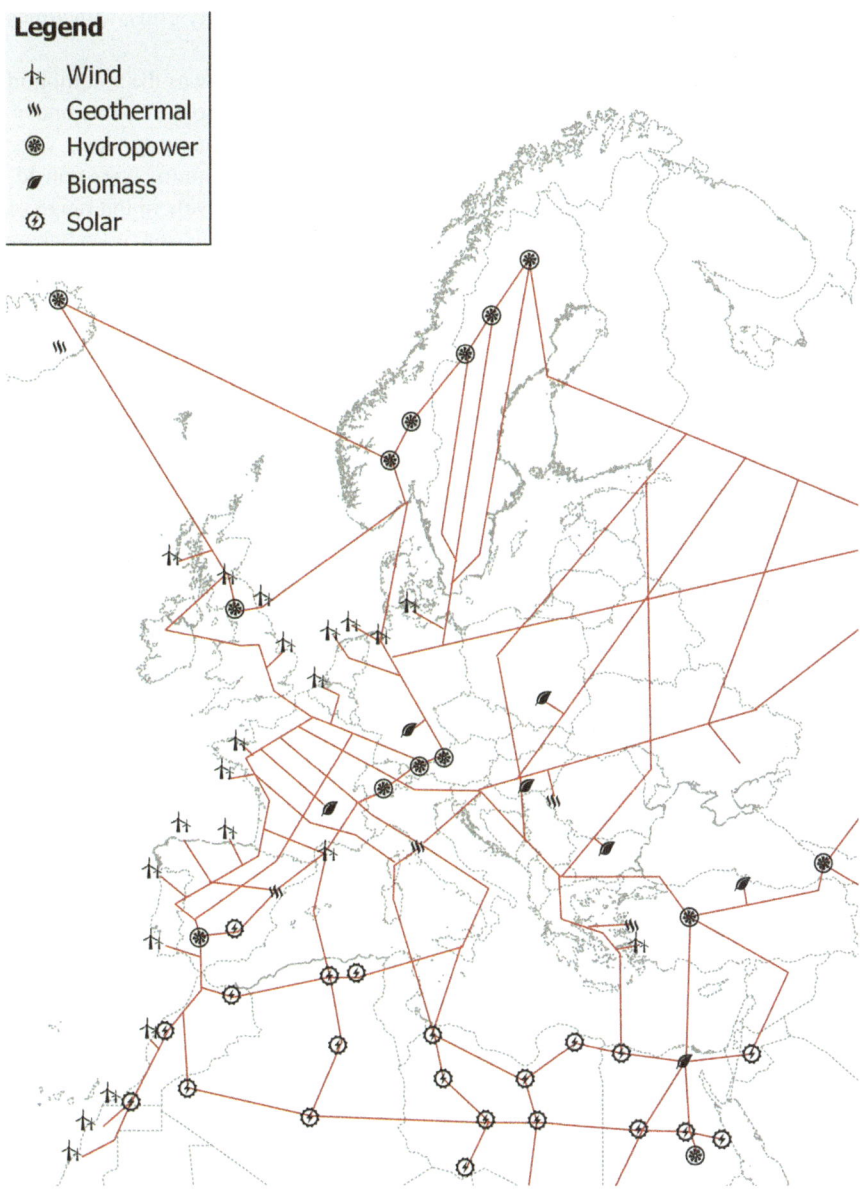

Fig. 11.2   Vision of a European electricity supergrid. Source: Own depiction, based on DLR (2006)

a project connecting the West European and the Russian electricity grids, sometimes referred to as the "Baltic Ring."[10]

An essential element of past studies on the supergrid concept was the assumption that transmission is cheap to construct and implement, leading to the (false) understanding of transmission expansion as a low-cost solution. Transaction costs were therefore entirely ignored, and only the costs of "steel and aluminum" were considered. As a result, these studies arrived at low investment cost levels in the range of €1.4 mn/km for a 2 GW circuit (Egerer et al. 2016). Transmission investments represented a very small share in the overall cost estimates, in particular when compared with the costs of low-carbon generation capacities. In a study by the European Climate Foundation, for instance, transmission requirements were estimated at €46 bn, below 5% of total costs of €2273 bn (ECF 2011).

There was broad consensus spanning numerous stakeholders including transmission engineers (e.g., the Friends of the Supergrid), think-tanks such as the European Climate Foundation, which estimated the needs for a decarbonized European electricity sector (ECF 2010, 2011), and environmental lobby groups such as Greenpeace Europe (see Tröster et al. 2011), which proposed a European backbone grid. While the transmission companies themselves have remained cautious about supergrid concepts, the idea has found its way into a number of European planning and legislative documents, and ended up being the preferred architecture for the European Energy Roadmap 2050 as well (EC 2011c, 75): the European Infrastructure Priorities (EC 2011a) assumed investment requirements of €142 bn for electricity transmission.

11.3.2 From Supergrid Projects Towards a More Regional and Local Approach

11.3.2.1 Supergrids Have Not Taken Off

Almost a decade after the first concrete supergrid ideas were proposed, there is an undeniable lack of progress to be seen, and the once widespread enthusiasm has dissipated. In fact, little has happened on the ground that would hint at the realization of some type of supergrid DC-overlay network in the near or medium-term future. Three prominent examples support this hypothesis. First, the EU-MENA cooperation has not progressed significantly: While a few joint projects in generation exist,

---

[10]In fact, the project of linking the ENTSO-E European grid and the Russian UPS network has occupied the European industry and policymakers for a long time. Since the opening of Central and Eastern Europe to the West European electricity grid in the early 1990s, several attempts have been made to connect Europe with the Russian grid as well, with a particular focus on the Baltic countries (Lithuania, Latvia, and Estonia), that remain physically integrated into the Russian electricity system to this day; a pilot project in this regard was the "Baltic Ring," a Transeuropean Project of the 1990s (Schrettl et al. 1998).

not a single large-scale transmission project has been built so far. The difficulties of the Desertec/DII and MedGrid projects provide further evidence of the gap between hopes and realities; whereas the future of Med Grid is rarely discussed, the Desertec Industrial Initiative (DII), once considered a groundbreaking project, was closed down in 2014.[11] Second, in the North Sea, the trend is not toward an integrated network structure but toward bilateral cable projects such as the NorGer between Norway and Germany (600 MW), set to begin operations by 2023, and between Norway and the UK (600–1000 MW), scheduled to open by 2022. In addition, the NorNed cable linking Norway and the Netherlands has been in operation since 2009 (700 MW, see Gerbaulet and Weber 2014). Third, in the Baltic Sea, the idea of a HVDC connecting in a meshed "Baltic Ring" has also not been pursued further.[12]

Among the reasons for the delay in developing a supergrid architecture are geopolitical changes in the partner countries (e.g., North Africa, Russia), the public debate about the need for additional infrastructure, and uncertainties about the future mix of electricity generation and demand, casting doubt on optimistic predictions of future infrastructure capacity requirements. In addition, some uncertainty remains around the technical feasibility of a pan-European DC grid, particularly the "DC breaker," an important element of multiterminal HVDC networks (although a patent for the DC breaker has been given to an equipment supplier). With respect to the design of future grids, Egerer et al. (2016) have proposed that a meshed AC network—that is, an upgrade of the existing structure—is more likely to emerge than an independent DC superstructure.

However, the largest obstacle to a pan-European DC supergrid lies in the institutional and regulatory requirements that would first need to be addressed, including issues of joint planning, financing, and operation of a multinational electricity grid. With respect to planning, a formal procedure has been enacted in Directive 2009/72/EC on the third internal electricity market to promote what are known as Ten-Year Network Development Plans (TYNDP). So far, however, these plans have been pieced together out of TSOs' national transmission plans, which have to be approved solely by national regulatory authorities. With respect to financing, no institutional vehicle has been proposed that would make it possible to share costs and benefits in a fair manner among the parties involved, let alone one that has been applied in practice. A European regulation developed for "Inter-TSO compensation" (ITC) exists on paper,[13] but is usually used only to compensate for system losses, not for investments (Ruester et al. 2012). With respect to regulation, access, and so on, national regulatory authorities (NRAs) remain the only entity to which TSOs are

---

[11]DII was dissolved on October 13, 2014, and some of the personnel was transferred to different previous members, such as ACWA Power (Saudi Arabia), RWE (Germany), and SGCC (China).

[12]One of the bilateral connections proposed in the "Baltic Ring" project in the 1990s has been realized: the back-to-back DC-linking between Poland and Lithuania (2015: 500 MW, to be expanded to 1000 MW later on).

[13]See EC. 2010. Commission Regulation (EU) No 838/2010 of 23 September 2010 on Laying down Guidelines Relating to the Inter-Transmission System Operator Compensation Mechanism and a Common Regulatory Approach to Transmission Charging.

accountable. Last but not least, it is still not clear under which regulatory structure a joint DC supergrid would be managed.[14]

11.3.2.2 Some Evidence from a Model Comparison

One other possible explanation for the lack of successful supergrid projects might be that the demand for transmission capacities was overestimated in the first place. In fact, a more realistic picture of transmission requirements emerges from our own modeling work, suggesting that the need for transmission expansion is lower than generally assumed—not only in Germany (see above Chap. 7) but also at the European level. In a Europe-wide model comparison ("EMF 28": The Effects of Technology Choices on EU Climate Policy), we applied our European ELectricity Model (ELMOD) to estimate transmission requirements for different low-carbon European scenarios.[15] This subsection outlines details of the model comparison in the search for a low-cost low-carbon approach to electricity generation for Europe (Egerer et al. 2016).

The model calculates a cost-minimizing electricity transmission path for a given set of low-carbon generation scenarios up to 2050. The EMF 28 model comparison suggested that three pre-defined, exogenous scenarios for generation should be used: first, a pathway with a 40% GHG-reduction target, based on conventional generation, including nuclear, coal, etc. (here called the 40% DEF scenario); second, a pathway with an 80% GHG emission reduction target with default conventional generation technologies (80% DEF); and third, a different 80% pathway with a focus on renewable generation (80% GREEN).

Table 11.1 and Fig. 11.3 summarize the main results of the modeling: Table 11.1 indicates the number of transmission line kilometers for which the model suggests an upgrade or newbuilds. The total number of line kilometers increases between the 40% scenario (27,978 km) and the 80% scenarios, but is almost identical among the 80% scenarios, and it is even slightly lower in the 80% GREEN scenario (50,993 km) than in the 80% DEF scenario (52,424 km). The two 80% scenarios differ; however, with respect to their distribution between DC and AC cross-border lines, as the 80% GREEN scenario has a higher share of DC cross-border lines. The results also reveal an interesting finding that is often ignored in aggregate analysis: domestic upgrades play an important role in all scenarios, and outweigh cross-border investments significantly (over 2:1 in the 40% DEF scenario, and over 3.5:1 in the 80% scenarios).

---

[14]See von Hirschhausen (2010) for an early account of these issues.

[15]ELMOD is a techno-economic model developed at Dresden University of Technology (Chair of Energy Economics), the Berlin University of Technology (Workgroup for Infrastructure Policy), and the German Institute for Economic Research (DIW Berlin) (see Leuthold et al. 2012; Egerer et al. 2014); it is a large-scale spatial model of the European electricity market including both generation and the physical transmission network (DC Load Flow Approach). The model optimizes line investments for specific years.

Table 11.1   Electricity transmission expansion in Europe up to 2050 (by line type)

In km	DC Cross-border	AC Cross-border	AC National	Total
40% DEF	4174	4611	19,194	27,979
80% DEF	5346	7173	39,905	52,424
80% GREEN	7057	4138	39,798	50,993

Source: Egerer et al. (2016)

          40%DEF   80%DEF   80%GREEN

Fig. 11.3   DC grid infrastructure investments in various 2050 scenarios. Source: Egerer et al. (2016)

AC (alternating current) transmission expansion clearly dominates DC (direct current) grid expansion. The 80% DEF and 80% GREEN scenarios generate slightly more investment in cross-border lines (+3700 km/+2400 km, respectively) and significantly more in the AC network within countries (+20,700 km/+20,600 km) than the 40% DEF scenario. In the 80% GREEN scenario, the higher renewable share results in higher DC cross-border investments in the North and Baltic Sea regions and additional AC lines as integration measures at the connection nodes of the DC lines with the AC network (e.g., in Sweden, France, and Germany, known as "hinterland connections"). On the contrary, the solar capacities in Southern Europe do not seem to generate a corresponding level of DC connections in the region. The 80% DEF scenario also requires some investments for the integration of increasing renewable generation; yet the renewable share is lower than in the 80% GREEN scenario as the scenario allows for more CCTS technology and an overall constant level of nuclear power in the European electricity system. This combination of a lower renewable share and a shift in the spatial allocation of nuclear and coal power plants results in lower investments in the North and Baltic Seas region and higher network development in Central and Eastern Europe.

Figure 11.3 depicts the investments in DC lines, realized among the 23 options provided by the backbone architecture. Curiously, and contrary to the common belief of "pan-European" electricity highways, the model only invests in the DC offshore cables between the non-synchronized networks of Ireland, Great Britain,

Scandinavia, and continental Europe, but not in the onshore DC cables or in any DC cable in the South of France (with one exception in the DEF 80% case).[16] Far from suggesting any form of a "supergrid," the analysis instead reveals more modest investments in HVDC technology altogether.

In terms of cross-border infrastructure development, as well, the ELMOD calculations identify significantly lower transmission requirements than those estimated by the European Commission. Thus, the figures used by the European Commission (calculated by the PRIMES model, approx. €142 bn) are about 3–5 times higher than the ELMOD investments (€31 bn for the scenario with 40% GHG emission reductions, and €51 bn for the 80% GHG reductions). The ELMOD estimates also show a different investment dynamic: While the European Commission has a flat, regular investment path of around €40–50 bn per decade, the ELMOD path flattens out after 2020 for the 40% GHG emission reduction scenario (€2–7 bn per decade only later); the same trend holds for the 80% GHG reduction scenario, except for the peak in the final decade (2050).

Table 11.2 shows the results of a comparison between a high-granularity model (ELMOD) and the more aggregated PRIMES model on the expected infrastructure needs, comparing the interconnector investments in the two models for comparable scenarios[17]: The Energy Roadmap requires more than five times higher investment than the disaggregated ELMOD model. The divergence increases over time: Whereas up to 2020, the amount was 2.5 times higher (€21,900 mn in PRIMES vs. €8652 mn in ELMOD), this factor increases to more than eight in the period 2031–2050 (€50,800 mn in PRIMES, vs. €6262 mn in ELMOD). As Egerer et al. (2016) show, this effect can partly be explained by the importance of domestic network upgrades in ELMOD, made possible by the high level of granularity, featuring not less than 3523 substations and 5145 lines, whereas the electricity module of PRIMES works with national aggregates (one country—one node).

Table 11.2 Comparison between PRIMES and ELMOD for investments in electricity interconnector capacities

Km	Before 2020	2021–2030	2031–2050	Total
Elmod	8652	2573	6262	17,488
Primes	21,900	21,200	50,800	93,900

Source: Egerer et al. (2016)

---

[16]Compared to the high mitigation scenarios, the 40% DEF scenario has one more cable connecting Great Britain to Germany but one less connecting it to Norway. Sweden is linked to continental Europe by one additional cable in the 40% DEF scenario, two in the 80% DEF scenario, and three in the 80% GREEN scenario. Overall higher DC investments in the 80% GREEN scenario also indicate a stronger integration of the non-synchronized transmission systems around the North and Baltic Seas.

[17]ELMOD: scenario 80%GREEN, PRIMES: scenario "high RES".

11.4   Natural Gas Infrastructure

11.4.1   Uncertain Perspectives for Natural Gas and Its Infrastructure Requirements

11.4.1.1   The "Dash for Gas" Is Over in Europe

Infrastructure developments in the European natural gas sector have little dynamics but reflect two major uncertainties in the sector. The main uncertainty about future infrastructure needs stems from the uncertain prospects for natural gas consumption, which have declined over the first half of this decade. Whereas experts once heralded the dawning of a "dash for gas" (IEA 2004), a fuel they thought would pave the way to a low-carbon transformation, the current thinking has shifted. Second, security of natural gas supplies continues to have shaping influence on the European natural gas infrastructure.

Natural gas was long considered to be an ideal partner in electricity generation for variable renewables due to its high flexibility, versatility, as well as the diversified supply sources. In 2012, natural gas was expected to become a "key for the energy future of Europe".[18] The International Energy Agency's idea of a "golden age" (IEA 2011) was also based on the understanding that natural gas is the natural complement to variable renewable energy: when the wind is not blowing and the sun is not shining, natural gas—a relatively low-carbon fuel—can take the lead in a low-carbon merit order. Some evidence favoring the possibility of a natural-gas-driven low-carbon transformation came from the USA and Japan (Neumann and von Hirschhausen 2015). With the idea that natural gas could play a key role, European transmission system operators conceived bold development plans for pan-European network development—similar to their plans for the electricity sector—in response to the perceived need for more natural gas supplies. These were supported by generous calculations of natural gas needs by the European Commission. At least €70 bn were expected to be invested in pipelines, LNG terminals, and the necessary connecting infrastructure up to 2020 (ENTSO-G 2013).

However, the situation has changed in Europe, and the prospects for natural gas have deteriorated. One reason is the continuously low CO_2 price: instead of increasing to levels where a fuel switch from coal to natural gas could occur, CO_2 prices have remained low and have not affected fuel choices at all. Thus, in a world of dirty—but cheap—coal plants with ample existing capacities, natural gas demand has not developed as expected. The rapid decrease in the cost of renewables has also contributed to the relatively poor outlook for natural gas: in combination with diverse storage technologies on the verge of becoming competitive (see Chap. 8),

---

[18]Speech of European Energy Commissioner G. Oettinger at the 10th Gas Infrastructure Europe Annual Conference in Krakow, Poland, May 24, 2012, quoted in the GIE article available at http://www.naturalgaseurope.com/oettinger-europe-gas-market (accessed January 23, 2013).

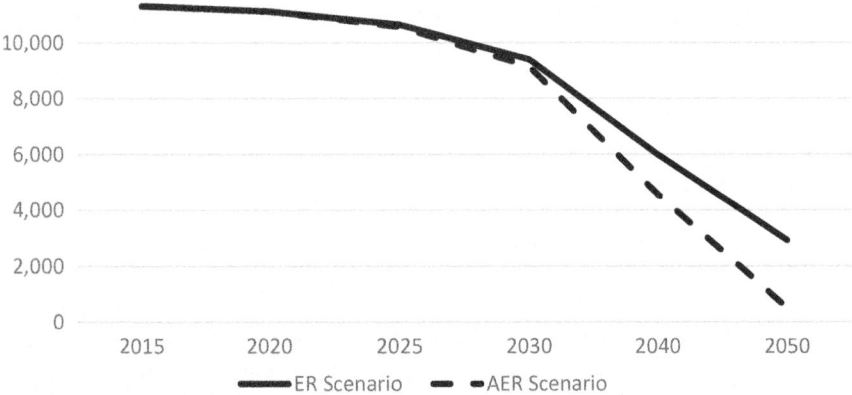

Fig. 11.4 The expected development of natural gas demand in the OECD Europe (PJ/a). Source: Greenpeace et al. (2015)

even variable renewables no longer require significant conventional backup capacity, thus depriving natural gas of its expected role as a "complement" to renewables.

Contrary to the expectations that prevailed around the turn of the century, natural gas no longer represents a cornerstone of the low-carbon transformation in Europe. With hindsight, the high expectations of natural gas demand were not based on solid energy economic analysis, but rather wishful thinking. In fact, when one takes the EU climate targets serious, natural gas is unlikely to play a major role in the 2050 modeling exercises: As laid out in detail in Chap. 13, natural gas use is likely to decline, both in electrification, industry, and heating. While the industry itself continues to hope for increased consumption, independent analysts do not share such expectations, see the analyses by Abrell et al. (2013), Aoun et al. (2015).[19]

Taking an NGO perspective to determine pathways of fossil fuel consumption under the climate constraint, Greenpeace et al. (2015) defines a world under which the energy-related carbon dioxide emissions are cut to zero by the year 2050 (Energy Revolution Scenario). In the more progressive scenario (Advanced Energy Revolution, AER) new technologies are implemented even faster leading to a complete decarbonisation of the power, heat and especially the transportation sector. As a result for the natural gas sector, the expected future demand is less promising than what the EU forecasts. As illustrated in Fig. 11.4, the natural gas demand in OECD Europe diminishes only slightly until 2030, whereas a rapid decline from then onwards is expected.

---

[19]In the EMF-28 model comparison, Abrell et al. (2013) expect natural gas consumption in the default case (-40% GHG emissions) to decrease from about 18 EJ (2010) to below 15 EJ (2050) in the EU-27.

11.4.1.2   Some Insights from Natural Gas Infrastructure Modeling

In order to gain some insights into current and future infrastructure needs, we refer to model exercises carried out with the "Global Gas Model" (GGM), a partial equilibrium model of the natural gas market that numerically simulates global natural gas production, consumption, and trade flows (Holz et al. 2015, 2016).[20] In particular, the model makes it possible to identify infrastructure bottlenecks in the existing European natural gas system and the potential need for upgrades. The results are summarized in a paper by Holz et al. (2016).

Contrary to the expectations of a need for significant pipeline expansion published by the pipeline transmission system operators (e.g., ENTSO-G 2013), the results from the GGM indicate a rather limited need for additional infrastructure. The model results clearly show that infrastructure investments are dominated by external forces. Figure 11.5 shows the expansions calculated, including both

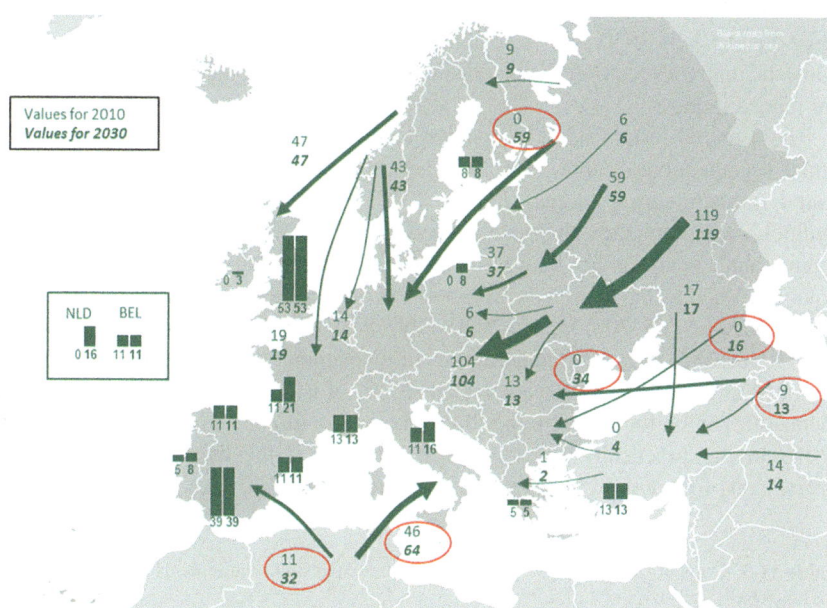

Fig. 11.5 Pipeline and regasification capacities to Europe in 2010 and 2030 (in bcm/year). Source: Holz et al. (2016)

---

[20]Egging (2013) provides a description of the main model setup and features. The model represents the supply chain structure of the sector, and allows a high level of detail, featuring demand seasonality, potential market power of trading agents, as well as endogenous investment in storage and transport capacity, both along the LNG supply chain and regarding pipeline connections. Whereas Egging (2013) presents a stochastic model, here we report results of deterministic versions with a particular focus on Europe and updated data sets. 25 of the EU-28 countries are incorporated individually in the data.

cumulative pipeline expansion within and to Europe up to 2050. The focus of—
economically rational—investments clearly lies on linking Europe with future supplies from the Caspian region and North Africa. The endogenous pipeline expansions serve primarily to accommodate a shift of import flows away from Russia and
LNG (which mainly serves the Asian market in the future). The pipeline expansions
take place along three main import routes: the first runs from Africa to Spain and
from there to France. The second runs from Africa to Italy and further on to Austria
and Germany. The third runs from the Caspian region, possibly via new pipelines
such as White Stream.

Most infrastructure expansions will be completed by 2020/2025 and are already
underway and/or are part of European investment plans (ENTSO-G 2013). These
plans include only small additions to existing LNG import (regasification) capacities
and several pipeline expansions that were decided in recent years. In contrast to the
expansions from the South, there seems to be no need to expand the pipeline
infrastructure from Scandinavia (Norway) to continental Europe. In total, the results
of the GGM point to investment needs in Europe of around €25 bn by 2050, of which
more than 65% by 2020, and more than 94% by 2025. Thus, if one abstracts from the
domestic upgrades required by increasing imports, the European infrastructure needs
are indeed modest.

These model results stand in contrast to the results reported by several international organizations. Table 11.3 provides estimates from the European Commission's infrastructure package and the IEA's (2014) "World Energy Investment
Outlook" and compares them with the Global Gas Model results. The absolute
figures and the dynamics between the two former and the latter estimates differ
notably: Whereas the European Commission and IEA estimate infrastructure needs
in the range of €65–70 bn up to 2020, Holz et al. (2016) only arrive at €23.6 bn—
until 2030. With regard to the dynamics of investment, the IEA predicts an increasing need in the following decade (2020–2030) of €100 bn, arriving at total investments of €165 bn for the entire period analyzed (up to 2030). In contrast, Holz et al.
(2016) predict practically no additional investments in the 2020s, and estimate total
investments for natural gas pipeline infrastructure until 2030 at €24.3 bn.

Table 11.3 Comparison of natural gas infrastructure expansion requirements within and into the
EU (in € bn)

	Until 2020	2020–2030	Until 2030 (sum)
European Commission (2011)	70	n.a.	n.a.
IEA (2014)	65	100	165
Holz et al. (2016)	23.6	0.7	24.3

Source: von Hirschhausen et al. (2014)

11.4.2   EU Natural Gas Sector Resilient Against Supply Shocks

Could Europe still suffer from supply shocks even if its infrastructure expansion needs are modest? In this subsection, we analyze the potential effects of specific supply interruptions from the East (Russia) and/or from the West (Netherlands). The results show that the European natural gas sector is indeed capable of resisting such supply shocks because it has a well-developed import infrastructure (LNG and pipelines) and access to a diverse set of suppliers.

11.4.2.1   Disruptions in the Supply of Russian Natural Gas Can Be Dealt With...

The situation in the late 2010s differs from that in the previous decade, when the low-carbon transformation in Europe had just begun, particularly due to the new political conflict between Russia and the EU. Russia has been a key supplier of natural gas to large parts of Europe for many decades. Hence, natural gas supply security is a particularly sensitive issue both in Germany and in the EU, and infrastructure links with Russia are key in this respect.[21] Previously considered a strategic partner, the EU weakened its strategic energy dialogue with Russia following the annexation of Crimea by Russia in 2014 and its support to the military separatist troops in the war in Ukraine. In contrast to the 1990s, a period characterized by "winds of change," the current decade has clearly been marked by a disturbance in European-Russian relations and the political destabilization of Ukraine. These shifts have rekindled concerns in the EU, the Eastern EU Member States, and Ukraine about energy supply security, particularly with regard to the potential threat of natural gas supply interruptions by Gazprom, the Russian natural gas export monopolist. This subsection therefore analyzes different aspects of the European natural gas supply and the role of Russia and Gazprom in it, with a focus on European policies designed to increase resilience against physical supply shocks.[22]

The share of natural gas imports from Russia in total primary energy supply in the European Union is relatively modest, below 25% on average (Holz et al. 2014, 3). Also, the resilience of the European natural gas infrastructure and supply diversification have significantly improved since the natural gas crises of 2006 and 2009 and even more so since 2014. In fact, since then, strategic efforts were undertaken to

---

[21]This subsection draws on Holz et al. (2014).

[22]Gazprom still controls the majority of natural gas production in Russia, and in 2013 it produced around 75% of total Russian natural gas of 600 bcm. Total exports have been fairly constant over the course of the current decade at slightly below 200 bcm/a, with 60% of exports going to non-CIS countries in 2013. Richter and Holz (2015) provide a detailed analysis of disruption scenarios of Russian natural gas supplies to Europe.

improve the interconnectivity of the cross-border European natural gas infrastructure, and as a result, diversification has become easier. As seen in Fig. 11.5, Europe is served by large and small export pipelines from a variety of suppliers: Russia and the Caspian region in the East, Norway in the North and North Africa (mostly Algeria, but also Libya) in the South. The figure also highlights Gazprom's ownership of Russian export pipelines in Europe.[23]

In order to analyze the resilience of the European natural gas supply to interruptions, we have simulated future patterns of natural gas production, consumption, and trade, with a specific focus on potential infrastructure expansion needs. The Global Gas Model (GGM) introduced above is used to analyze counterfactual scenarios such as the disruption of pipeline capacity between Russia and Europe: The GGM *Base Case* is set up in line with projections of the New Policies Scenario (NPS) of the World Energy Outlook 2013 (IEA 2013), a moderate climate policy scenario. In addition, two scenarios have been constructed around the disruption of Gazprom majority-owned infrastructure (for a detailed description of the scenarios and results, see Holz et al. (2014).

- In the first scenario, "UKR disruption," it is assumed that all pipeline connections to Ukraine, which serve to deliver Russian natural gas to Europe, are interrupted; hence, no transit via Ukraine can take place.
- In the second scenario, "Gazprom," all infrastructure that is majority-owned by Gazprom or its subsidiaries, is interrupted.

Figure 11.6 summarizes the effects of the two disruption scenarios and their relation to the base case, both with respect to total consumption and the supply structure. In the UKR disruption scenario, the EU is only slightly affected, with small average reductions of consumption levels (by 2%, or 11 bcm), but the deviation across countries is large, with Eastern Europe being most severely affected.[24] By contrast, the "Gazprom" disruption scenario has a stronger impact on EU countries: At the aggregate level, EU consumption in 2015 is reduced by 10%, or 53 bcm.

However, modeling results also indicate that the European gas supply can be further diversified, and that infrastructure is available to accommodate this diversification. The shortfall of Russian supply to some countries ("UKR Disruption") or all European countries ("Gazprom") in 2015 is compensated by an increase in domestic European production as well as by imports from other world regions. While domestic EU production is only marginally increased in the UKR disruption scenario, production is larger in the Gazprom scenario, namely by 5%, or 8 bcm relative to the

---

[23]Russia's project to expand its pipeline capacities through the Baltic Sea ("Nord Stream 2") has sparked debates in Europe, particularly in the context of growing geopolitical disputes. See Holz et al. (2014, 26) for a detailed list of the export pipelines from Russia to Europe, and Neumann et al. (2018) for a critical assessment of the investment project.

[24]In particular, in Croatia, Hungary, and Romania, consumption is reduced substantially by more than 20% but also in Austria, the transit disruption effect is notable (-4% consumption in UKR Disruption relative to the Base Case).

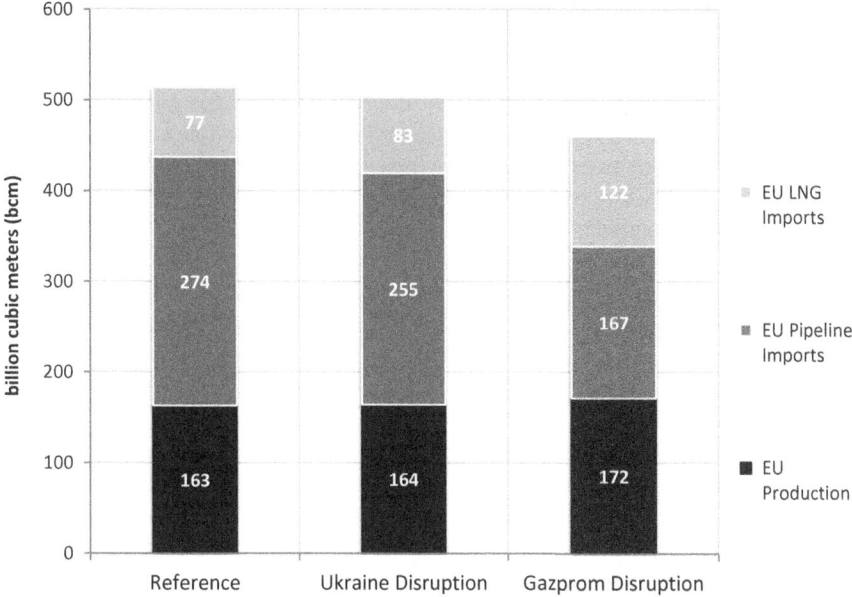

Fig. 11.6 EU supply structure in 2015 across scenarios (bcm). Source: Holz et al. (2014, 57)

base case. In the Gazprom scenario, the shortfall of 110 bcm (14 bcm) imports from Russia (and the Caspian region) relative to the base case is compensated by the increase of 8 bcm in domestic production and by additional imports from other suppliers amounting to 62 bcm.[25] The share of LNG imports is substantially increased (+45 bcm, or almost 60% higher in the Gazprom scenario than in the base case), while total pipeline imports drop significantly, despite small increases from North Africa and Norway.[26]

In essence, the model-based analysis of two supply disruption scenarios confirms that the real potential threat of Gazprom arises for Ukraine (and Belarus) and Eastern Europe, and much less in Central and Western Europe. Mainly Russia's East European neighbors are severely affected by the Ukraine-disruption scenario: Romania, Croatia, Hungary, and above all Ukraine. In contrast, West European countries have multiple options for diversification and are, therefore, less affected: They can compensate for reduced imports from Russia through domestic production, and higher pipeline and LNG imports from other suppliers.

---

[25]Concretely, imports from Africa +18 bcm; Middle East +19 bcm; South America +15 bcm, and from Rest of Europe +10 bcm; the remaining 53 bcm reflect the reduction in EU consumption.

[26]In the Gazprom scenario, the increase of LNG imports to the EU comes mainly from Qatar, African countries like Nigeria, Algeria and Egypt, and Trinidad and Tobago.

11.4.2.2 ... And Earthquakes Disrupting Dutch Natural Gas Supplies are Manageable as Well

In the 2010s, another major shock to the European natural gas supply occurred, coming from earthquakes in the natural gas production region around Groningen, Netherlands. However, an extension of the modeling work described above indicates that it is possible to compensate for this supply shock as well and that European supply security is ensured.[27]

Since the 1960s, the Netherlands have nevertheless been one of the leading European natural gas producers, with 86 bcm of natural gas in 2013, corresponding to 20% of EU consumption. Approximately one third of the natural gas produced in the Netherlands is consumed directly in the country—mainly for power and heat generation—and two thirds is exported to neighboring countries in northwestern Europe, namely to Germany, Belgium, Luxembourg, and France. However, the role of the Netherlands as Europe's natural gas supplier will have to decrease considerably in the future. Intensive natural gas extraction in the northeast of the country has triggered an increasing number of earthquakes since 2010, in particular in the Groningen province. Under public pressure and growing safety concerns, the cap on natural gas production at the giant Groningen field has been lowered repeatedly and reached 30 bcm in 2015, whereas real production was only 25 bcm that year. In early 2018, it was decided to completely end natural gas extraction from the Groningen field until 2030.

Similar to the Gazprom disruption case, Holz et al. (2017) simulated a scenario of low Dutch natural gas production to check the effects on European gas supplies for the period 2015–2040. In the first scenario, the impact of reduced natural gas production in the Netherlands on the European natural gas market is analyzed. In the second scenario, this lower production rate is combined with a scenario of Russia disrupting its supply of natural gas to Europe. The lower production at the Groningen field is envisaged to be 33 bcm in 2015 instead of the 39 bcm originally planned, which is followed by further cuts in subsequent years.

Figure 11.7 summarizes the results: the lower line shows the shortfall of Dutch natural gas output, amounting to almost 30 bcm in 2040 compared to previous forests. Due to a slight price increase (+0.7%), European consumption is modestly reduced compared to the base case, but only by 3 bcm in 2030 and 2040. Imports of natural gas into the EU take care of the rest: the large majority of additional imports comes in the form of LNG from North America (+25%), South America (+17%), and the Middle East (+10%). One reason for the relatively low impact on the European natural gas market is the oligopolistic market structure combined with the availability of diverse import infrastructure, since the market is attractive for a large number of suppliers due to high prices.

Additional model results indicate that the EU natural gas sector is even resilient if the Dutch supply shock adds up to a disruption of Russian supplies. In this case, the

---

[27]This section draws on Holz et al. (2015, 2017).

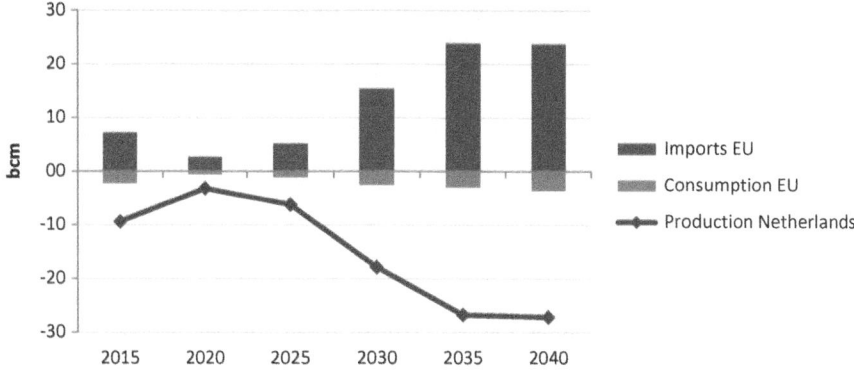

Fig. 11.7 Differences to earlier assumptions on EU natural gas consumption and imports as well as Dutch production in case of a Dutch supply shock (in bcm). Source: Holz et al. (2017)

traditional importers of Dutch gas (mainly Germany, Belgium, and the UK) would need to diversify further. For example, Germany would import about 9 bcm less gas from the Netherlands in 2040 but would increase its imports from other suppliers (Holz et al. 2017).

11.4.2.3 Nord Stream 2 Pipeline Not Needed for EU Supply Security

In the context of European infrastructure projects, a controversial debate emerged about the economic and political arguments of another large-scale pipeline connecting Russia directly to North Germany, called "Nord Stream 2", thus preventing Ukraine from its traditional transit revenues.[28] The project consists of the extension and new construction of inlet and outlet natural gas pipelines in Russia and Germany and the main line of two parallel offshore pipelines through the Baltic Sea. The offshore pipeline is largely parallel to the Nord Stream pipeline (approx. 1200 km), and the investment needs for the entire Nord Stream 2 project are estimated at 17 billion USD.[29]

In a model-based study of this question, we concluded that the planned pipeline project Nord Stream 2 is not necessary to secure natural gas supplies for Germany and Europe. The energy consumption forecasts on which the project is based, especially the EU Reference Scenario, significantly overestimate natural gas demand in Germany and Europe. On the supply side, there will be no supply gap if Nord

---

[28]See for details Neumann et al. (2018).

[29]On the Russian side, a new pipeline from Ukhta to Gryazovets (970 km) and the extension of the Gryazovets-Volkhov connection to the Slavyanskaya compressor station, the entry point to the Nord Stream 2 offshore pipeline, are required. See for details (Sberbank Investment Research 2018).

Stream 2 is not built. Different profitability studies suggest that high losses up to the billions can be expected from the project (Neumann et al. 2018).

The optimistic assessment of natural gas is now generally broadly criticized: The EU Reference Scenario used to plan Nord Stream 2 postulates a roughly constant demand for natural gas, but its assumptions and methodology are controversial: The energy system model used for the Reference Scenario calculations, PRIMES, systematically favors fossil fuels, especially coal and natural gas (as well as nuclear power, which is not discussed here), whose significance is structurally overestimated, especially in the energy sector. The systematic use of a technology that does not exist, CO_2 capture technology (Carbon Capture, Transport, and Storage, CCTS), strengthens the bias in favor of fossil natural gas: The costs for CCTS are erroneously set so low that this technology would be used starting in 2020 for economic reasons alone; this is not plausible, neither in any EU member state nor worldwide. In contrast, the importance of renewable energies is systematically underestimated by ignoring technical improvements and by overestimating costs. The rapid developments in storage technology are also ignored in the PRIMES model by using inflated cost values (Neumann et al. 2018).

European natural gas supply is already diversified, and can increase its robustness further. The Nord Stream 2 pipeline is not necessary for European supply; on the contrary, Europe could assure future consumption needs even without *any* Russian natural gas export. In addition to the regional supply via natural gas pipelines, highly diversified LNG deliveries guarantee long-term supply security. The possibility of landing LNG at numerous import terminals along the European coasts and subsequently implementing efficient distribution through the existing pipeline system strengthens supply security. Currently, the capacity utilization of existing LNG import terminals is very low: in 2016, only 25% of existing import capacities in Europe were used. This also indicates that there will not be an infrastructure shortage (Neumann et al. 2018).

Neither is Nord Stream 2 a lucrative business. Building the pipeline will not increase Russian natural gas sales in Germany or the EU, and the additional low revenue Nord Stream 2 would bring is offset by very high costs; as a result, no profit can be made from the construction of Nord Stream 2 (Roar Aune et al. 2017). An analysis from the Russian investment bank Sberbank concludes that Nord Stream 2 destroys rather than creates value; the costs of Nord Stream 2 (17 billion USD) is compared with the savings of approximately 700 million USD per year from avoiding transit through Ukraine. Additionally, it is assumed that natural gas sales in Europe will not increase and that the pipeline is operating at 60% capacity. Based on these assumptions, the present value of the investment will be negative at six billion USD (approximately five billion EUR) (Sberbank Investment Research 2018).

11.5 CO$_2$ Pipelines

11.5.1 Early Expectations for a European-wide CO$_2$ Pipeline Network

The case of CO$_2$ pipelines is unique because this type of network does not exist yet in Europe, but was conceived to become an integral part of the European low-carbon transformation. The fact that it still has not fulfilled this role again demonstrates the difficulty of developing large infrastructure systems of the supergrid type, and the discrepancy between planning and the more difficult implementation on the ground. In fact, the CO$_2$ pipeline case study is extreme because, as of today, not a single kilometer of CO$_2$ infrastructure of the planned supergrid has been built, and it seems unlikely that this will change in the foreseeable future.

Ideas of a Europe-wide CO$_2$ pipeline network gained popularity in the 2000s, in accordance with the vision of large-scale carbon capture roll-out (see also the previous Chap. 8). At least since the IPCC (2005) Special Report on Carbon Capture and Storage, this technology was considered essential to reducing CO$_2$ emissions from the coal sector and fossil fuels more generally. The technology indeed had a particular appeal because it was supported by stakeholder groups that were at odds under other circumstances, including the traditional fossil fuel industry and climate activists (von Hirschhausen et al. 2013). As a result, not only carbon capture but also carbon transport received significant attention at the European level and in some Member States. All the major top-down models assumed that large-scale CO$_2$ abatement would be possible through wide availability of CCTS as of 2020. Since the latter half of the 2000s, all infrastructure development plans for Europe, as well as the official EU Reference Scenarios and other planning documents, have included substantial CO$_2$ pipeline capacities: The 2011 Reference Scenario of the EU Energy Roadmap 2050 forecasted CCTS capacity in the electricity sector of 11 GW by 2030 and of 100 GW by 2050 (EC 2011b).

Consequently, the European public policy approach long favored a pan-European solution centered around a European Directive to impose CCTS standards and options on the Member States (EC 2009). With respect to the underlying infrastructure, the idea of a pan-European CO$_2$ pipeline network rapidly gained credence, reinforced by engineering studies such as Ainger et al. (2010), Neele (2012), and by the Joint Research Centre (JRC) of the European Commission itself (Morbee et al. 2010). Following the traditional "the-more-the-better" philosophy, CO$_2$ pipelines came to be seen as a precondition for the success of CCTS. Hence, a proactive approach appeared most suitable, whereby the European level should share responsibility for the gradual development of a pan-European CO$_2$ pipeline network. The JRC model suggested a network of 20,000 km of CO$_2$ pipelines by 2050, for instance, from Slovakia to the North Sea, from Lithuania across Finland to the West coast of Norway, and from the Italian Alps to the French Atlantic (see Fig. 11.8).

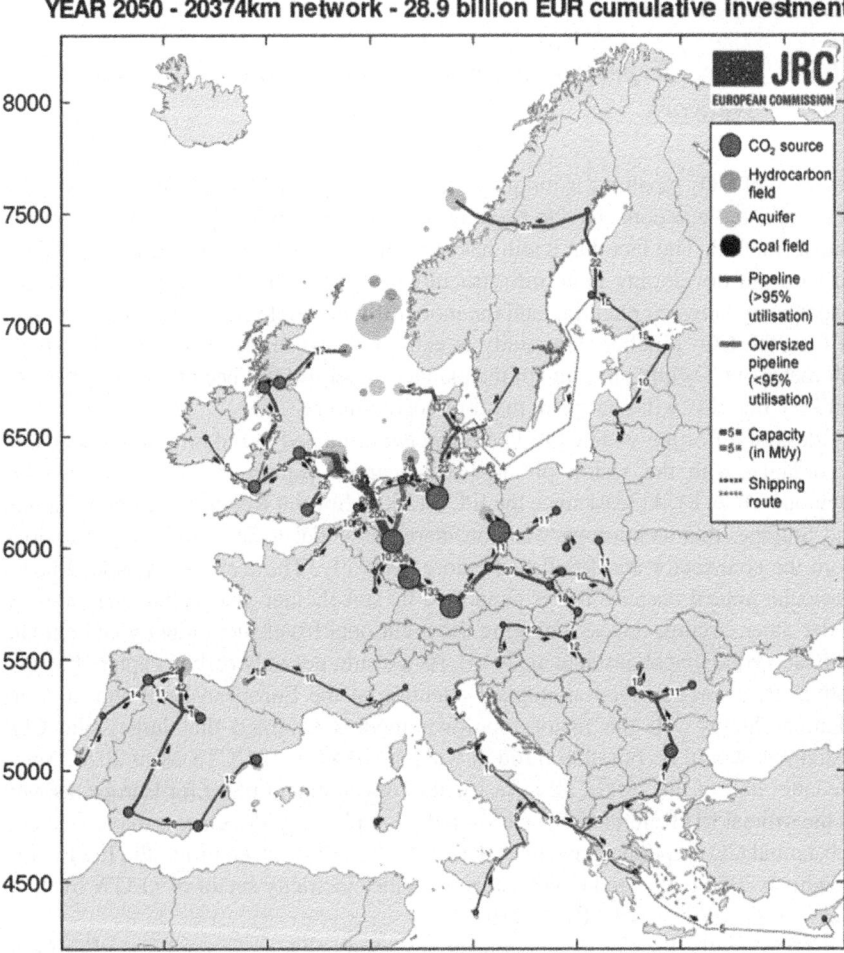

Fig. 11.8 Design of a Europe-wide CO_2 pipeline network. Source: Morbee et al. (2010, 10)

While studies differ with respect to the concrete figures, the general trends they have identified are similar. Table 11.4 summarizes the key estimates of various studies, including one by ourselves (Oei et al. 2014) on the assumed CCTS technology development. As seen in the table, there was a certain consensus among these studies that by 2050, CCTS would represent around 100 GW of electricity generation. Annual storage was considered to be between 500 and 1000 mn t per year, and the length of the pipeline network would lie in the range of 20,000 km. Assuming the commercial availability of CCTS technologies, the long-term financing instruments for a trans-national CO_2 pipeline system would require a substantial role of European

Table 11.4 Estimates of CCTS technology deployment in 2020 and 2050

Technology deployment	Source model	Year 2020	2050
Power Generation in GW	OECD/IEA[a]	5.5	140
	IEA[b]	4.9	77
	PRIMES[c]	3	108
Storage in Mt CO_2 per year	OECD/IEA	37	990
	IEA	52	550
	JRC[d]	36	900
	PRIMES	18	347
	CCTS-Mod[e]	0/0	450/750
Pipeline length in km	OECD/IEA	1400	27,500
	JRC	2005	20,374
	CCTS-Mod	0/0	9800/20,400

Source: Mendelevitch (2014)
[a]IEA. 2010. Energy Technology Perspective 2010. Paris, France: International Energy Agency. Blue Map Scenario; values for transport infrastructure are averages of spans given in the study
[b]IEA. 2012. Energy Technology Perspective 2012. Pathways to a Clean Energy System. Paris, France: International Energy Agency
[c]EC (2011a): Values are averages from scenarios for Energy Roadmap for 2050
[d]Morbee et al. (2010): InfraCCS model with input data from PRIMES Baseline Mitigation 1 Scenario009. Model used in assessment of European CO_2 transport infrastructure requirements
[e]Oei et al. (2014): Modeling results of CCTSMOD runs: Scenario BAU (ON 75)/OFF 75

financing, both for further R&D as well as for commercial roll-out of the technology after 2020.

11.5.2   Current Status: The Network That Did Not Become a Reality

Somewhat similar to the European electricity supergrid, the European CO_2 pipeline supergrid has not taken off and it appears unlikely to do so in the foreseeable future. Chapter 10 has provided a detailed account of pilot projects in CCTS at the European and global levels, and discussed why this idea has not materialized as an option for a low-carbon energy transformation. Today, around a decade since the unveiling of ambitious plans for CCTS, there is very little progress to be seen. And more than ten years after the IPCC (2005) report and more than a dozen failed pilot projects—many funded substantially by the EU and its Member States—not a single EU demonstration project has been completed either in the energy or in the industry sector. Plans for a pan-European pipeline network have accordingly been shelved.

It is too early to judge whether future CCTS projects planned in Europe will fail or whether modest progress may be made on some in the coming years. If pilot projects are implemented, they will most likely be around the North Sea, a region with direct access to offshore storage sites. Using the captured CO_2 for enhanced oil recovery (CO_2 EOR) creates an additional revenue stream at some of these offshore locations.[30] Mendelevitch (2014) and Mendelevitch and Oei (2017) have developed scenarios of potential CCTS development for the North Sea region including the option of CO_2 EOR. In the case of purely national developments, pipelines could extend from potential onshore power plant sites into the North Sea. In other scenarios, a small CO_2 pipeline network around the North Sea, perhaps connecting sites in the UK, Norway, and the Netherlands, also remains a possibility. However, the times of pan-European CO_2 infrastructure planning are certainly over.

11.6   Conclusions

In this chapter we have analyzed the European approach to infrastructure development as part of its low-carbon transformation. As in the German energiewende, large-scale infrastructures, often called "supergrids," have been a key element of reform proposals. However, real developments have run somewhat counter to this hypothesis, as little progress has been achieved on large corridors, whereas bilateral and regional infrastructure initiatives have been more successful. The three previous sections have shown that large-scale supergrid infrastructures, once considered to be essential for the low-carbon transformation in Europe, have not developed as expected over the last decade, or, in the case of CO_2 pipelines, have not developed at all. However, the fact that supergrid corridors have not emerged as expected does not mean that "nothing" is going on or that infrastructure has no role to play in the European low-carbon transformation. It does, however, indicate that the initial vision of continental-wide corridors may have been unrealistic, and that infrastructure will have a more important role to play at the regional or the national level.

The literature on the role of infrastructure yields two interesting lessons: on the one hand, climate policy and infrastructure finance can complement each other usefully to reap a "triple dividend", i.e. climate, infrastructure, and economic growth (Edenhofer 2013). On the other hand, the Hirschman (1958) reasoning reminds us of the fact that infrastructure investments themselves are not the sole necessary ingredient in the low-carbon energy transformation. Rather, a low-carbon energy system requires low-carbon energy generation, which is largely unrelated to the issue of transmission infrastructure (see Chap.10 on generation). The "cables for carbon" argument even runs in the opposite direction: in a very CO_2-intensive energy system, the construction of more infrastructure may imply even higher CO_2 emissions.

The chapter reviewed a decade of developments in three important energy infrastructure sectors: electricity, natural gas, and CO_2 pipelines. Empirical evidence can be found in infrastructure planning by European grid operators and international

---

[30]CO_2-EOR is not an abatement technology, because a large part of the CO_2 pumped into the ground resurfaces later on.

organizations during the first decade of the low-carbon transformation: Whereas pan-continental infrastructure programs were heralded in the early days of the transformation by international organizations and independent think-tanks, there has been a trend away from these catch-all solutions towards more regional approaches in recent years. The three big electricity projects (EU-MENA, North Sea Grid, Baltic Ring) have not progressed substantially; natural gas pipeline developments have advanced smoothly, but were only partially able to reduce the dependency on Russia. And with respect to CO_2 pipelines, not a single kilometer of the proposed European-wide grid has been constructed. Overall, our own model-based analyses indicate that the infrastructure challenges of the low-carbon transformation are modest. European initiatives to foster cross-country infrastructure cooperation are important, as a large part of the identified requirements stretch across borders. But supergrid solutions seem not only infeasible but also unnecessary to master the transformation challenges.

Evidence reported in this chapter also suggests that more attention is currently being paid to regional and local infrastructure initiatives: whereas the focus in the first period of the low-carbon transformation was on larger-scale, pan-European infrastructure corridors, the current tendency favors hands-on, regional solutions with a much higher probability of realization. Therefore, in recent years, there has been a shift away from the "pan-continental" infrastructure models toward more regional approaches, supported by both modeling work as well as the industry itself. The Associations of European electricity and natural network operators, ENTSO-E and ENTSO-G, for instance, have decentralized its transmission planning from a centralized approach to six "Regional Groups" that work independently of each other (while of course sharing information). In the gas sector, a "natural" regional focus is provided by the existing pipeline infrastructure: As the Russian-Ukrainian natural gas crisis has shown, solutions such as reverse flows or cross-border storage use need to be sought regionally, and pan-European corridors are of little help in such situations. Model results show that even a disruption of Russian and Dutch gas supplies could be handled by a resilient EU natural gas infrastructure. CO_2 pipeline plans have also narrowed in scope from the European to the regional level, focusing in particular on the North Sea region.

References

Abrell, Jan, and Sebastian Rausch. 2015. Cross-country electricity trade, renewable energy and transmission infrastructure policy. ETH Zurich Working Paper WP3-2015/06. Zurich, Switzerland.

Abrell, Jan, Clemens Gerbaulet, Franziska Holz, Casimir Lorenz, and Hannes Weigt. 2013. Combining energy networks: the impact of Europe's natural gas network on electricity markets until 2050. Discussion Paper 1317. Berlin, Germany: DIW Berlin.

Ainger, D., S. Argent, and S. Haszeldine. 2010. Feasibility study for Europe-Wide CO_2 infrastructures. Study for the European Commission Directorate-General Energy Issue Rev.00 TREN/372-1/C3/2009. Leeds, UK: Ove Arup & Partners Limited.

Aoun, Marie-Claire, and Sylvie Cornot-Gandolphe. 2015. *The European gas market looking for its golden age? Les Etudes*. Paris/Brussels: Ifri.

Aschauer, David Alan. 1989. Is public expenditure productive? *Journal of Monetary Economics* 23 (2): 177–200.

Brancucci Martínez-Anido, Carlo. 2013. Electricity without borders – the need for cross-border transmission investment in Europe. Proefschrift/Dissertation, The Netherlands: Technische Universiteit Delft.

Buchanan, James M. 1965. An economic theory of clubs. *Economica, New Series* 32 (125): 1–14.

Czisch, Gregor. 2006. Low cost but totally renewable electricity supply for a huge supply area – a European/Trans-European Example. Kassel, Germany.

Davis, Steven J., Ken Caldeira, and H. Damon Matthews. 2010. Future CO_2 emissions and climate change from existing energy infrastructure. *Science* 329 (5997): 1330–1333.

DLR. 2006. Trans-mediterranean interconnection for concentrating solar power. Final Report. Stuttgart, Germany: Deutsches Zentrum für Luft- und Raumfahrt e.V. (DLR).

EC. 2009. Directive 2009/31/EC on the Geological Storage of Carbon Dioxide (CCS Directive).

———. 2011a. Energy roadmap 2050: impact assessment, Part 1/2. SEC(2011) 1565/2. Commission staff working paper. Brussels, Belgium: European Commission.

———. 2011b. Energy roadmap 2050: impact assessment, Part 2/2. SEC(2011) 1565. Commission staff working paper. Brussels, Belgium: European Commission.

———. 2011c. Energy roadmap 2050. Communication from the Commission to the European Parliament, the Council, the European Economic and Social Committee and the Committee of the Regions. Brussels, Belgium: European Commission.

ECF. 2010. Roadmap 2050 – a practical guide to a prosperous, low-carbon Europe. Technical analysis. Brussels, Belgium: European Climate Foundation.

———. 2011. *Power Perspectives 2030 – On the Road to a Decarbonised Power Sector*. Brussels: European Climate Foundation.

Edenhofer, Ottmar. 2013. The economics of uranium, fossil fuels and climate change stabilization – trade-offs, synergies and solutions. Keynote at the IAEE 2013 European Conference. Presented at the 13th European IAEE Conference, Düsseldorf, Germany, August 19.

Egerer, Jonas. 2016. Open source electricity model for Germany (ELMOD-DE). DIW Berlin Data Documentation 83. Berlin, Germany: DIW Berlin.

Egerer, Jonas, and Clemens Gerbaulet. 2009. European electricity transmission in 2050 – an engineering economic analysis of a combined AC-DC super grid based on renewable energy sources (RES). Vienna, Austria.

Egerer, Jonas, Lucas Bückers, Gregor Drondorf, Clemens Gerbaulet, Paul Hörnicke, Rüdiger Säurich, Claudia Schmidt, et al. 2009. Sustainable energy networks for Europe – the integration of large-scale renewable energy sources until 2050. Electricity Market Working Papers WP-EM-35. Dresden, Germany.

Egerer, Jonas, Clemens Gerbaulet, Richard Ihlenburg, Friedrich Kunz, Benjamin Reinhard, Christian von Hirschhausen, Alexander Weber, and Jens Weibezahn. 2014. Electricity sector data for policy-relevant modeling: data documentation and applications to the German and European electricity markets. DIW Data Documentation 72. Berlin.

Egerer, Jonas, Clemens Gerbaulet, and Casimir Lorenz. 2016. European electricity grid infrastructure expansion in a 2050 context. *The Energy Journal* 37 (1): 101–124.

Egging, Ruud. 2013. Benders decomposition for multi-stage stochastic mixed complementarity problems – applied to a global natural gas market model. *European Journal of Operational Research* 226 (2): 341–353.

Egging, Ruud, Steven A. Gabriel, Franziska Holz, and Jifang Zhuang. 2008. A complementarity model for the European natural gas market. *Energy Policy* 36 (7): 2385–2414.

Ehrenmann, Andreas, and Yves Smeers. 2005. Inefficiencies in European Congestion Management Proposals. *Utilities Policy* 13 (2): 135–152.

ENTSO-G. 2013. Ten-Year Network Development Plan (TYNDP) 2013–2022. Brussels, Belgium.

EUROGAS. 2010. *Long term outlook for gas demand and supply 2007–2030*. Brussels: The European Union of the Natural Gas Industry.

European Commission. 2011. *European energy sector: large investments required for sustainability and supply security*. Impact Assessment Energy Roadmap 2050, SEC (2011).

FOSG. 2011. The challenge of evolving from a member state planning investment methodology to the 1st phase of a European supergrid. Report. Brussels, Belgium.

———. 2015. New market design needs strong transmission grid. Press Release. Brussels, Belgium.

Gerbaulet, Clemens, and Alexander Weber. 2014. Is there still a case for merchant interconnectors? Insights from an analysis of welfare and distributional aspects of options for network expansion in the Baltic Sea Region. Discussion Paper 1404. DIW Berlin: Berlin, Germany.

Gramlich, Edward M. 1994. Infrastructure investment: a review essay. *Journal of Economic Literature* 32 (3): 1176–1196.

Greenpeace, GWEC, and SPE. 2015. *Energy [r]Evolution – A Sustainable Energy Outlook*. Hamburg: Greenpeace International, Global Wind Energy Council, SolarPowerEurope.

Hirschman, Albert O. 1958. *The Strategy of Economic Development*. Vol. 10. New Haven: Yale University Press.

Holz, Franziska, and Christian von Hirschhausen. 2013. The infrastructure implications of the energy transformation in Europe until 2030 – lessons from the EMF28 modeling exercise. *Climate Change Economics* 4 (1).

Holz, Franziska, Philipp M. Richter, and Ruud Egging. 2016. The role of natural gas in a low-carbon Europe: infrastructure and regional supply security in the global gas model. *The Energy Journal* 37 (Special Issue 3): 33–59.

Holz, Franziska, Hella Engerer, Claudia Kemfert, Philipp M. Richter, and Christian von Hirschhausen. 2014. *European Natural Gas Infrastructure: The Role of Gazprom in European Natural Gas Supplies; Study Commissioned by The Greens/European Free Alliance in the European Parliament*, Politikberatung kompakt. Vol. 81. Berlin: DIW Berlin.

Holz, Franziska, Phillipp M. Richter, and Ruud Egging. 2015. A global perspective on the future of natural gas: resources, trade, and climate constraints. *Review of Environmental Economics and Policy* 9 (1): 85–106.

Holz, Franziska, Hanna Brauers, Philipp M. Richter, and Thorsten Roobeek. 2017. Shaking Dutch grounds won't shatter the European gas market. *Energy Economics* 64: 520–529.

IEA. 2004. *Energy Policies of IEA Countries: 2004 Review*. Paris: OECD.

———. 2011. *Are We Entering a Golden Age of Gas? World Energy Outlook 2011 – Special Report*. Paris: OECD.

———. 2013. *World Energy Outlook 2013*. Paris: OECD.

———. 2014. World Energy Investment Outlook – Special Report. In *OECD Publishing*. Paris: International Energy Agency.

IPCC. 2005. IPCC special report on carbon dioxide capture and storage. Prepared by Working Group III of the Intergovernmental Panel on Climate Change. Cambridge, UK: Cambridge University Press.

Leuthold, Florian, Hannes Weigt, and Christian von Hirschhausen. 2012. A large-scale spatial optimization model of the European electricity market. *Networks and Spatial Economics* 12 (1): 75–107.

Makholm, J.D. 2015. Regulation of natural gas in the United States, Canada, and Europe: prospects for a low carbon fuel. *Review of Environmental Economics and Policy* 9 (1): 107–127.

Mendelevitch, Roman. 2014. The role of CO_2-EOR for the development of a CCTS infrastructure in the North Sea Region: a techno-economic model and applications. *International Journal of Greenhouse Gas Control* 20 (January): 132–159.

Mendelevitch, Roman, and Pao-Yu Oei. 2017. The Impact of Policy Measures on Future Power Generation Portfolio and Infrastructure: A Combined Electricity and CCTS Investment and Dispatch Model (ELCO). *Energy Systems* 9: 1–30.

Midwest ISO. 2010. RGOS: Regional Generation Outlet Study. In *Saint Paul, MN*. Midwests ISO.

Mieth, Robert, Richard Weinhold, Clemens Gerbaulet, Christian von Hirschhausen, and Claudia Kemfert. 2015. Electricity grids and climate targets: new approaches to grid planning. DIW Economic Bulletin 5 75–80. Berlin, Germany.

Morbee, Joris, Joana Serpa, and Evangelos Tzimas. 2010. The evolution of the extent and the investment requirements of a trans-European CO_2 transport network. JRC Scientific and Technical Reports JRC61201. Brussels, Belgium: European Commission, Joint Research Centre.

Neele, Filip. 2012. Towards a transport infrastructure for large-scale CCS in Europe. Final Report to the European Commission 226317. TNO.

Neumann, Anne, and Christian von Hirschhausen. 2015. Natural gas: an overview of a lower-carbon transformation fuel. *Review of Environmental Economics and Policy* 9 (1): 64–84.

Neumann, Anne, Leonard Göke, Franziska Holz, Claudia Kemfert, and Christian von Hirschhausen. 2018. Natural gas supply: another Baltic Sea pipeline is not necessary, DIW Berlin Weekly Report, no. 27–2018: 241–248.

Oei, Pao-Yu, and Roman Mendelevitch. 2016. European scenarios of CO_2 infrastructure investment until 2050. *The Energy Journal* 37: 171–194.

Oei, Pao-Yu, Johannes Herold, and Roman Mendelevitch. 2014. Modeling a carbon capture, transport, and storage infrastructure for Europe. *Environmental Modeling & Assessment* 19 (6): 515–531.

Rausch, Sebastian, and John Reilly. 2015. Carbon taxes, deficits, and energy policy interactions. *National Tax Journal* 68 (1): 157–178.

Richter, Philipp M., and Franziska Holz. 2015. *All Quiet on the Eastern Front? Disruption Scenarios of Russian Natural Gas Supply to Europe.* Berlin: DIW.

Roar Aune, Finn, Rolf Golombek, Arild Moe, Knut Einar Rosendahl, and Hilde Hallre Le Tissier. 2017. The future of Russian gas exports. *Economics of Energy & Environmental Policy* 6 (2).

Ruester, Sophia, Claudio Marcantonini, Xian He, Jonas Egerer, Christian von Hirschhausen, and Jean-Michel Glachant. 2012. EU involvement in electricity and natural gas transmission grid tarification. Think Research Report. Florence, Italy: Florence School of Regulation.

Sberbank Investment Research. 2018. *Russian Oil and Gas – Tickling Giants.* Moscow: Sberbank Investment Research.

Scheer, Hermann. 2010. *Der energethische Imperativ: 100% jetzt: Wie der vollständige Wechsel zu erneuerbaren Energien zu realisieren ist.* 1st ed. Munich: Kunstmann, A.

Schrettl, Wolfram, Christian von Hirschhausen, and Herbert Brücker. 1998. Socio-economic effects of the baltic ring projects, Expertise for the Baltic Ring Study Group. Subcontractor study for the EU-TEN Project Baltic Ring. Berlin, Germany; Brussels, Belgium.

Smith, Adam. 1776. *An Inquiry into the Nature and Causes of the Wealth of Nations.* 1st ed. - London: Methuen & Co.

Stern, Jonathan. 2017. Challenges to the future of gas: unburnable and unaffordable. Oxford Institute for Energy Studies (OIES) Paper NG 116. Oxford, UK: Oxford Institute for Energy Studies (OIES).

Tröster, Eckehard, Rena Kuwahata, and Thomas Ackermann. 2011. European grid study 2030/2050. Commissioned by Greenpeace International. Darmstadt/Langen, Germany: Energynautics GmbH.

Turmes, Claude. 2010. Cables for carbon? Connecting the European electricity grid to non-EU mediterranean countries: The Bigger Picture. Report to the European Parliament. Brussels, Belgium: European Parliament.

VDE. 2015. Der Zellulare Ansatz – Grundlage einer erfolgreichen, regionenübergreifenden Energiewende. VDE-Studie. Frankfurt am Main, Germany: VDE Verband der Elektrotechnik Elektronik Informationstechnik e.V. – Energietechnische Gesellschaft im VDE (ETG).

von Hirschhausen, Christian. 2010. Developing a 'supergrid': conceptual issues, selected examples, and a case study for the EEA-MENA Region by 2050 ('Desertec'). In *Harnessing Renewable Energy in Electric Power Systems: Theory, Practice, Policy*, ed. Boaz Moselle, Jorge Padilla, and Richard Schmalensee . Washington, DC: RFF Press.Chapter 10

————. 2011. Financing trans-European energy infrastructures: past, present, and perspectives. Policy Paper 48. Notre Europe.

————. 2012. Green electricity investment in Europe: development scenarios for generation and transmission investments – study for the European Investment Bank. EIB Working Papers 2012/ 04. Luxembourg: European Investment Bank.

von Hirschhausen, Christian, Clemens Gerbaulet, Franziska Holz, and Pao-Yu Oei. 2013. European energy infrastructure integration Quo Vadis? Sectoral analyses and policy implications. In *Europa am Scheideweg*, ed. Theresia Theurl, vol. 338. Berlin: Duncker & Humblot.

von Hirschhausen, Christian, Franziska Holz, Clemens Gerbaulet, and Casimir Lorenz. 2014. European energy sector: large investments required for sustainability and supply security. *DIW Economic Bulletin* 9 Berlin, Germany.

Zweibel, Ken, James Mason, and Vasilis Fthenakis. 2008. A Solar Grand Plan. *Scientific American* 298 (1): 64–73.

Chapter 12
Cross-Border Cooperation in the European Context: Evidence from Regional Cooperation Initiatives

Casimir Lorenz, Jonas Egerer, and Clemens Gerbaulet

> *"We are convinced that an intensified regional cooperation is an important step towards further EU market integration, that it will increase energy security, reduce energy prices and costs and promote further integration of renewable energy."* Joint Declaration for Regional Cooperation on Security of Electricity Supply in the Framework of the Internal Energy Market, Signed in Luxembourg on June 8, 2015, by Germany and its 12 "electrical neighbors" (p. 1).

12.1 Introduction

Cross-border cooperation on energy policies is crucial for achieving the ambitious goals of the low-carbon transformation in Europe and the energiewende in Germany. Because the European electricity system is so densely interconnected, reform processes in one country affect the broader European market, whether through price effects, cross-border flows, or the sharing of backup capacity. "Electrical neighbors"

This chapter is based on previous research by the authors on cross-border cooperation, amongst them Gerbaulet and Weber (2018), Egerer et al. (2013, 2015, 2016), and Lorenz and Gerbaulet (2014); the usual disclaimer applies.

C. Lorenz (✉)
DIW Berlin, Berlin, Germany

TU Berlin, Berlin, Germany
e-mail: cl@wip.tu-berlin.de

J. Egerer
Friedrich-Alexander-Universität (FAU), Erlangen, Nürnberg, Germany
e-mail: jonas.egerer@fau.de

C. Gerbaulet
TU Berlin, Berlin, Germany

DIW Berlin, Berlin, Germany

© Springer Nature Switzerland AG 2018
C. von Hirschhausen et al. (eds.), *Energiewende "Made in Germany"*,
https://doi.org/10.1007/978-3-319-95126-3_12

of the EU play a role in this process, in particular Norway and Switzerland, with their significant (de facto already integrated) hydro capacities but also other neighboring regions in the East (Russia, etc.) and the South (Mediterranean). Intensified cooperation can help to stabilize the market and regulatory environment, leading to greater long-term stability and welfare gains. The coordination of electricity market segments between countries or regions could potentially also reduce the required investment costs by reducing the total need for infrastructure and allocating resources more efficiently, as suggested by the European Climate Foundation (ECF 2011) and Newbery et al. (2013), among others.

Countries engaged in cross-border cooperation face the transaction costs of implementing new regulatory regimes and sometimes significant distributional effects. Coordination of cross-border policies on generation, renewables, transmission, and other aspects of energy policy entails the challenge of equitable decision making over the distribution of costs and benefits. Spillover effects of investments in one country can be either positive or negative for neighboring countries. For example, support for renewables in one country can lead to lower spot market prices that benefit customers in other countries, but it may also hurt producers in these countries.

Even within a given region, different countries often exhibit different levels of cooperation. At the beginning of the low-carbon transformation process, *Europe-wide* coordination was the main driver of development, whereas today, *regional* cooperation among several neighboring countries plays an important role, and there are also cases of bilateral cooperation between countries over *national* energy policies (von Hirschhausen 2012). The European electricity sector has traditionally emphasized the European single market and the integration of national electricity systems, from Portugal to the Baltic States, and from Greece to Ireland. In the EU's third energy package, the European Commission aimed to create a Europe-wide internal energy market connecting all of the Member States by expanding the existing transmission infrastructure.[1] As described in Chap. 11, the implementation of Europe-wide coordination, e.g., supergrid infrastructures, has proven to be complex. This led to discussion over regional cooperation schemes focusing primarily on existing system operations as well as investments in cross-border transmission infrastructure and capacity instruments. Regional cooperation can function as an intermediate step towards broader integration, yet it can sometimes lead to a more permanent state of cooperation, as with the Nordic electricity market in Scandinavia.

In this chapter, we analyze different forms of cooperation in the context of the European low-carbon transformation process and provide empirical evidence on some concrete cooperation schemes. The chapter focuses on *regional* cooperation schemes, as these provide plentiful evidence of developments and progress to date. The next section discusses potential fields of cooperation and specifies our classification of cooperation types. Section 12.3 focuses on the potential scope of regional

---

[1]European Commission. 2009. Directive 2009/72/EC. Brussels, and European Commission. 2009. EC Regulation No 714/2009. Brussels.

cooperation in the electricity sector, and describes existing examples of cooperation, such as the Pentalateral Energy Forum (PLEF) and the North and Baltic Sea Grid Initiatives. Sections 12.4 and 12.5 provide model-based analysis of the concrete effects of regional cooperation: joint balancing markets in the Alpine region, and transmission expansion in the North and Baltic Sea Region. Section 12.6 draws lessons from the analysis and concludes.

12.2 Different Cooperation Schemes

Cooperation may be established to achieve a number of different goals, such as the creation of a common market to increase consumer welfare by reducing prices, the low-carbon transformation of the energy system, energy efficiency, security of supply, etc. Two types of policies can be put in place to attain these goals: those that merely modify the existing technical system and its operational policies, for instance, by harmonizing product definitions between countries in the balancing market, and those targeting the capital stock, for instance, joint investments in renewable and/or conventional generation capacities and transmission infrastructure.

Different types of cross-border cooperation differ in their effects and burdens of implementation. In general, there is an inverse relation between the scope of a coordinating action and the transaction costs required to implement it. Two neighboring countries can agree relatively easily on the joint use of existing capacities such as electricity generation, while the coordination of large-scale, capital-intensive investments at the European level is obviously more complicated.

In a study for the European Investment Bank (EIB), von Hirschhausen (2012) put forward a stylized classification of cross-border cooperation patterns in Europe (Table 12.1): From a geographical perspective, one can distinguish between a Europe-wide approach, encompassing *all* countries of the European Union, and a focused approach, limited to a certain geographical region, including bi- or multi-lateral coordination. From an institutional perspective, these partnerships may rely on a Europe-side coordinating framework, such as EU Directives, or they may not, in which case the national institutions play a stronger role. As Europe-wide coordination without European coordinating institutions is unlikely, the upper right quadrant of Table 12.1 may be left aside, resulting in three patterns that we characterize as (1) "Europe centralized," (2) "Regional," and (3) "National."

Table 12.1 Cross-border cooperation patterns in Europe

		European coordinating institutions	
		...in place	...not in place
Geographical scope...	...Europe-wide	(1) Europe centralized	/
	...Focused	(2) Regional	(3) National

Source: von Hirschhausen (2012), based on an idea of Thorsten Beckers and Albert Hoffrichter

Reform options were, until recently, considered either in a pan-European (quadrant 1) or a purely national context (quadrant 3), whereas the "regional" level hardly existed at all, either in the political sphere or in the institutional setting. The EU single market initiative therefore addresses all Member States, whereas bilateral cooperation initiatives such as the renewables schemes between Sweden and Norway are based on national initiatives. In the previous chapter on infrastructure, we have seen that pan-European coordination on capital-intensive infrastructure can be quite challenging, thus hampering the expected speed of integration. On the other hand, it is evident that the targets of the low-carbon transformation, and most other goals, cannot be achieved by bilateral cooperation alone (Clemens Gerbaulet et al. 2014).

The third approach, *regional* cooperation, is currently thriving, and seems to constitute an appropriate level for integration policies that do not overstretch the institutional requirements. Thus, a regional approach may represent a compromise and a pragmatic solution for an industry based on physical infrastructure connections. De Jong and Egenhofer (2014), who examine a regional approach to EU energy policies, note some advantages of regional cooperation, such as the close technical interrelatedness of neighboring countries and easier (local) negotiation of distributional effects. The following sections of this chapter therefore focus on available evidence of regional cooperation taking place in the framework of the European electricity market.

12.3 Regional Cooperation Patterns in European Electricity

12.3.1 "Regional Groups" and Similar Ongoing Activities

The regionalization of a substantial share of energy policies is a relatively recent process, but it is now in full swing. The request by the EU and the European network of transmission system operators for electricity (ENTSO-E) is representative of this development: It requires that "member states should coordinate energy policies starting at regional level".[2] Such a regional approach can be a "fast track" towards reaching a fully integrated European market, and at the same time acknowledge the national right and responsibility of each Member State to determine their own energy mix and security of supply. Since this approach requires a stable regulatory framework, the European Commission aims to foster "regional co-ordination of national energy policy decisions". ENTSO-E also argues that regional groups should not be mandated top-down by the EU but should instead take shape organically, based on

---

[2]ENTSO-E. 2015. "Member States to Coordinate Energy Policies Starting at Regional Level with ENTSO-E Commitment to Contribute." Press Release. Brussels, Belgium: European Network of Transmission System Operators for Electricity.

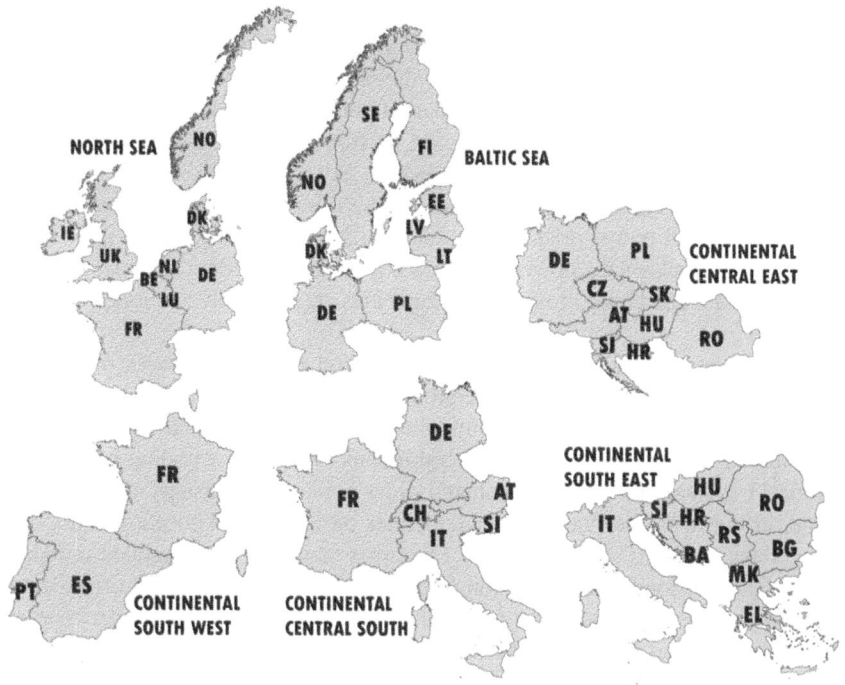

Fig. 12.1 Regional cooperation in Europe: Specific regional groups for transmission coordination. Source: Own depiction based on ENTSO-E (2016b)

the participants' interest in cooperation. Therefore, ENTSO-E establishes the term "energy policy regions," in which regional cooperation should be used to address political and regulatory issues. These regions should be based on a "common history, trust," and "geographic proximity," the participant's "interconnection and energy mix," and similar challenges, and should be a "manageable size" (ENTSO-E 2016a).

Figure 12.1 shows the structure of regional cooperation that has been formalized within the ten-year network development planning (TYNDP) process to formulate specific regional investment plans, but also serves other fields of cooperation, e.g., balancing markets (see below). Central countries like France and Germany can be part of more than just one regional group.

Other regional cooperation initiatives have developed as well: with regard to market coupling, the Nord Pool Spot Market allows for coordinated trading in the wholesale markets of Estonia, Finland, Latvia, Lithuania, Norway, and Sweden.[3] In November 2011, the Central-Western Region (called CWE), consisting of the

---

[3]NordPoolSpot. 2014. Europe's leading power markets. Presentation. http://www.nordpoolspot.com/Global/Download%20Center/Annual-report/Nord-Pool-Spot_Europe's-leading-power-markets.pdf.

Benelux countries, Germany, and France, introduced market coupling using the concept of available transfer capacity (ATC) to make grid capacity calculations. In May 2015, the CWE region changed their day-ahead market coupling mechanism from the ATC approach to flow-based market coupling (FBMC).[4] In conjunction with market coupling in the CWE region, a form of price coupling has been introduced in the North-Western Europe (NWE) day-ahead markets. The so-called Price Coupling of Regions (PCR) solution allows the participating countries to use a common day-ahead power price calculation. Furthermore, the common price coupling mechanism allows the power exchanges to use cross-border capacities directly in their auctions, and hence cross-border capacities are implicitly auctioned (Weber et al. 2010).[5]

Following an approach similar to PCR, different power exchanges of Europe together with transmission system operators (TSOs) from 12 European countries[6] decided to create a joint integrated continuous intraday cross-border market (called XBID Market Project). It will enable a continuous cross-border trading based on a common IT system that provides a shared order book and a capacity management module, in line with the Framework Guidelines on Capacity Allocation and Congestion Management (CA-CM) and the EU target model for an integrated intraday market. Hence, as long as transmission capacity is available, orders and offers from distinct countries can be directly matched. Furthermore, the cross-border market supports explicit and implicit cross-border capacity allocation.[7] After a delay, the European Cross-Border Intraday Solution XBID started operation in 2018.[8]

Regional cooperation is in place not only in wholesale markets but also in the balancing market. Within the international grid control cooperation (IGCC), eight countries are cooperating to reduce balancing energy costs through imbalance netting.[9] Regional cooperation is also envisaged by the guideline on Electricity

---

[4]Here the allocated capacity on the interconnectors is not static, but flows on adjacent markets' interconnectors influence the transfer capacity available to the market to fully utilize available energy transfer capacity. Therefore, FBMC is likely to increase cross-border electricity transfer and reduce the price spread between markets while maintaining the same level of security of supply.

[5]For details on the coupling see APX Group (2014). North-Western European Power Markets Successfully Coupled – A landmark in the integration of the European power market. Press release, Amsterdam, The Netherlands, last accessed September 14, 2016 at http://www.apxgroup.com/press-releases/north-western-european-power-markets-successfully-coupled/.

[6]Austria, Belgium, Denmark, Finland, France, Germany, Great Britain, Italy, Luxembourg, Norway, Portugal, Spain, Sweden, Switzerland, and The Netherlands.

[7]See also Pickles, Mark (2016): XBID: Cross-Border Intraday Market Project—Third User Group Meeting; Brussels.

[8]Epex Spot SE 2018. Exchange council supports migration of products from local trading systems to xbid; https://www.epexspot.com/de/presse/pressarchive/details/press/Exchange_Council_supports_migration_of_products_from_Local_Trading_Systems_to_XBID.

[9]ENTSO-E 2017: Update on imbalance netting. Presentation at the Balancing Stakeholder Group on the 28.09.2017, Brussels. https://www.entsoe.eu/Documents/MC%20documents/balancing_ancillary/2017-09-28/170928_BSG_Imbalance_netting.pdf.

Balancing (NC EB)[10] that aims at increasing cooperation on balancing markets. All countries should form so-called coordinated balancing areas (CoBAs) that must consist of at least two different control areas and can be seen as a starting point for emerging regional cooperation.[11]

The German Ministry of Energy contributed to a regional initiative in which 12 ministers of the countries that are Germany's "electrical neighbors" signed key political declarations on enhanced regional cooperation in the field of electricity markets. These countries[12] agreed to continue working to strengthen the single energy market and to allow free trading even in times of high prices on the electricity grid as the most important tool to deliver energy security. Based on the idea that market signals with price spikes would increase the flexibility of supply and demand, the signatories agreed to refrain from introducing legal price caps. Furthermore, under the agreement, cross-border exchanges are not limited, even in times of scarcity. More cross-border grid expansion should be fostered. And as a final but important stipulation, the generation capacity adequacy assessment should not be done from a national standpoint but from a regional standpoint using a common approach. Regarding the energy mix, no common goal was formulated; rather, the national sovereignty to decide upon the national energy mix was confirmed (BMWi 2015).

By contrast, note that so far the coordination of renewables policies has *not* been an important topic of debate. European Renewables Directive No. 2009/28/EC provides cooperation mechanisms, but they have not been used to date. The new European Energy and Climate package to be implemented in the coming years may alter this state of affairs by providing incentives for regional markets, following the example of Sweden and Norway, which merged their renewables certificate trading. A decision by the European Court of Justice has strengthened national governments' renewables targets, although more integration of regional renewables markets, or even a pan-European market, is still on the agenda. In the European Court of Justice's Åland decision, it rejected the request of a Finnish producer of renewable electricity to take part in the Swedish renewables support program. The Court argued that national renewables targets and the pursuit of national environmental objectives were higher priorities than the general internal market.[13] In the future, national regulators are likely to auction some of their renewable capacities in neighboring countries as well.

---

[10]By publishing in the Official Journal of the European Union the Network Codes became the EU regulation: Commission Regulation (EU) 2017/2195 of 23 November 2017 establishing a guideline on electricity balancing.

[11]ENTSO-E. 2014. "ENTSO-E Network Code on Electricity Balancing – Version 3.0." Brussels, Belgium: European Network of Transmission System Operators for Electricity.

[12]In addition to Germany: Poland, the Czech Republic, Austria, Switzerland, France, Luxemburg, Belgium, the Netherlands, Denmark, Norway, Sweden.

[13]European Court of Justice. 2014. "Ålands Vindkraft AB v Energimyndigheten." Judgment of the Court (Grand Chamber) of 1 July 2014 Case C-573/12. Luxemburg.

12.3.2   Pentalateral Energy Forum (PLEF)

One of the oldest regional cooperation schemes in Europe is the Pentalateral Energy Forum (PLEF), initially initiated by the five ("penta") neighboring countries of Belgium, Netherlands, Luxemburg, France, and Germany, in 2007. In addition to the respective energy ministries, the Forum also includes the TSOs and the power exchanges. The goal of the initiative is to "enhance the cooperation between all relevant partners to create a regional Northwest-European electricity market as an intermediate step towards one common European electricity market" (PLEF 2007). This has been confirmed by its second political declaration of June 2015.[14] Although it stressed its adherence to the goals of the common European electricity market, the five countries had been pursuing their own, specific agendas since the beginning: the first concrete project was market integration, whereby the operation of the electricity markets, including the use of interconnector capacity, was merged into flow-based market coupling. Initially planned for implementation as early as 2011, flow-based market coupling was implemented in 2015 (PLEF 2015). For testing and quantification purposes, the flow-based market coupling was run in parallel to NTC market coupling. The results of the parallel run show positive welfare effects for flow-based market coupling. Negative welfare effects were only observed if internal remedial actions as redispatch were assumed in the ATC mechanism but explicitly excluded from the flow-based mechanism.[15]

Another field of cooperation in the PLEF is generation adequacy and security of supply (SoS). Since 2015, Working Group 2 of the PLEF has been coordinating a joint report on system adequacy in the region involving the participation of all TSOs, using a common data set and a joint model. In this exercise, it is assumed that capacities in one country could be traded to be available in another country, even in times of scarcity. Under this assumption, and based on national generation adequacy plans that are taken as exogenous, the current thinking is that there will be no adequacy risk in the early 2020s. As seen in Fig. 12.2, demand can be satisfied in *all* hours where in all cases, remaining generation capacity is available within the PLEF region. In January 2018 the Second Pentalateral Generation Adequacy Assessment was released[16] which includes a better representation of the electricity grid by using a Flow-Based approach and an improved model for taking into account flexibilities on the demand side.

---

[14]PLEF. 2015b. "Second Political Declaration of the Pentalateral Energy Forum of 8 June 2015." Luxembourg.

[15]CASC. 2014. "CWE Flow Based Market-Coupling Project: Parallel Run Performance Report." Luxembourg.

[16]PLEF 2018: Generation Adequacy Assessment. Support Group 2. Brussels, Belgium. http://www.benelux.int/files/1615/1749/6861/2018-01-31_-_2nd_PLEF_GAA_report.pdf.

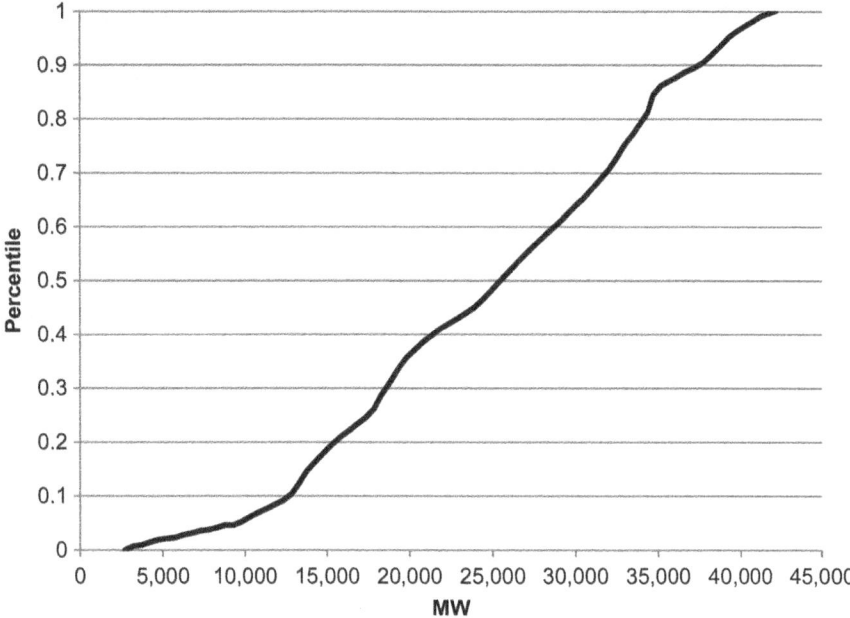

Fig. 12.2   Remaining regional capacity for year 2015/2016. Source: (PLEF 2015)

12.3.3   North Sea Countries Offshore Grid Initiative (NSCOGI)

The North Sea Countries Offshore Grid Initiative (NSCOGI) is a regional cooperation of ten riparian countries around the North Sea with the target of coordinating the development of an offshore grid: Belgium, Denmark, France, Germany, Ireland, Luxembourg, Netherlands, Norway, Sweden, and the UK. The objectives are a sustainable low-carbon economy, cost-effective security of energy supply, and a strategic, coordinated, and cost-effective development of the offshore grid. NSCOGI was formed in 2010 after a memorandum of understanding was signed by the energy ministries, regulators, and transmission system operators; it is subdivided into three working groups on grid configuration, regulatory issues, and planning and permitting. A special focus lies on investments in hybrid offshore structures (interconnectors with offshore wind farms connected) and facilitating trading among these. Therefore, different studies on market coupling and cost allocation have been published, indicating the beneficial welfare effects.

12.3.4 Baltic Energy Market Interconnection Plan (BEMIP)

The Baltic Energy Market Interconnection Plan (BEMIP) was formed in 2009 by the High Level Group on Baltic interconnections. This group was set up by the President of the European Commission at the time, José Barroso, following the agreement of the Member States of the Baltic Sea Region, and is chaired by the Commission on "Baltic Interconnections." Initial participants were Denmark, Germany, Estonia, Latvia, Lithuania, Poland, Finland, and Sweden; Norway acts as an observer.[17] The goal of the BEMIP is to form an integrated energy market in the Baltic region, supported by the expansion of the necessary infrastructure; BEMIP is also part of the EU Strategy for the Baltic Sea Region. This should lead to higher energy supply security and more efficient market outcomes. The BEMIP's aims include reducing electricity congestion, removing regulated energy tariffs, establishing common energy reserves, and opening the retail market completely. One important objective, the inclusion of the Baltic States in the Nord Pool market, has already been achieved. Most of the infrastructure projects of the BEMIP are either part of the so-called Projects of Common Interest (PCI) or part of the European Energy Security Strategy (EESS).

In mid-2015, the participating countries signed a memorandum of understanding to further strengthen and modernize the BEMIP. The memorandum's objectives are to further integrate the markets for electricity and gas to enhance competition, to accelerate the coordination of energy infrastructure projects, and to emphasize the targets for energy efficiency and renewable energy. BEMIP is considered a political success, with progress reports regularly reporting that the implementation of the BEMIP is broadly on track and proceeds accordant to schedule.

12.3.5 Trilateral Cooperation for the Expansion of Pumped Storage Plants: Switzerland, Austria, Germany

At a more regional level, the Energy Ministries of Germany, Austria, and Switzerland committed to joint initiatives to promote the expansion of pumped storage power plants (PSP) in the Alpine region. They see PSP as crucial for the integration of renewable energy sources (RES) and as the only mature technology for electricity storage. Therefore, it was decided that PSP capacities should be expanded in all three countries and the resulting necessary line expansions should be coordinated trilaterally.[18]

---

[17]See EC (2014). Baltic Energy Market Interconnection Plan – 6th Progress Report. last accessed September 14, 2016 at https://ec.europa.eu/energy/sites/ener/files/documents/20142711_6th_bemip_progress_report.pdf.

[18]UVEK, BMWFJ, and BMWi. 2012. "Erklärung von Deutschland, Österreich und der Schweiz zu gemeinsamen Initiativen für den Ausbau von Pumpspeicherkraftwerken." Berlin, Germany; Bern, Switzerland; Vienna, Austria.

According to German Minister for Economic Affairs, this should lead to cross-border utilization of existing storage capacity and joint activities to foster the expansion of capacities. Austrian Minister of Economic Affairs Reinhold Mitterlehner, who sees potential benefits for the security of supply and a more efficient utilization of the Austrian potentials, confirms this (BMWi 2012).

The first result of this cooperation is a trilateral study on the potentials and the economic situation of PSP in the region.[19] It concludes that new investments into PSP are currently not economically feasible, but are crucial for the success of the energiewende. The future role of PSP appears to depend on the development of the European electricity system overall. To avoid an inefficient national solution, regional cooperation would be useful, and harmonization of national regulatory frameworks would be necessary to avoid market distortions. In-depth discussion on cooperation in the fields of market design, security of supply, transmission expansion, and ancillary services are crucial for the evaluation of PSP development (Weber et al. 2014).

12.4 Case Study (I): A Joint Balancing Market in the Alpine Region

12.4.1 An Important Market Segment . . .

This case study reports progress of a regional cooperation in a specific market segment, the "balancing market," the market for very short-term reserves, following the closure of the day-ahead and spot markets. Balancing energy markets help to stabilize systems that encounter deviations from nominal frequency due to unexpected fluctuations in demand or generation. Increased cooperation on balancing markets is important, as an increasing share of fluctuating renewables is normally accompanied by higher balancing needs.[20] Besides introducing changes in market design and offering opportunities for fluctuating renewables to provide balancing reserves, a common (regional) market for balancing reserves could significantly reduce costs.

The large potential for cross-border cooperation was also recognized in the latest regulation ("network codes"), which is designed to foster cross-border exchange of balancing services with the objective of lowering overall costs and increasing social

---

[19]See Hildmann, M. et al. (2014). Pumpspeicher im trilateralen Umfeld Deutschland, Österreich und Schweiz. Report. Zurich, Switzerland, last accessed September 15, 2016 at http://www.bmwi. de/BMWi/Redaktion/PDF/Publikationen/Studien/trilaterale-studie-zu-pumpspeicherkraftwerken-deutschland-oesterreich-schweiz-zusammenfassung,property=pdf,bereich=bmwi2012, sprache=de,rwb=true.pdf.

[20]Dena. 2014. "dena-Studie Systemdienstleistungen 2030 – Sicherheit und Zuverlässigkeit einer Stromversorgung mit hohem Anteil erneuerbarer Energien." Endbericht. Berlin, Germany.

welfare.[21] The relevant technical and regulatory framework for balancing markets and possible cooperation is described in the NC EB and in the Network Code on Load-Frequency Control and Reserves (NC LFCR).[22] Different studies have shown that cross-border exchanges of balancing services are in general beneficial and lead to significant cost savings. An Impact assessment by Mott MacDonald for the EU Directorate-General for Energy concludes that for the analyzed cases of France and the UK, the gains from increased cooperation outweigh the costs of implementation.[23] Similarly, a model-based analysis by Gebrekiros et al. (2015) shows that transmission capacity reservation for the exchange of balancing services is beneficial in Northern Europe.

Currently, there are different projects for regional cooperation in balancing markets underway. These projects aim at different degrees of cooperation ranging from (1) imbalance netting[24] to (2) joint activation[25] all the way to (3) joint reservation.[26] When allowing for joint activation or even joint reservation, a joint dimensioning of the needed reserves can be conducted, which could additionally lower the total cost as less overall capacity needs to be reserved. The ENTSO-E lists eight cross-border pilot projects on electricity balancing throughout Europe with different degrees of cooperation for different balancing products. Most relevant for our case study is the international grid control cooperation (IGCC) between Austria, Belgium, Czech Republic, Denmark, France, Germany, The Netherlands, and Switzerland (see Fig. 12.3). It allows for imbalance netting with regard to secondary control reserves (SC)[27] (comparable the definition of automatic Frequency Restoration Reserve (aFRR) by the ENTSO-E) and accumulated savings of up to €350 mn in the period 2011–2018[28]. Further studies by Fattler and Pellinger (2015)[29] and Sprey et al. (2015) confirm the benefits of the IGCC.

---

[21]ENTSO-E. 2014. "ENTSO-E Network Code on Electricity Balancing – Version 3.0." Brussels, Belgium: European Network of Transmission System Operators for Electricity.

[22]The NC LFCR will be merged into the Singe System Operation Guideline. A first draft of the merged guideline was published on May 4, 2016. ENTSO-E. 2013. "Network Code on Load-Frequency Control and Reserves." Brussels, Belgium: European Network of Transmission System Operators for Electricity.

[23]Mott MacDonald. 2013. "Impact Assessment on European Electricity Balancing Market." Contract EC DG ENER/B2/524/2011. Brighton, UK: European Commission, Directorate General for Energy.

[24]Imbalance netting describes the process of netting positive and negative imbalances in different control areas and thereby reducing the total imbalance of both control areas.

[25]Joint activation describes the usage of a common merit order list for the activation of reserves across two or more control areas.

[26]Joint reservation describes the joint determination of reserve capacities across two or more control areas.

[27]Secondary control reserves (*Sekundärregelleistungen*) are short-term reserves activated in case of imbalances within 5–10 min by the TSOs.

[28]IGCC (2017): Regular report on social welfare Q3/2017. Brussels, Belgium.

[29]Fattler, Steffen, and Christoph Pellinger. 2015. "Auswertungen und Analysen zur International Grid Control Cooperation." In. Vienna, Austria.

Fig. 12.3 Development of balancing market integration within the IGCC. Source: Own depiction

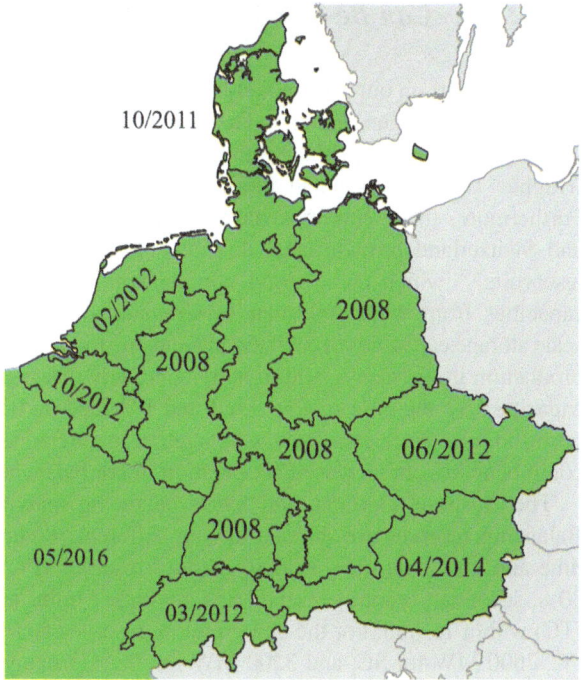

Apart from the IGCC, currently there are eight important balancing pilot projects in Europe. The two most important ones are TERRE and EXPLORE, which represent the ideas of CoBAs formulated in the NC-EB. Each of the initiatives has different countries participating and different product specifications and balancing concepts. The TERRE project consists of TSOs from France, Italy, Switzerland, Spain, Portugal, and the UK. Its main characteristics are the exchange of replacement reserves (RR) for which a common European RR platform will be built. Furthermore it is characterized by a proactive balancing concept, where imbalances are forecasted and replacement reserves are activated when the forecasted imbalance occurs. The EXPLORE project consists of TSOs from Austria, Belgium, Germany, and the Netherlands. It investigates how a FRR balancing market design could look like, considering the interdependencies between aFRR and mFRR. As most EXPLORE TSOs largely use aFRR with a reactive balancing approach, it is concluded that a joint activation of aFRR and imbalance netting will further reduce the need for cross-border mFRR balancing energy exchange.[30]

---

[30]50 Hz, Amprion, TransnetBW, APG & Elia. 2015. EXPLORE Status Update. Presented at Balancing Stakeholder Group on November 27, 2015. ENTSO-E. 2014. ENTSO-E Network Code on Electricity Balancing – Version 3.0. Brussels, Belgium.

12.4.2   ... Can Be Beneficial for Regional Cooperation

We now report on our own model-based analysis of the benefits of a joint balancing market in the Alpine region.[31] The partners in cooperation are Germany, Austria, and Switzerland, which share a long history of cooperation in the electricity sector (Hughes 1993; Schnug and Fleischer 1999; Horstmann and Kleinekorte 2003). Furthermore, these countries have complimentary generation portfolios: Austria and Switzerland provide large dispatchable hydro capacities for run-of-river and reservoirs,[32] which are a good complement to the large fluctuating renewable capacities from wind and solar in Germany.[33] Furthermore, the pumped large-scale storage capacities of Switzerland and Austria are able to store excess renewable production in Germany. Additionally, Germany offers comparatively large thermal capacities for the case of low wind and PV feed and low reservoir levels. These portfolios also work extremely well together in the traditional electricity markets and could benefit from increased cooperation in the balancing market as well.

The Austrian, German, and Swiss balancing markets are similarly organized. Balancing reserves are divided among different products dependent on response time and provision time. Primary control reserves (PC) must but activated within 30 s, secondary control reserves (SC) within 5 min, and tertiary control reserves (TC) within 15 min. For the entire region, reserve capacities are about 700 MW for PC, 2600 MW for SC, and 3200 MW for TC. Germany accounts for up to 85% of those reserve requirements.

In our analysis, we go a step further and introduce the highest level of *cooperation* and allow for joint reservation of reserves between Austria, Germany, and Switzerland. We apply a cost-minimizing model with mixed integer constraints that represents the spot and balancing market. We assume a social planner who is

---

[31]The model and the results described in this section are based on the DIW Discussion Paper 1400 (Lorenz and Gerbaulet 2014).

[32]The electricity system in Austria is based mainly on renewable energy sources: in 2016, hydro power contributed about 50.3% to the total net generating capacity of 21.9 GW, followed by fossil fuel based power plants (25%), wind (10.5%), solar (about 3,7%) and other renewables, see APG. 2016. "Installierte Kraftwerksleistung 2016." Austrian Power Grid. *Swiss* electricity generation relies mostly on hydro; more than two thirds of its installed capacity consists of hydro plants with 9.8 GW of storage, and pumped storage plants and run of river plants with 3.7 GW. Nuclear power plants with an installed capacity of 3.2 GW are the second-largest source of generation. Ongoing development of the Swiss electricity system is affected by the nuclear phase-out planned for 2034, which will be compensated by further expansion of hydro capacities, combined-cycle gas turbine plants and other renewable sources (see BFE. 2012. "Erläuternder Bericht zur Energiestrategie 2050." Bern, Switzerland).

[33]Germany's renewable energy sources provided 27% of electricity generation in 2012. Lignite represents 25.7%, hard coal 19.1% and nuclear 16.1% of the total power generation. The goals of the energiewende include phasing out nuclear by 2022, increasing of the share of renewables beyond 80% by 2050 (about 37% in 2017), and reducing greenhouse gas emissions (-40% by 2020, -80 to 95% by 2050). Germany remains a large net exporter of electricity (50 TWh in 2015), but overall consumption is supposed to be reduced by 50% by 2050 (compared to 2008).

minimizing total system cost, taking into account constraints due to generation restrictions, reserve restrictions, and flow limitations between countries. A detailed description can be found in Lorenz and Gerbaulet (2014).

We analyze four different degrees of cooperation within our scenarios:

- In the *no cooperation* scenario, cross-border exchanges are limited to the spot-market activity;
- in the *imbalance netting* scenario, reservation and activation take place nationally, but imbalances are netted when possible;
- in the *joint activation* scenario, reservation takes place nationally, but reserves are activated jointly if available cross-border capacity allows;
- in the *full cooperation* scenario, reservation and activation are jointly coordinated.

Our results confirm that regional cooperation in the balancing market is beneficial (although the reductions are relatively small compared to the total market volume of the spot market). As could be expected, the full cooperation scenario shows the highest possible cost reductions for the reservation of balancing capacity, amounting to €36 mn per year. Implementing joint activation or imbalance netting alone shows cost reductions for the activation of balancing energy of up to €11 mn annually. Only minor differences for activation costs are observable under the *joint activation* scenario in comparison to the *imbalance netting* scenario, since joint activation is only possible when sufficient interconnector capacity is available, but in this scenario, no prior reservation of interconnector capacity takes place.

Apart from cost savings, several distributional effects can be observed with regional cooperation. Especially in the *full cooperation* scenario, we see changes in the reserved capacities from Germany to Austria and Switzerland (Figs. 12.4 and 12.5). For positive SC and TC, 20% of capacity shifts out of Germany on average, which results for Austria in a 140% increase of reserved capacity for secondary and

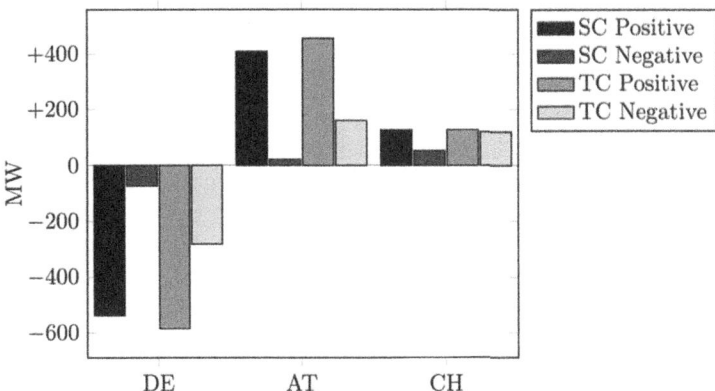

Fig. 12.4 Average change of reserved capacity in *full cooperation* compared to *no cooperation*. Source: Gerbaulet et al. (2014, 56)

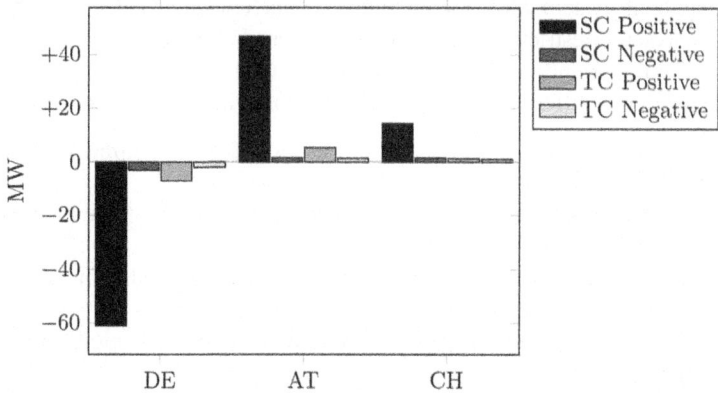

Fig. 12.5 Change of called energy in *full cooperation* compared to *imbalance netting*. Source: Gerbaulet et al. (2014, 56)

200% for tertiary on average. These large changes can be explained by the different cost structures of the national generation portfolios. The possible cooperation also has significant effects on the distribution of the activated balancing energy: on average, about 25% of the activated energy volumes for Germany shift towards Austria and Switzerland. Our analysis confirms that balancing market integration, e.g., in the framework of the IGCC, is beneficial, and that there is still unused potential for further cooperation. Integration results in additional benefits if the degree of cooperation was deepened in the direction of joint reservation of reserves, especially in light of the high market prices and market power that exist on the Austrian balancing market. The NC EB together with the successful ongoing project of the IGCC offer a starting point and framework for further cooperation.

12.5   Case Study (II): North and Baltic Sea Cooperation

More quantitative evidence on regional activities can be drawn from the cooperation among the North Sea and Baltic Sea riparian countries. Due to the large number of countries involved and to the different energy mixes in each of them, the potential benefits of this cooperation are large, making this region an important driver of the low-carbon transformation in Europe. The available evidence includes a welfare economic analysis of different development scenarios in the region and a detailed analysis of the effects of closer cooperation in the Baltic Sea.

12.5.1 Welfare Economic Analysis of Development Scenarios

12.5.1.1 Development Scenarios

Within the regions likely to benefit from intensified cooperation, the North and Baltic Seas Grid features very prominently on the agenda for European energy infrastructure development. The North Sea Grid was an early initiative that received significant political support: It was declared a priority of the "European energy infrastructures for 2020 and beyond," whose overarching objective is to develop an "Offshore Grid in the Northern Seas and a Connection to Northern as well as Central Europe" (EC 2010, 10). The aim is "to integrate and connect energy production capacities in the Northern Seas with consumption centers in Northern and Central Europe and hydro storage facilities in the Alpine region and in Nordic countries" (EC 2010, 10). This project is important since it would enable continental Europe to accommodate large volumes of wind and hydropower surplus electricity generation in and around the Northern and Baltic Seas, while connecting these new generation hubs, as well as major storage capacities in Northern Countries and the Alps with the major consumption centers in Continental Europe.[34]

Our research group is one of several to have analyzed the potentials and difficulties of closer integration in the region. In Egerer et al. (2013), we provided a survey of the existing literature, as well as a model-based analysis of allocative and distributional effects of potential projects. The TradeWind study (EWEA 2009) was in fact the first analysis of the proposed offshore grid designs with a flow-based model, the objective function of which was to minimize the total operating costs of the system. In a high wind case, annual cost reductions for generation amount to €326 mn; investment costs were estimated in the range of €300–400 mn/year. The analysis also suggested that a meshed solution yields a better benefit-cost ration than "only" radial connections. The OffshoreGrid study (De Decker and Kreutzkamp 2011) concluded that the connection costs of offshore wind farms could be reduced by €14 bn over the next 25 years through offshore clustering, and that the optimal offshore grid design should include meshed network elements instead of only radial connections. Meeus et al. (2011) provided a general assessment of engineering and economic analyses of offshore grids.

The study by Egerer et al. (2013) complements the aggregate analysis with distributional aspects, in terms of both nation-wide effects and the distribution of benefits between electricity generators and consumers for different design configurations of a potential North and Baltic Seas Grid. It uses the electricity model ELMOD (Leuthold et al. 2012), a techno-economic model of the European electricity market. In the following subsection, we provide the key findings of this research; for details, see Egerer et al. (2013, 123 sq).

Figure 12.6 shows the scenarios for network development that are used to derive the effects of integration on trade flows, prices, and welfare. In fact, the scenarios

---

[34]This section is based on Egerer et al. (2013).

	2009 Case	Wind+ Case
Status Quo Scenario		
Trade Scenario		
Meshed Network Scenario		

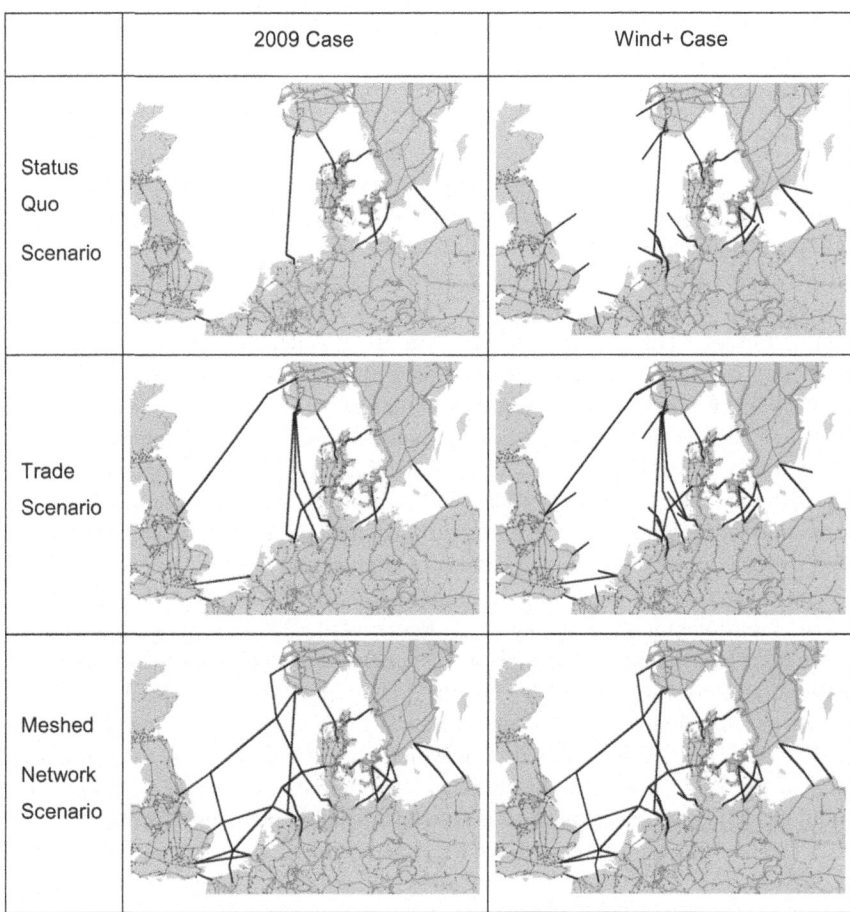

Fig. 12.6 Offshore grid and wind scenarios for the model runs. Source: Egerer et al. (2013, 127)

deployed in the study follow the above differentiation into "European," "regional," and "national" scenarios. The horizontal axis of Fig. 12.6 shows two settings for renewables development: (1) the status quo of 2009: 74 GW wind onshore, no wind offshore, and (2) an aggressive wind expansion case, called "Wind+": 177 GW onshore, 47 GW wind offshore. With respect to the nature of cooperation, we use the same classification as von Hirschhausen (2012), translating it into three different patterns of network development (vertical axis of Fig. 12.6):

1. The status quo scenario largely corresponds to the "national" scenario sketched out above, i.e., Great Britain harnesses wind off its coast, Norway and Sweden use their storage potential for domestic balancing, and Germany, the Benelux countries, France, Poland, etc. develop wind parks and connect them to their national territory. This corresponds roughly to the early period of analysis (~2009): besides very few bilateral connections, all offshore wind parks are

connected "only" to the next (national) shore; there are no bilateral connections or multilateral connections or a meshed network.

2. The "regional" scenario consists mainly of bilateral connections throughout the region, i.e., point-to-point trade cables that connect two countries. In this context, "regional" includes both bilateral cooperation and the potential trade connection between three or more countries (e.g., Sweden, Denmark, and Germany). In this scenario, about five new interconnections are added to the network, and some existing lines are expanded; the overall expansion corresponds to 5300 GW/km (see Egerer et al. (2013), for details).

3. By further extending the lines, one arrives at a meshed integrated network structure that can be interpreted as "European-wide": It is fully integrated between the different regional markets involved (last row in Fig. 12.6). In particular, some hubs are developed, such as the Kriegers Flak connector in the Baltic Sea, which acts like a multi-connection hub.[35]

12.5.1.2 Welfare Results

The model runs indicate, first, changes in the electricity trading in the region, and second, welfare effects, both at the aggregate and at the national level. Cross-border flows increase from 40 TWh/year (2009) in the status quo scenario to 110 TWh/year in the trade scenario and 140 TWh/year in the meshed network scenario. Clearly a higher level of integration leads to more electricity exchange between the participating countries, and a higher utilization of the wind mills.

The welfare effects are significant, as well: in the meshed scenario, one observes an overall social welfare gain of €210 mn/year, and the trade scenario still yields a benefit of €100 mn/year vis-à-vis the status quo. Figure 12.7 presents the major trends in the trade and meshed network scenarios when compared to the status quo scenario that is maintained as reference.[36] Previously isolated markets, such as Great Britain, generally benefit from interconnection. The high-price-producer countries, such as Germany and France, lose welfare. The effects are even stronger in the meshed network scenario, with additional supplies from Great Britain and Scandinavia.

12.5.1.3 Distributional Consequences

Each cross-border cooperation partnership entails distributive effects, making information at a more disaggregated level useful. Welfare gains can be distributed among

---

[35]In the case study by Egerer et al. (2013), there is only slightly more expansion (5500 GW/km), but the structure of the network is more interconnected than in the "regional" scenario.

[36]Solid areas indicate an improvement of national welfare, whereas areas with horizontal line shading indicate a deterioration of national welfare.

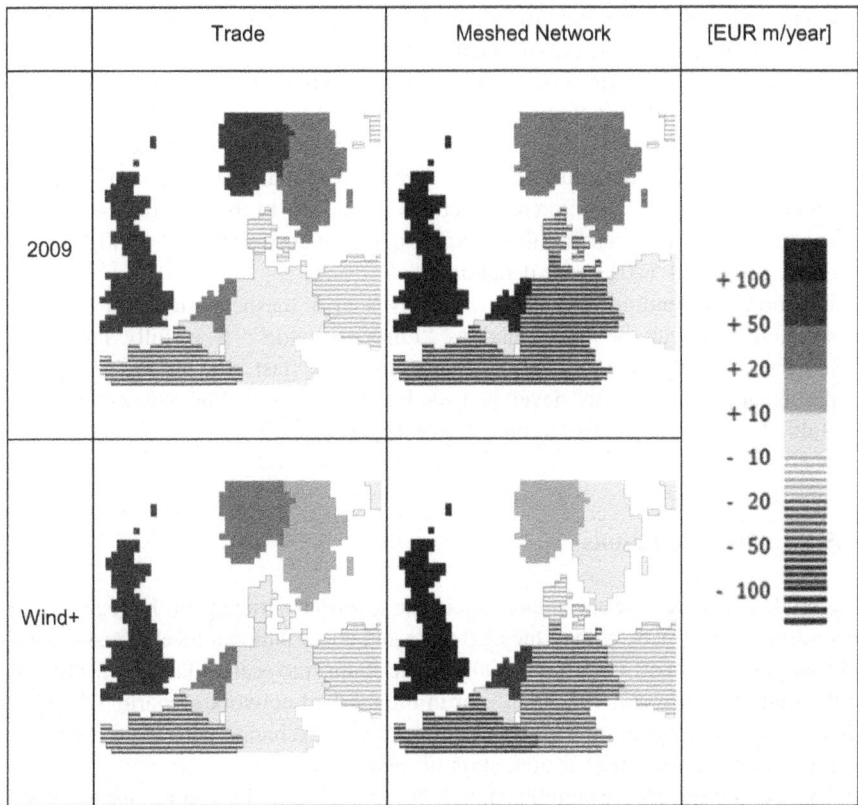

Fig. 12.7 Development of national welfare compared to no offshore extensions. Source: Egerer et al. (2013, 129)

countries, consumers, producers, and network operators. Figure 12.8 provides a disaggregation of the welfare effects outlined above, ordered by country but also by participating stakeholder. In addition to the effects at the national level, discussed above, for most countries, the changes in consumer and producer surplus are also significant. Overall, average prices fall due to higher wind integration and more efficient sharing of the fossil resources; this benefits consumers and hurts expensive producers. Thus, consumers in continental Europe and producers in the UK and Scandinavia benefit; by contrast, expensive German and French producers lose surplus. Clearly the individual welfare effects are stronger in the case of meshed network development ("Europe-wide," right column in Fig. 12.8).

Figure 12.8 also shows that the individual welfare effects are much more important than the aggregated ones. In fact, the aggregated welfare effects by country are modest, and are below €200 mn./year even for the highest beneficiaries (UK, Norway). Even if the national welfare increases for a certain network expansion scenario, higher electricity prices can trigger public and political opposition. This seems most likely to occur in the Scandinavian countries. The national welfare gain

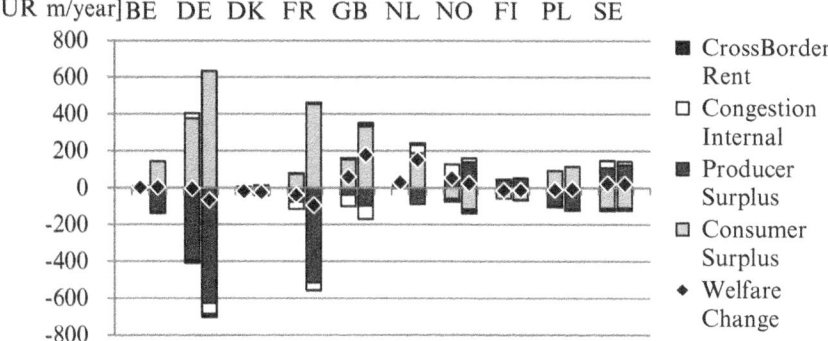

Fig. 12.8 Distributional effects of the implementation of the trade (left) and meshed (right) network design. Source: Egerer et al. (2013, 130)

in Norway and Sweden is positive, but consumers collectively must pay an additional €100–150 mn/year. This argument is discussed in detail for the Norwegian case by Midttun et al. (2012) on the national debate about the future of offshore connectors.

12.5.2 Focus on Transmission Cooperation Among the Baltic Countries

Additional evidence on the effects of bilateral and/or regional cooperation is provided by a study coordinated by Agora Energiewende (Germany) and the Swedish think-tank Global Utmaning. Within the framework of the Agora study, extending the work of Egerer et al. (2013) and focusing particularly on the Baltic countries, Ea et al. (2015) and Egerer et al. (2015) worked together with other stakeholders to examine the distributional effects of system integration in the region, with a particular focus on the Baltic riparian countries, and some considerations of the implication for stakeholders, mainly the energy-intensive industry, and household consumers. The studies also produced differentiated scenarios for moderate or higher integration of transmission grids, and for moderate or higher development of variable renewable sources (Ea et al. 2015). While the findings of this study with respect to transmission expansion were similar to the case study reported above, there are interesting implications for the energy mix: in fact, the results of wholesale prices and overall welfare depend significantly on how fast renewable energies are able to expand. This is shown in the following for wholesale prices and the distributional effects of integration. Table 12.2 explains the scenario setup for different levels of grid integration and renewables.

Figure 12.9 shows the change of annual wholesale electricity prices for the countries in question, in three scenarios with respect to the base scenario,

Table 12.2 Scenario definition in the Baltic Sea study

	Moderate RES-E	High RES-E
Moderate integration of grids	ModRE_ModTrans	HighRE_ModTrans
High integration of grids	ModRE_HighTrans	HighRE_HighTrans

Source: Own depiction, based on Ea et al. (2015, 93)

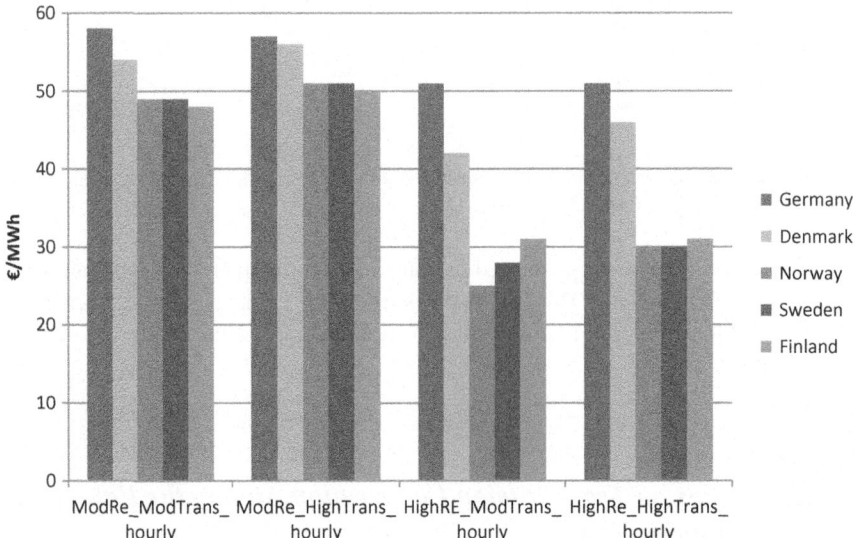

Fig. 12.9 Electricity prices for different renewable and transportation scenarios. Source: Ea et al. (2015)

Table 12.3 Distributional effects of integration on national socioeconomic welfare

		Norway	Sweden	Finland	Denmark	Germany	Total
Moderate renewables	[m €/ year]	+35	+12	−1	−5	+34	+75
High renewables	[m €/ year]	+54	+115	+10	+1	+10	+191

Source: Ea et al. (2015, 24)

ModRe_ModTrans. The price effects are much stronger for the case of "high renewables," in particular for the Nordic countries: wholesale prices in Norway, Sweden, and Finland fall from €50/MWh to almost €30/MWh. The price drop for Germany and Finland is also strong, although relatively less significant.

The distributional effects of the integration process are also stronger in the case of high renewables expansion in the Baltic region. Table 12.3 provides an overview of the effects, summarizing the shifts in national gains and losses while taking into account consumer rents, generator profits, and congestion rents, but excluding additional infrastructure costs for generation and transmission ("additional value of increased integration [HighTrans] as compared to moderate integration [ModTrans] scenario"). For the region as a whole, rents increase in the HighTrans scenarios and even more in the HighRES scenario. As price changes are asymmetric, integration triggers an uneven redistribution. The biggest beneficiaries in terms of market rents are Norway and Germany under the moderate scenario and Sweden and Norway under high deployment scenario. In other countries, the effects on the whole are moderate and non-uniform, which might seem surprising, especially in the case of Denmark, in view of its central location as a transit country.

It is evident that the deployment of renewables has a strong effect on regional industry electricity prices, and that this will weigh on the political feasibility of the integration policy. Therefore the distributional effects of integration are presented in detail for the different stakeholder groups in Fig. 12.10. The effects between the stakeholders within one country are much greater than the welfare effects between countries. For Norway and Sweden, the internal redistribution effects from consumer to producers are the strongest (€900 mn and €300 mn), as increased average electricity prices reduce the consumer and congestion rent while the rent for hydro and wind producers is increased. For Germany, the results are the opposite, as average prices are declining. Accordingly, consumer rent is increased, while renewable and conventional producers rent is reduced. However, these redistributions are on a lower level of €182 mn, which is additionally split among more stakeholders.

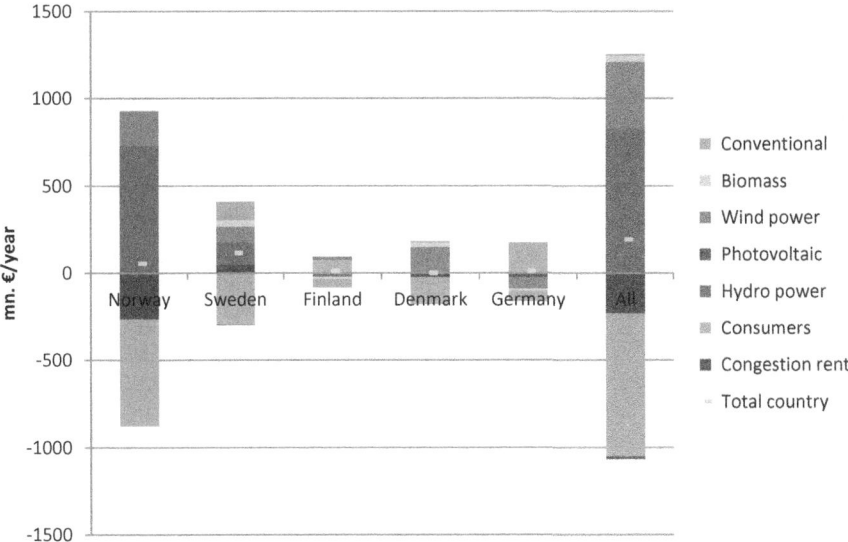

Fig. 12.10 Distributional effects of integration in the HighRE_HighTrans compared to the HighRE_ModTrans in mn €/year. Source: Ea et al. (2015, 28)

12.6 Conclusions

The success of the low-carbon transformation of the electricity sector depends, among other things, on successful market integration and increased cooperation at the bilateral, regional, and the European levels. This chapter has provided a framework for the analysis of different integration schemes, and sketched out evidence on ongoing cooperation, mainly at the regional level. In fact, the old dichotomy between "European vs. national" policies has given way to a third level, the *regional* level, acting under a joint European institutional umbrella, but pursuing very specific objectives.

We observe that the scope of regional cooperation schemes is broadening: the "Regional Groups," established for joint transmission planning by ENTSO-E, have set a standard for regionalization. Among the more advanced forums for cooperation are the Pentalateral Energy Forum (PLEF), the North Sea Countries Offshore Grid Initiative (NSCOGI), and the Baltic Energy Market Interconnection Plan (BEMIP). Smaller successful cooperation projects also exist, such as the Swedish-Norwegian joint renewables scheme.

Two empirical case studies highlight the further potential for cooperation: In a case study on the secondary and tertiary balancing markets in Germany, Austria, and Switzerland, we show that coordinated procurement and dispatch can lead to overall cost decreases compared to the current national settings. Increased cooperation leads to a shift in reservation from Germany to Austria and Switzerland, as well as to monetary distributional effects. The second case study on North and Baltic Sea Grid cooperation shows that the potential effects of binational or even meshed cooperation can be significant, but that cooperation also leads to distributional effects, thus tempering the appetite of certain interest groups such as large electricity consumers. The distributional effects between stakeholders within countries are significantly larger than the distributional effects between countries; thus, some market participants might strongly prefer or disapprove of new interconnections. Increased cross-border integration and deployment of renewable energy sources complement each other. With increasing renewable shares, the value of additional interconnection between Germany and the Nordic countries increases. At the same time, increased interconnection also leads to better utilization of these renewable energy sources as renewable curtailment is reduced, thus also reducing CO_2 emissions.

Looking back over the last two decades since the implementation of the first Directive on the Internal Electricity Market (1996), significant progress can be observed. Without this process, as explained in Chap. 10, the low-carbon transformation of the European electricity markets might not even have occurred. However, the last two decades also show that reforms always take much more time than expected, and by far not all the integration benefits calculated on paper can be reaped in reality. While much progress has been achieved in better coordinating existing assets, agreeing on capital-intensive investments in generation or transmission capacities has proven significantly more difficult.

References

BMWi. 2012. Germany, Austria and Switzerland start an initiative to promote pumped storage power plants. Press Release. Berlin, Germany.

———. 2015. Joint declaration for regional cooperation on security of electricity supply in the framework of the internal energy market. Berlin.

De Decker, Jan, and Paul Kreutzkamp. 2011. Offshore electricity grid infrastructure in Europe – Final Report. Brussels, Belgium.

De Jong, Jacques, and Christian Egenhofer. 2014. Exploring a regional approach to EU energy policies. CEPS Special Report 84. Brussels, Belgium.

Ea, DTU, and DIW. 2015. Increased integration of the nordic and German electricity systems (full version). Study on behalf of Agora Energiewende and Global Utmaning. Berlin, Germany: Agora Energiewende.

EC. 2010. Energy infrastructure priorities for 2020 and beyond – a blueprint for an integrated European energy network. Communication from the Commission to the European Parliament, the Council, the European Economic and Social Committee and the Committee of the Regions COM(2010) 677/4. Brussels, Belgium: European Commission.

ECF. 2011. *Power Perspectives 2030 – On the Road to a Decarbonised Power Sector.* Brussels, Belgium: European Climate Foundation.

Egerer, Jonas, Friedrich Kunz, and Christian von Hirschhausen. 2013. Development scenarios for the north and baltic seas grid – a welfare economic analysis. *Utilities Policy* 27 (December): 123–134.

Egerer, Jonas, Christian von Hirschhausen, and Alexander Zerrahn. 2015. Increased integration of the nordic and German electricity systems. Distributional effects of system integration and qualitative discussion of implications for stakeholders – work package 2. Study on behalf of Agora Energiewende and Global Utmaning. Berlin, Germany.

Egerer, Jonas, Clemens Gerbaulet, and Casimir Lorenz. 2016. European electricity grid infrastructure expansion in a 2050 context. *The Energy Journal* 37 (1): 101–124.

ENTSO-E. 2016a. Regional cooperation and governance in the electricity sector. Policy Paper. Brussels, Belgium.

———. 2016b. Specific regional groups for the TYNDP. The ten-year network development plan – general. May 20, 2016.

EWEA. 2009. Integrating wind – developing Europe's power market for the large-scale integration of wind power. Final Report. Brussels, Belgium: European Wind Energy Association (EWEA).

Fattler, Steffen, and Christoph Pellinger. 2015. *Auswertungen und Analysen zur International Grid Control Cooperation.* Vienna, Austria. https://www.ffe.de/download/article/549/20150211_Langfassung_IGCC_IEWT.pdf

Gebrekiros, Yonas, Gerard Doorman, Stefan Jaehnert, and Hossein Farahmand. 2015. Reserve procurement and transmission capacity reservation in the Northern European power market. *International Journal of Electrical Power & Energy Systems* 67 (May): 546–559.

Gerbaulet, C., and A. Weber. 2018. When regulators do not agree: are merchant interconnectors an option? Insights from an analysis of options for network expansion in the Baltic Sea Region. *Energy Policy* 117 (June): 228–246.

Gerbaulet, Clemens, Casimir Lorenz, Julia Rechlitz, and Tim Hainbach. 2014. Regional cooperation potentials in the European context: survey and case study evidence from the Alpine Region. *Economics of Energy & Environmental Policy* 3 (2): 45–60.

Horstmann, Theo, and Klaus Kleinekorte. 2003. Strom für Europa: 75 Jahre RWE-Hauptschaltleitung Brauweiler 1928–2003/Power for Europe: 75 Years of the Brauweiler Control Centre 1928–2003. Essen, Germany: Klartext.

Hughes, Thomas Parke. 1993. *Networks of POWER: Electrification in Western Society, 1880–1930.* Baltimore, USA: Johns Hopkins University Press.

Leuthold, Florian, Hannes Weigt, and Christian von Hirschhausen. 2012. A large-scale spatial optimization model of the European electricity market. *Networks and Spatial Economics* 12 (1): 75–107.

Lorenz, Casimir, and Clemens Gerbaulet. 2014. New cross-border electricity balancing arrangements in Europe. Discussion Paper 1400. Berlin, Germany: DIW Berlin.

Meeus, Leonardo, Manfred Hafner, Isabel Azevedo, Claudio Marcantonini, and Jean-Michel Glachant. 2011. Transition towards a low carbon energy system by 2050: what role for the EU? Final Report 3. THINK Project. European University Institute.

Midttun, Atle, Tiina Ruohonen, and Raffaele Piria. 2012. Norway and the North Sea Grid. Key positions and players in Norway, from a Norwegian perspective. SEFEP Working Paper. Berlin, Germany: SEFEP, the Smart Energy for Europe Platform.

Newbery, David, Goran Strbac, D. Pudjianto, and Pierre Noël. 2013. Benefits of an integrated European energy market. Final Report, Prepared for: Directorate – General Energy, European Commission.

PLEF. 2007. Memorandum of understanding of the pentalateral energy forum on market coupling and security of supply in Central Western Europe. Luxembourg.

———. 2015. Generation adequacy assessment. Pentalateral energy forum support group 2. - Brussels, Belgium: Pentalateral Energy Forum.

Schnug, Artur, and Lutz Fleischer. 1999. *Bausteine für Stromeuropa: Eine Chronik des elektrischen Verbunds in Deutschland; 50 Jahre Deutsche Verbundgesellschaft.* Heidelberg, Germany: Deutsche Verbundgesellschaft.

Sprey, Jens, Tim Drees, Denis vom Stein, and Albert Moser. 2015. Potential einer Harmonisierung der europäischen Regelleistungsmärkte. In. Vienna, Austria: 9. Internationale Energiewirtschaftstagung an der TU Wien.

von Hirschhausen, Christian. 2012. Green electricity investment in Europe: development scenarios for generation and transmission investments – study for the European Investment Bank. EIB Working Papers 2012/04. Luxembourg: European Investment Bank.

Weber, Alexander, Dietmar Graeber, and Andreas Semmig. 2010. Market coupling and the CWE project. *Zeitschrift Für Energiewirtschaft* 34 (4): 303–309.

Weber, Alexander, Thorsten Beckers, Sebastian Feuß, Christian von Hirschhausen, Albert Hoffrichter, and Daniel Weber. 2014. Potentiale zur Erzielung von Deckungsbeiträgen für Pumpspeicherkraftwerke in der Schweiz, Österreich und Deutschland. Study commissioned by the Swiss Federal Office of Energy SFOE. Berlin, Germany: TU Berlin, IAEW RWTH Aachen.

Chapter 13
Modeling the Low-Carbon Transformation in Europe: Developing Paths for the European Energy System Until 2050

Konstantin Löffler, Thorsten Burandt, Karlo Hainsch, Claudia Kemfert, Pao-Yu Oei, and Christian von Hirschhausen

> *To truly transform our economy, protect our security, and save our planet from the ravages of climate change, we need to ultimately make clean, renewable energy the profitable kind of energy.*
> U.S. President Barack Obama (New York Times, February 25, 2009)

The chapter is an update of the discussion paper by Hainsch et al. (2018); the usual disclaimer applies.

K. Löffler (✉) · T. Burandt · P.-Y. Oei
Junior Research Group "CoalExit", Berlin, Germany

TU Berlin, Berlin, Germany

DIW Berlin, Berlin, Germany
e-mail: kl@wip.tu-berlin.de

K. Hainsch
TU Berlin, Berlin, Germany

C. Kemfert
DIW Berlin, Berlin, Germany

Hertie School of Governance, Berlin, Germany

German Advisory Council on the Environment (SRU), Berlin, Germany

C. von Hirschhausen
TU Berlin, Berlin, Germany

DIW Berlin, Berlin, Germany

13.1   Introduction

One of the biggest contributors of greenhouse gas (GHG) emissions is the energy sector, accounting for more than two thirds of the global emissions (IEA 2016). The most important greenhouse gas is CO_2, which is responsible for more than 80% of the emissions in the energy sector (Foster and Bedrosyan 2014). Therefore, various challenges arise for different countries when it comes to decarbonizing their energy systems. The European Union (EU), being a major economic force, has set several climate goal targets, which should lead to an energy system with almost no GHG emissions. In recent years, the focus was heavily set on decarbonizing the electricity sector. However, in a fully decarbonized energy system the heating and transportation sector deserve just as much, if not more attention, due to the challenges of phasing out fossil fuels in these areas. A high degree of electrification in these sectors is predicted in future scenarios, which implicitly affects the power sector.

As discussed in the previous chapters, in particular Chap. 10, long-term scenarios of the low-carbon energy transformation in Europe are quite diverse. In this chapter, we provide a detailed discussion of various scenarios leading to a far-reaching decarbonization of the European energy system to 2050. We use an updated version of the Global Energy System Model (GENeSYS-MOD), developed by our group to study various low-carbon transformation processes at global, continental, or national level. The modeling results suggest that a largely renewables-based energy mix is the lowest cost solution to the decarbonization challenge, and that the distribution of the carbon budget has a strong impact on the results. Our top-down model calculations thus confirm bottom-up results obtained for the electricity sector, in Chap. 10, suggesting that the solution to the carbon challenge is the increased use of renewable energy sources, mainly solar and wind.

The power sector is by far the most wide-spread sector of choice when it comes to analyzing energy system transitions towards less GHG emissions. Some studies focus solely on the electricity sector on a European scale and analyze impacts of high renewable penetration (Czisch 2007; PwC 2011; Scholz 2012; Plessmann and Blechinger 2016; Gerbaulet et al. 2017). Gerbaulet, et al. (2017) analyze different scenarios for the European electricity sector with high amounts of renewables, showcasing that neither high shares of carbon capture, transport, and storage (CCTS) nor nuclear power are necessary for such a system to be feasible. Scholz (2012) shows that most European countries will be able to cover their domestic power demand on their own, with countries like Belgium or Luxembourg relying on grid interconnections with other countries. Czisch (2007) comes to similar results, concluding that given enough grid capabilities between European and North African countries, renewable electricity could be produced and distributed at costs similar to today's. If cross-border transmission is restricted, however, significant increases in capacity can be observed. Plessmann and Blechinger (2016) suggest that the 2050 EU emission reduction target can be met with investments of 403 billion Euro (EUR). The result is an electricity sector based on mainly wind and photovoltaic (PV), with hydro power and natural gas as complements.

In addition to production and distribution, electricity storages and their incorporation into the power sector are the focus of many other studies. While all of the above-described authors mention storages as an element of future energy systems, others take a closer look. In general, a positive correlation between high shares of renewables and storage capacities can be found across the literature. Zerrahn and Schill (2016) additionally highlight that the relevance of power storages is even higher if other flexibility options are less developed. Bussar et al. (2015) suggest storage capacities of 804,300 Gigawatt hours (GWh) in 2050, most of which consist of gas storages. Also, a negative correlation between storage and trade capacities can be observed, showcasing the power grid as another form of storage. In contrast, Rasmussen et al. (2012) find that without additional balancing storage capacities of 320 Terawatt hours (TWh) are required. However, they acknowledge that hydrogen storages would increase this number substantially. Even more optimistic results are presented by Tröndle (2015), suggesting 150 TWh of storage capacities. He also highlights the fact that excess capacity can substantially reduce the need for storages. Comparing storage options to conventional generation coupled with CCTS technology, Bogdanov and Breyer (2016) show that the former outperform the CO_2 heavy alternatives. Similar to the studies in the previous paragraphs, a decrease in cost can be observed if cross-border electricity trade is enabled.

On a global scale, Jacobson et al. (2017a) published one of the most comprehensive studies, showcasing 100% renewable energy roadmaps for 139 countries of the world. Electricity is produced by wind, water, and solar technologies and a significantly more aggressive pathway is projected than what the Paris agreement calls for. However, his methodology and results of a different paper were origin to a controversy between researchers (Clack et al. 2017; Jacobson et al. 2017b). This not only showcases the prevalence of the topic, but it also highlights the various paradigms within the field.

The rest of this chapter is structured in the following way: The next section provides a non-technical description of the model, the Global Energy System Model (GENeSYS-MOD); it is an energy system model developed recently for scenario analysis, providing a high level of technical detail, and the integrated coverage of all sectors and fuels. Section 13.3 presents different GHG emissions pathways, related to a 1.5° increase of the global mean temperature, a 2° increase, and a business-as-usual (BAU) case with a much larger emission budget. For each scenario, we distributed the emission budget to countries according to different criteria, i.e. free distribution, share of European GDP, share of current emissions, or share of population. Section 13.4 presents model results, suggesting that renewable technologies gradually replace fossil-fuel generation, starting in the power sector: By 2040, almost all electricity generation is provided by a combination of PV, wind, and hydropower, in addition some storage. The pathways for transportation and heat are more diverse, but they follow a similar general trend. The commitment for a 2 °C target only comes with a cost increase of about 1–2% (dependent on the emission share) compared to a business-as-usual-pathway, while yielding reduced emissions of about 25%. The different regions and demand sectors each experience different decarbonization pathways, depending on their potentials, political settings, and technology options. Section 13.5 concludes that with already known technologies, even ambitions climate targets can be met in Europe, at moderate costs, as long as strict carbon constraints are applied.

13.2 Model and Data

13.2.1 General Model Description

The model is based on the formulation of the Global Energy System Model (GENeSYS-MOD), as described by Löffler et al. (2017a, b). In order to overcome some of the key shortcomings of the aforementioned model, especially when it comes to renewables, the model has been revised and improved to a new version. The model version described by Löffler et al. (2017a) will be referenced as GENeSYS-MOD v1.0 from here on, whereas the new version being presented here is named GENeSYS-MOD v2.0.

In essence, GENeSYS-MOD can be illustrated as an integrated, a flow-based cost-optimization model. The different nodes are represented as "Technologies", which are connected by "Fuels". Examples for technologies are production entities like wind or solar power, conversion technologies like heat pumps, storages, or vehicles. Fuels serve as connections between these technologies and can be interpreted as the arcs of the network. In general, Fuels represent energy carriers like electricity or fossil fuels, but also more abstract units like demands of a specific energy carrier or areas of land are classified as Fuels. Also, technologies might require multiple different Fuels or can have more than one output fuel, e.g., a combined heat and power plant could use coal as an input fuel and produce electricity and heating energy as an output fuel. Efficiencies of the technologies are being accounted for in this exact process, which would allow to model energy losses due to conversion. Energy demands are classified into three main categories: electricity, heating, and transportation. They are exogenously defined for every region and each year. The model then seeks to meet these demands through a combination of technologies and trade between the different regions. Figure 13.1 provides a general overview of the different technologies and the connections between them.

GENeSYS-MOD v2.0 offers a fully revised data set for all global parameters, such as fuel prices, general cost assumptions, and emissions data. Furthermore, the list of available technologies has been revised and extended, now including more options in the transportation sector, as well as a representation of CCTS plants. Additionally, the model has been upgraded with new equations and revised formulations that offer more and new functionalities (see Fig. 13.2):

- The trade system (especially with respect to power trade) has further been improved. It introduces transmission capacities and the option for the model to

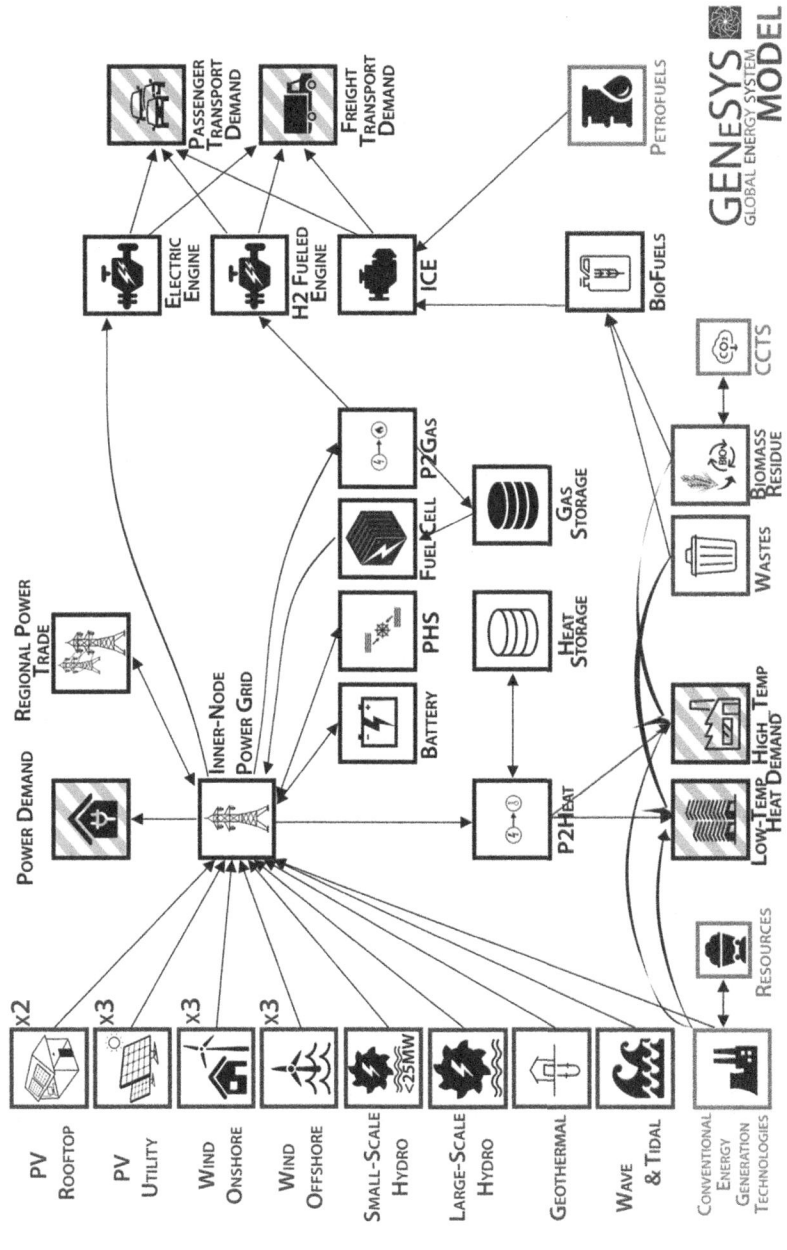

Fig. 13.1 Model structure of GENeSYS-MOD v2.0. Source: Own illustration

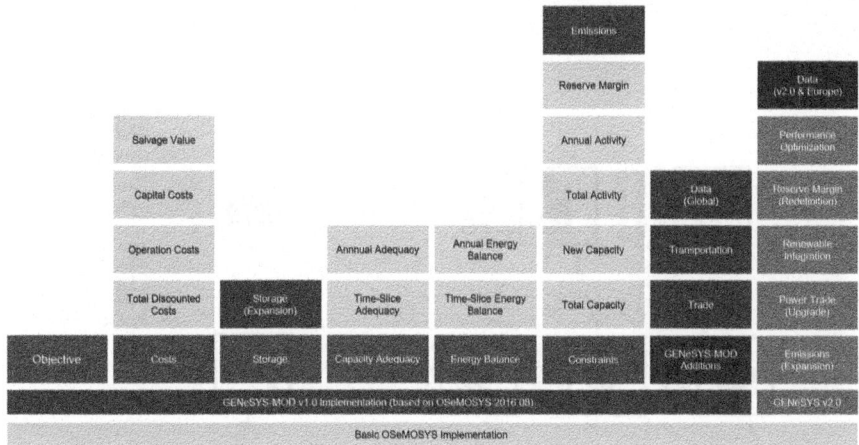

Fig. 13.2 Block structure of GENeSYS-MOD v2.0. Source: Own illustration, based on Howells et al. (2011)

endogenously expand them. The approach for the endogenous grid expansion is the same as described by Hosenfeld et al. (2017).

• New constraints limit the phase-in and phase-out of new technologies, as well as renewable electricity growth. The new equations make sure that new future technologies are not being used in one year and then completely disregarded in the next, as well as old technologies being constructed and then ending as a stranded asset.

• Emission targets can be set globally as well as for individual regions.

• The efficiency of technologies depends now on the year of construction, rather than on the current date, avoiding an overestimation of potentials.

• The ReserveMargin has been redefined to better fit the flexibility requirements of a largely decarbonized system. The new formulation requires the model to produce a certain share of its production of a selected Fuel (e.g., power) with selected technologies that offer the necessary flexibility when it comes to load balancing (e.g., technologies with fast ramp-up times, such as electric storages or gas-fired plants).

• Implemented performance optimization reduce the necessary memory resources and calculation time.

The general data foundation as described in Löffler et al. (2017a) has been revised and updated with a new spatial and temporal resolution. Also, new regional data for Europe has been researched and was added to the model. The list of available technologies has been updated with their respective cost assumptions, potentials, and efficiency parameters. Emission data and fuel costs for fossil energy carriers have been updated. If not stated otherwise, data is adopted from GENeSYS-MOD v1.0 (see Löffler et al. (2017a) and Burandt et al. (2016)).

13.2.2   Model Setup and Data

13.2.2.1   Spatial Resolution and Grid Data

Since the focus of this work is the European region, a new geographical resolution had to be found. The broad world region "Europe" was split up into multiple smaller nodes to fit the scope of the study. The same approach has been used in a case study for India with GENeSYS-MOD v1.0 (see Löffler et al. (2017b)).

The model for the European energy system consists of 15 nodes. A focus has been placed on Germany and its central role, both geographically and politically. Hence, Germany and all its neighboring countries are modeled as single regions (with Luxembourg being the exception), whereas the resolution gets broader moving to the edges of the European region. There, multiple countries are aggregated into one region, based on matching regional potentials and conditions. The chosen regional disaggregation of Europe with 15 nodes in total leads to a stylized version of the European electricity grid. The resulting grid structure with its possible connections between nodes can be found in Fig. 13.3. Grid capacities for Europe have been taken from Gerbaulet and Lorenz (2017).

13.2.2.2   Temporal Resolution

This temporal disaggregation has been revised and updated, now featuring four quarters of a year, and four daily time brackets, to a total of 16 time slices per year. A similar approach can be found in Welsch et al. (2012), where they show that

Region
- Austria
- Baltic States
- Belgium & Luxembourg
- Czech Republic
- Denmark
- Europe South-East
- France
- Germany
- Great Britain
- Italy
- Netherlands
- Poland
- Portugal & Spain
- Scandinavia
- Switzerland

Fig. 13.3 Grid structure and node set-up for Europe. Source: Own illustration

an energy system model using an enhanced version of Open Source Energy Modelling System (OSeMOSYS), utilizing 16 time slices, can achieve almost the same results as a full hourly dispatch model.

For each quarter, the daily time brackets which determine the time slices are slightly different. This approach was chosen to facilitate a better match of solar load profiles, since sun availability vastly differs between seasons. Each day is split up into (a) morning, (b) peak, (c) afternoon, and (d) night. The daily sunlight-hours for Germany (taken as representative for Europe, as its central geographical location gives a good mean value for the region) have been used as outlines for the daily time brackets.

13.2.2.3 Technology Representation in the Model

The list of technologies has been taken from the prior model version and has then been revised. Some technologies did receive updates in their implementation, others were added, and some have been removed. Heating technologies no longer have centralized and decentralized counterparts. This simplification is due to our rough regional disaggregation, which does not profit from such a distinct analysis of heating technologies. Instead, centralized and decentralized heating technologies for each type (e.g., low-temperature gas heating) have been combined into one unified technology. A total of 15 centralized heating technologies (including the area technologies) have been omitted from the model.

New technologies for the import of fossil fuels outside of the modeled region have been implemented. This is important when conducting case studies, where the rest of the world is not being calculated endogenously. Since resources might be scarce in the modeled region (such as crude oil reserves in Europe, for example), the model now has the option to import fossil fuels at world market prices. The model now distinguishes between hard coal and lignite—a change that was necessary considering the strong usage of lignite, as well as large amounts of existing capacities, in some parts of the European region. For this purpose, new technologies for the use and production of lignite have been implemented (but no import technology, since lignite is inefficient to transport and used in close proximity to the mining site instead).

The list of transportation technologies has been revised and expanded by new technologies not previously considered in GENeSYS-MOD v1.0. Electric options for road-based freight transport were added to the model.[1] Passenger road-based transport was expanded by plug-in hybrid electric vehicles and air-based transport has a new technology using biokerosene. These new technologies have been added to offer a broader variety of options for the model to choose from, including more possible future solutions for the transportation sector.

---

[1]New technologies for electricity-based road freight transport: overhead-powered trucks, battery-electric trucks, plug-in hybrid electric trucks.

Furthermore, two new storage technologies for electricity storage were added to the model: reduction–oxidation (redox)-flow-batteries and compressed air energy storage (CAES). Also, a new methanation technology has been put in place, which is able to transform biomass and synthetic hydrogen to methane, thus enabling more options for sector coupling inside the model. The produced methane is treated the same way as methane out of natural gas.

Carbon capture, transport, and storage (CCTS) for biomass-based power plants is now added as a technology into GENeSYS-MOD v2.0. While our stance on CCTS remains critical, using it in conjunction with biomass enables negative net emissions, which are a common basis for climate-focused model results. In order to be able to better compare our results with such models, the option for bio-energy with carbon capture, transport, and storage (BECCTS) has been added.

Residual capacities for 2015 for the power production of all European countries have been taken from Farfan and Breyer (2017). The future capacities were then projected based on the construction years and the respective Operational Life. For the heating sector, capacities described by Fraunhofer ISI et al. (2016)[2] were considered.

13.2.2.4 Potentials of Renewable Energy Sources

The total potential for renewable technologies is often disputed, even among experts, with heavily varying values. The choice of maximum land usage, as well as the underlying weather data (e.g., choice of the base year), strongly impact these numbers and quickly lead to an over- or underestimation of actually available potentials. The renewable potential data for the European region presented in this study stem from the model dynamic Electricity Model (dynELMOD) (Gerbaulet and Lorenz 2017), which, in turn, is based on an expert assessment by the Potsdam Climate Institute.

Sensitivity calculations with own assessments based on suitable land usage and solar radiation maps have been conducted to test the robustness of model and data, and will be discussed in Sect. 13.4.5.2.

13.2.2.5 Capacity Factors

Capacity factors for renewable generation have been taken from Pfenninger and Staffell (2016), given as an hourly time series for the year 2014. For each region, multiple samples have been taken, placed into a category, and then taken as average

---

[2]Fraunhofer ISI, Fraunhofer ISE, Institute for Resource Efficiency and Energy Strategies, Observ'ER, Technical University Vienna, Energy Economics Group, and TEP Energy. 2016. "Mapping and Analyses of the Current and Future (2020–2030) Heating/Cooling Fuel Deployment (Fossil/Renewables)."

for each region, category and time slice. Solar PV and onshore wind are divided up into the three categories (a) optimal, (b) average, and (c) inferior, while offshore wind has been categorized as (a) shallow, (b) transitional, and (c) deep.[3] The categories of PV and onshore wind only differ in the capacity factors whereas the particular types of offshore wind parks additionally have different capital and Operation and Maintenance (O&M) costs. Therefore, we decided to use another kind of categorization for offshore wind.

The hourly data for each quarter of the year has been aggregated with the corresponding time slice definition, as described in Sect. 13.2.2.2. This leads to the final capacity factors for each Timeslice. Because of the high dependency of the capacity factors on the selected year from Pfenninger and Staffell (2016) for aggregating the modeled timeslices, further sensitivity analyses have to be underdone. Especially extreme weather conditions in individual years are flattened out or not included in this aggregation. Hence, we have the plan to add additional daytypes to reflect different possible weather conditions for the power generation of renewable energy sources (RES).

13.2.2.6   Cost Data

For utility-scale PV and onshore wind, expenses have been assumed to be the same across all three categories. For offshore wind, the placement of turbines influences the resulting construction costs a lot more (e.g., near-shore vs. deep-water placement) with cost estimate ranges of up to more than double the price. Hence, offshore wind has its capital costs given separately for each category. The capital costs for fossil-based plants are assumed to be constant over the years, while renewables experience decreasing costs over the modeled time frame. Fixed costs are assumed as a percentage of capital costs, as in GENeSYS-MOD v1.0. Variable costs for renewable technologies are still considered to be negligible.

The prices for fossil fuels in the second version of GENeSYS-MOD have been split up into local and global prices. These global prices are tied to the global market price of each fuel and have been updated from the 2015 version of the World Energy Outlook of the International Energy Agency (IEA) to the 2016 version (IEA 2016). This means a drastic reduction in the price forecast, especially for oil (where the difference results in nearly a 50% reduction of future oil prices compared to the forecast from 2015 (IEA 2015).

Because of the regionally dependent availability and usage of lignite, local prices have been applied, where available.[4] For hard coal, natural gas, and crude

---

[3]The regional potential has been assumed to be evenly distributed across the categories, as per Gerbaulet and Lorenz (2017).

[4]The value for Portugal & Spain is the average of the other values, since no reliable source for a specific value was found.

oil, it has been assumed that local production is 5% cheaper than the global market price.[5]

13.2.2.7   Reserve Margin

The modifications made to the implementation of the calculation of the reserve margin require a change of the underlying parameter values. Fuels and technologies are tagged to indicate whether they need a reserve margin, or can contribute to the reserves, respectively.[6] The parameter ReserveMargin then sets the required relative amount of energy that has to come out of flexible supply technologies.

For this model set-up, only the Fuel 'Power' requires this form of load balancing. Technologies that are able to fulfill these flexibility requirements are gas- and oil-based power plants, batteries, and pumped hydro storages, as well as fuel cells.

ENTSO-E (2013) suggests a ReserveMargin between 5 and 10% on a country level, acknowledging that high shares of renewables might require higher percentages of additional capacity. Hence, our assumed values increase at the beginning to reflect that development. Simultaneously, they mention that a high degree of interconnection between different regions lower the need for such measures. Hence, we opted to reduce the necessary reserve margin for the later model periods.

13.2.2.8   Emissions Budget

The emissions budget available for the model has been reevaluated in GENeSYS-MOD v2.0. Additionally, a regional, European, limit was obtained from the given global emission budgets that are provided in the most recent literature.

In the modeled scenarios, keeping the temperature well below 2 °C is the primary goal, and the corresponding available CO_2 budgets provided by the IPCC (2014) are used. For the calculation of the total CO_2 budget for Europe, the data provided by the Stockholm Environment Institute was used.[7] This discussion briefly assesses the pathways that were released in the Fifth Assessment Report of the Intergovernmental Panel on Climate Change (IPCC 2014) and further elaborates different budgets for the various types of greenhouse gases. Accordingly, a global CO_2 budget of 890 $GtCO_2$ for the years 2012 to 2050 is accessible. Based on the yearly CO_2

---

[5]Only countries that currently mine hard coal are assumed to have this price advantage. Countries that have reserves, but do not currently mine hard coal, have their price increased by 5% compared to the market price to avoid the unrealistic domestic production in such cases.

[6]The functionality of these tags has not been changed from the original OSeMOSYS version and is documented in Howells et al. (2011).

[7]Kartha, Sivan. 2013. "The Three Salient Global Mitigation Pathways Assessed in Light of the IPCC Carbon Budgets." Discussion Brief. Stockholm Environment Institute.

emissions of around 36 Gt, as found in the Global Carbon Atlas,[8] the global budget is reduced to 782 $GtCO_2$ for the modeled base year 2015. Because GENeSYS-MOD does not include exogenous CO_2 emissions from specific industrial branches (e.g., cement manufacturing), we reduce the limit by 2 $GtCO_2$ for all years from 2015 to 2050 (Boden et al. 2017; UNFCC 2017; BP 2017). This leaves a final global CO_2 budget of 712 $GtCO_2$ available until 2050.

13.2.2.9 CO_2 Storage Potential

The available CO_2 storage potentials for CCTS are given on a regional basis. As the current political framework prohibits the transport of pollutants and waste, CO_2 must be captured within each country. Thus, countries without any CO_2 storage capacities cannot utilize CCTS technologies. Based on the calculations and data available from Oei et al. (2014), only offshore storage capacities in aquifers, and depleted gas fields are included.

13.2.2.10 Carbon Pricing

While the global implementation of GENeSYS-MOD v1.0 (Löffler et al. 2017a) opted for a strict emissions budget and a 100% renewable energy target, the constraint of a fixed RES target for 2050 has been lifted. Before, no carbon price was set, since the much stricter target for renewable energies and perfect foresight of the model showed that the difference in terms of model results was negligible (Burandt et al. 2016). With the removal of these limitations, the introduction of a carbon price to the model was necessary. The carbon prices for Europe have been taken from the IEA (2016).

13.3 Scenario Definition

A comparison of a single, European, limit (which is optimally allocated by the model), and different regional allocations, is done, in order to identify the optimal distribution of the available CO_2 limit. This problem is of specific relevance to the present situation in Europe, as the strong importance of decarbonization in the political debate of energy transformation is generally accepted. Nevertheless, the question of distributing the remaining available budgets and the country-specific allocations, has to be clarified. Without any joint measures against climate change, and agreements from the individual national governments, reaching the target of keeping the rise of the global mean

[8]See http://www.globalcarbonatlas.org/en/CO2-emissions for further information. Data is based on Boden et al. (2017), UNFCC (2017), and BP (2017).

temperature below 2 °C is getting more and more difficult. Therefore, this paper tries to find answers to the question of national distributions of the available CO_2 limit and the fairest distribution for the European region.

The WBGU (2009) promoted an emission per-capita approach of distributing the CO_2 budget. Hereby, a differentiation between a "historical responsibility" and a "responsibility for the future" concept has to be made. Whereas in the "historical responsibility" case, the total emissions from 1990 are used to determine the share for each country, the "responsibility for the future" utilizes only the current (2010) emissions per capita for this calculation. Considering the relatively homogeneous historical development regarding CO_2 emissions, both approaches would only differ in small amounts. Therefore, we use the values from our base year 2015 as key-indicators. Staying in the definitions by the German Advisory Counsil on Global Change (WGBU), we look at scenarios within the "responsibility for the future" approach.

To define these distributions, several national key indicators were used. Considering the possible combinations of these scenario types, a total of 12 different scenarios is set up.

The following three emission pathway scenarios were implemented:

- **1.5°**: The model gets a strict CO_2 limit of 24.15 $GtCO_2$ for Europe. Considering the current yearly CO_2 emissions of around 5.6 $GtCO_2$, this budget would be exhausted within the next four to five years. Therefore, immediate action would be required. This pathway serves as a probability study if, and under what conditions the target of keeping the global mean temperature rise below 1.5 °C is possible.
- **2°**: The scenario of keeping the temperature below 2 °C is used to compare the different decarbonization pathways of the modeled European regions. It has a carbon budget of 49.27 $GtCO_2$. This emission pathway, coupled with a free distribution of the European CO_2 limit, is further referenced as the base scenario.
- **BAU**: When using the current yearly emissions and possible efficiency additions of 30% as a base-line for the future years, we get a total budget 137.39 $GtCO_2$. This scenario serves to analyse if a decarbonization would still happen, even with a relaxed emissions budget.

Furthermore, we consider four different emission distribution scenarios as follows:

- **Regional Limit/Free Distribution**: No fixed share of the European CO_2 budget is included in the model run, and therefore the model can endogenously decide for the cost-optimal allocation of the emissions.
- **Share by GDP**: In this scenario, the 2015 gross domestic product (GDP) of each country is used as a key indicator to distribute the available budget.
- **Share by current emissions**: The emissions from the base year 2015 are used to define the share for the available budget.
- **Share by population**: Here, the available budget is shared between the modeled regions with respect to their population in the year 2015.

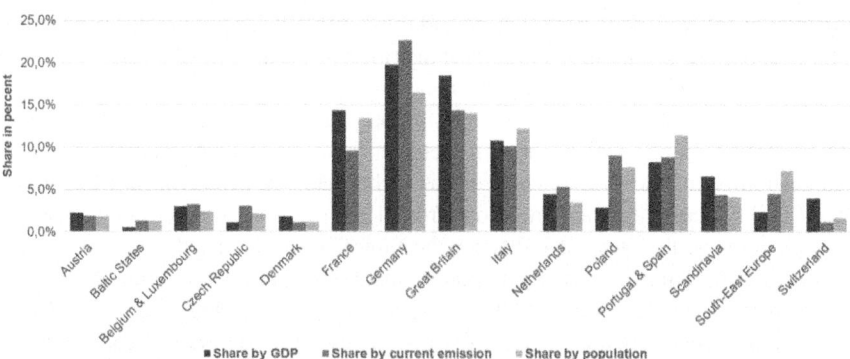

Fig. 13.4 Calculated emission shares in the different scenarios. Source: Own illustration, based on The World Bank (2017)

For distributing the emissions to our model regions, data available from The World Bank (2017) was used.

Using data and assumptions, the regional shares, as seen in Fig. 13.4, were calculated.

13.4   Results and Discussion

13.4.1   Emission Pathway: 2 Degrees

This section analyzes the results for the scenario, where the emission budget is derived from the 2°pathway. Also, the allocation of these emissions is not constrained, showcasing the ideal case, where a centralized planner is able to optimize.

Starting with the power sector, Fig. 13.5 shows the electricity generation pathway, summed up over all modeled regions. As a general trend, starting in the year 2020, renewable technologies continuously replace fossil-fueled generation. By 2040, almost all electricity generation is provided by the combination of PV, onshore wind, and hydropower.

When examining fossil fuels in depth, some interesting developments can be observed. Both hard coal and lignite are facing a constant phase-out across all regions. The emission budget is tight enough to force a rapid phase-out of these CO_2-intensive technologies. Natural gas, on the other hand, experiences a slight increase of importance in the power sector between 2015 and 2020, only to be phased out afterwards at a similar pace as the opposing coal technologies. The early growth is tied to a substantial rise in production, which originates from demand increases and the beginning of the electrification of the other sectors. By 2040, both natural gas and coal are almost nonexistent in the power sector. Nuclear energy is the

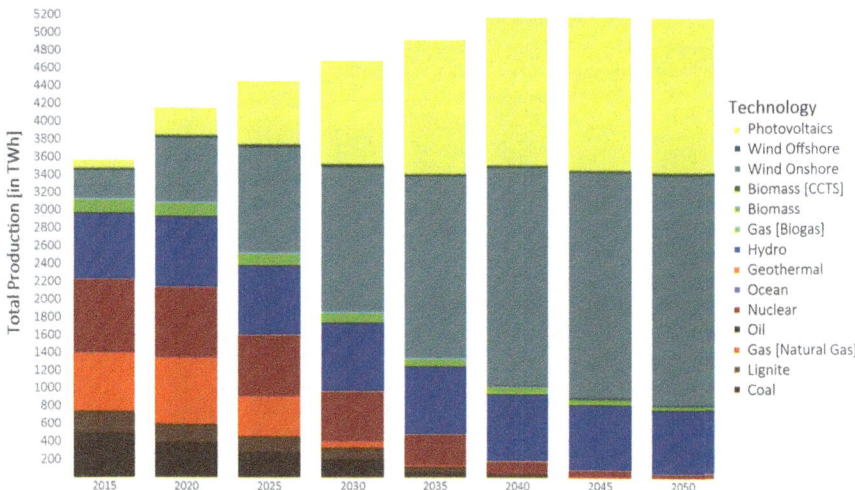

Fig. 13.5 Power production in the base scenario. Source: Own illustration

only conventional generation technology that survives until 2050, although its share by then is substantially lower than today.

As for renewable energy sources, onshore wind, PV, and hydro power are the predominant technologies. PV and wind experience rapid increases in generation capacities between the modeled periods. Onshore wind appears to be superior, where high potentials, an already very mature technology, and favorable cost developments enable high shares. In the final electricity mix of 2050, it accounts for about 47% of the total generation. Solar PV offers a similar development, the only notable difference being the lower potential, leading to upper limits being reached faster. Hydropower behaves slightly different than the two other technologies, since potentials are already quite used up, without much room for growth. Other renewable generation technologies, such as offshore wind, biogas or -mass, and geothermal energy are produced in small amounts compared to the aforementioned three technologies.

The overall electricity production increases by about 44%, which is a result of higher degrees of sector coupling and electrification of the other sectors. Taking a look at one of these sectors, the heating sector, and analyzing its development, we observe an increase in electricity use. In Fig. 13.6 the pathway of the low-temperature heating energy is shown. As before, conventional technologies are grouped in the bottom of the figure, while new, "green", technologies are shown in the top part of the graph.

Currently, natural gas is the most significant energy carrier in the low-temperature heating sector, accounting for more than 65% of the total production. This share, however, starts to decrease rapidly when the decarbonization of the energy system is taken seriously. Within the first ten years, the amount of heating provided by natural gas is more than halved, and, by 2040 natural gas has vanished. Coal, while being

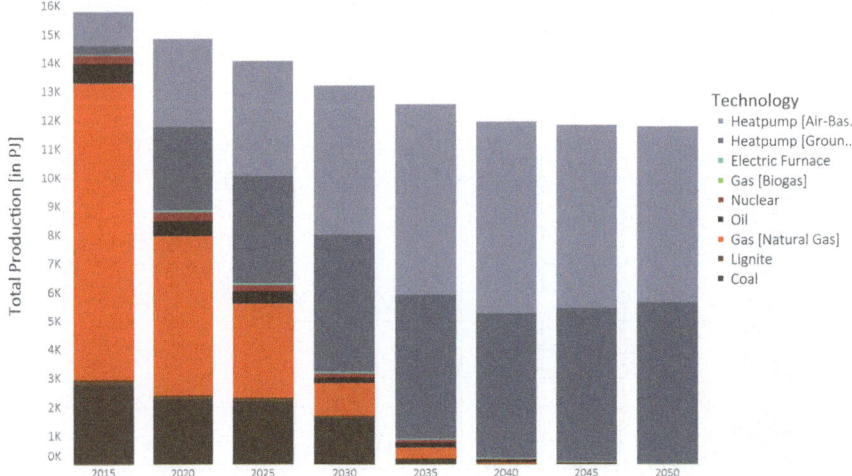

Fig. 13.6 Yearly low-temperature heat production in the base scenario. Source: Own illustration

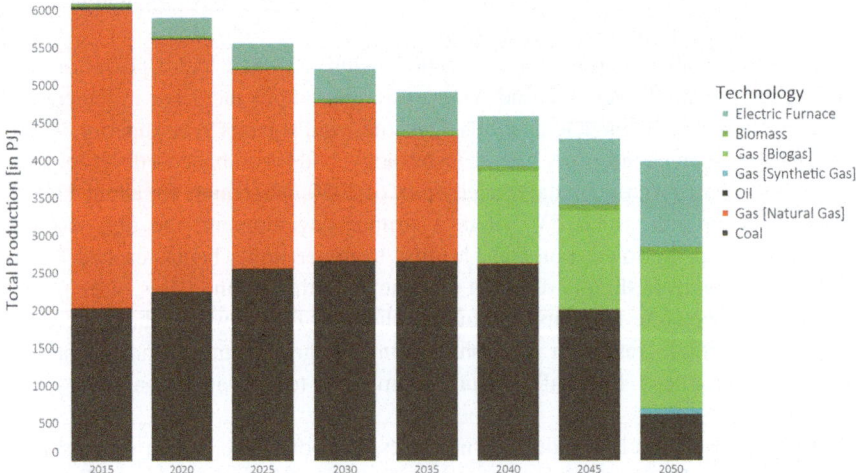

Fig. 13.7 Yearly high-temperature heat production in the base scenario. Source: Own illustration

less significant than natural gas, does not show such a drastic decrease in importance. Still, by 2040, even coal is phased out of the low-temperature heating sector.

The high-temperature heating sector is the most challenging to decarbonize. Figure 13.7 illustrates that this sector relies heavily on conventional energy sources such as natural gas and coal. Most of the gas-based heating is replaced with biomass between 2035 and 2040. A steady replacement of fossil fuels with biogas-based generation and electric furnaces results in decreased emissions, although coal stays the predominant technology until as late as 2050. Regarding efficiency and costs, high-temperature heating with hydrogen (H_2) becomes a viable option only in 2050. In the years from 2050 on, a shift towards H_2 could likely be observed. Also, with decreasing costs of power generation, electric furnaces could become an even more prominent technology.

Figures 13.8 and 13.9 show the resulting modal shares from 2015 until 2050 for both transportation sectors. In the passenger transportation sector, an early adoption of plug-in

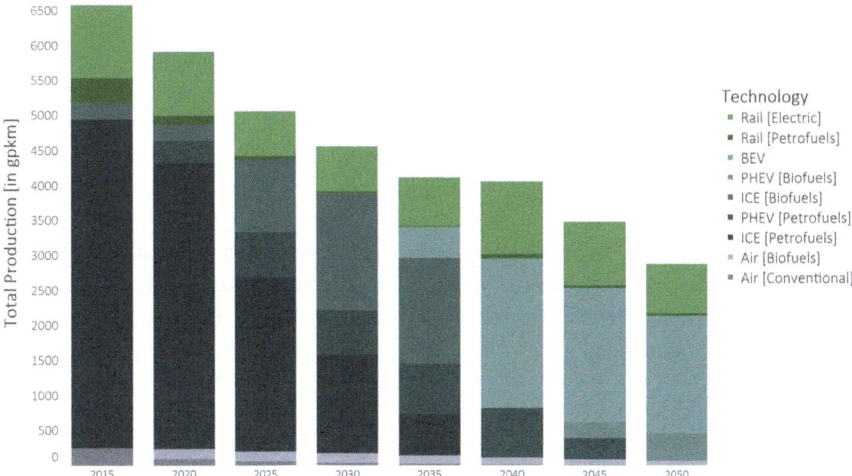

Fig. 13.8 Passenger transportation services in the base scenario. Source: Own illustration

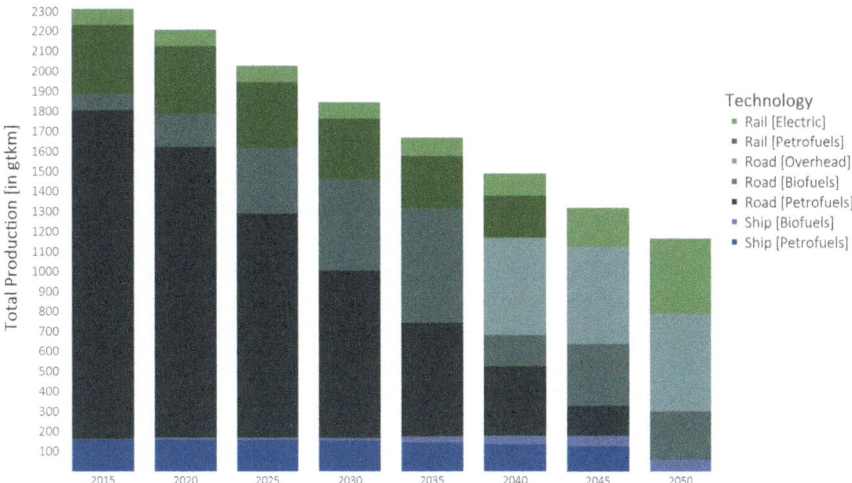

Fig. 13.9 Freight transportation services in the base scenario. Source: Own illustration

hybrid electric vehicles (PHEVs), fueled with conventional petrofuels can be observed in 2025. Furthermore, most existing diesel-electric trains are phased out in the 2020s and replaced by fully electric trains. In the second quarter of the century, biofuels gain in importance, becoming the main fuel for internal combustion engine (ICE) vehicles and PHEVs. This leads to substantial reductions of GHG emissions in the passenger transportation sector by 2035. Only in later time periods, fully electric battery electric vehicles (BEVs) start to replace conventional vehicles, whereas the newer PHEVs switch from petro- to biofuels. Due to the decreasing costs of electricity, BEV become the primary provider of passenger transportation services from 2040 on. Additionally, air transport faces a steady shift towards biofuels, coupled with a decreasing share of passenger transportation via airplanes. Thus, the passenger transportation sector is nearly decarbonized by 2050, with only small shares of diesel-electric trains remaining.

In contrast, a high reliance on fossil fuel-based ICEs can be expected in the freight transportation sector, even in the 2030s. Trains, which are currently mostly diesel-electric, stay fossil fuel-driven until 2040, facing a rapid shift towards cleaner alternatives only in the last decade. Road-based transportation experiences a steady phase-in of biofuels, which peak in 2035, with a percentage of 50% of all heavy goods vehicle (HGV) transports. Afterwards, a fast introduction of trolley-trucks can be observed. Those are powered by electric overhead lines and are thus a fully electric transportation technology, becoming the dominant technology in 2050. Conventional fuels are the main fuel for water-based freight transportation until 2045, but are entirely phased-out in 2050 and replaced with other means of transportation. While the main reason for that lies in the set-up and nature of the model (i.e. a linear cost optimization constraint by capacity expansion limits), further analysis or limitations of this rapid phase-out will have to be conducted in future work. In conclusion, similar to the passenger transportation sector, freight transportation is fully decarbonized by 2050, due to high shares of electric HGV and biofuels.

On a regional level, it can be seen that especially northern countries rapidly integrate capacities of onshore wind into their power generation, whereas the central and southern regions utilize higher shares of solar PV in 2030 (Fig. 13.10). This trend is continued between 2030 and 2050, but limited PV potentials lead to an increasing share of wind generation in some regions.

With high shares of RES in the electricity mix of 2050, their variability and flexibility have to be considered. Therefore, storages play an important role of balancing these loads. Figure 13.11 shows the charging and discharging profiles of electric storages in the different time slices for the year 2050. The backbone of the European storage capacities are lithium-ion batteries that are capable of providing intra-day storage possibilities. Energy stored in the peak-time of a day will therefore be used as an auxiliary energy-source in the night, to provide a stable energy generation. CAESs and pumped hydro storages (PHSs) are used as seasonal energy storages. Their stored energy is mostly discharged in the winter months to compensate the inferior capacity factors of RES. The maximal peak-amount of charging or discharging storages in Europe is around 250 TWh and thus less than 5% of the yearly power production.

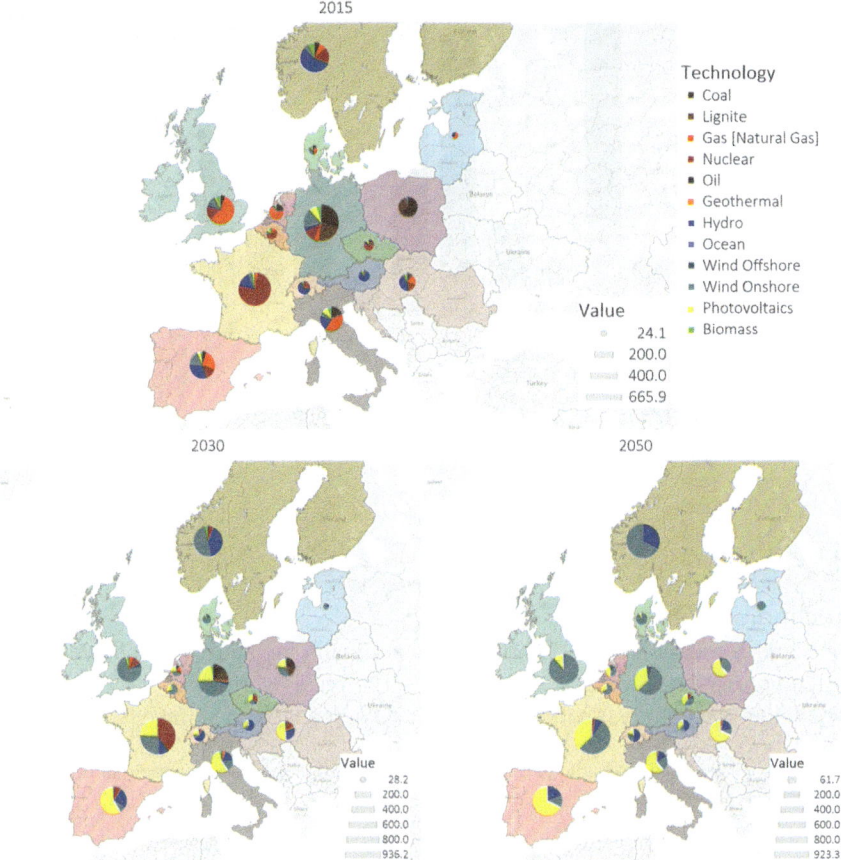

Fig. 13.10 Regional power production in 2015, 2030, and 2050 in the base scenario. Source: Own illustration

Countries with a high base level of emissions like Germany, the UK, or Italy experience a quicker phase-out of fossil fuels than other regions. The reason for that is the tight emission budget, forcing a rapid phase-out of fossil fuels in the early stages of the modeling period in order to achieve the 2° goal. Figure 13.12 graphs the cumulative emissions over all regions per year. The red bars show the total emissions in this period, while the gray line is the sum of all emissions during the modeling period. First, the yearly emissions show a steady decline in total emissions, which by 2030 are more than halved compared to 2015 levels. This reduction is considerably lower than current emission reduction targets from the EU or the respective countries. Second, following this path, more than 90% of the total emissions are being produced until 2035, which showcases the steep decarbonization pathway at later years. Another remarkable observation is that this emission trajectory would surpass the 1.5° budget as soon as 2020.

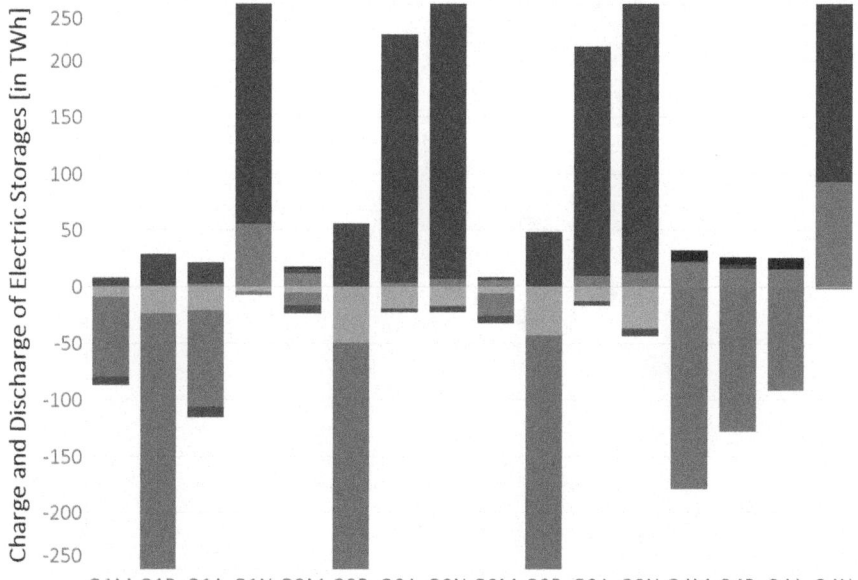

Fig. 13.11 Charge and discharge of electric storages in 2050 in the base scenario (per time slice).
Source: Own illustration

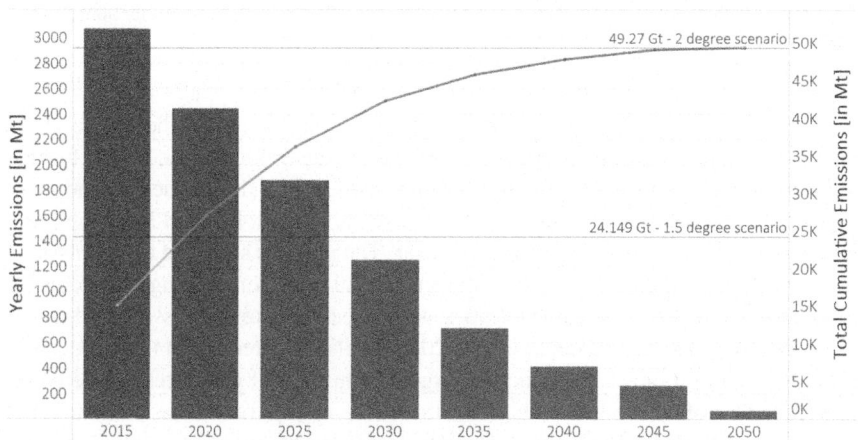

Fig. 13.12 Cumulative CO$_2$ emissions in the 2°pathway. Source: Own illustration

13.4.2   Emission Pathway: Business as Usual

In the business as usual (BAU) pathway, the model faces a carbon budget of 137.39 $GtCO_2$, almost triple the amount of the baseline 2°pathway. This implies a relaxed constraint, enabling many regions to use more of their fossil fuels for a longer period. The total carbon budget is not reached in any of the distribution scenarios, as renewable energy sources still beat fossil-fueled power plants in terms of future costs. Even though the model would, in theory, be able to emit more CO_2 and thus construct new fossil power plants, it decides against it on a cost basis, resulting in a total carbon amount of only 60.76 $GtCO_2$.

Figure 13.13 shows the development of electricity generation over the modeling period. The results are close to the 2°pathway, indicating a substantial shift towards RES by 2045, where more than 95% of power production is decarbonized. The main difference is the usage of lignite, which is phased out later than hard coal, contrary to the base scenario.

A much more prominent difference is the generation of high-temperature heat (Fig. 13.14) which, in the BAU scenario, is mainly based on coal as an energy carrier. While gas capacities see the same fuel switch from natural gas to biogas as seen in the 2°pathway, coal is not phased out, and instead actually being used more up until the year 2040. Only then does the usage of hard coal as the preferred source for process heat decline, with biogas, some synthetic gas and a small share of electric furnaces entering the fuel mix.

Another major difference between the emission pathways lies in the freight transportation sector, which can be seen in Fig. 13.15. While overhead-powered trucks were the backbone of freight transportation services in the 2°pathway, the BAU scenario opts for bio-fueled combustion-based trucks instead.

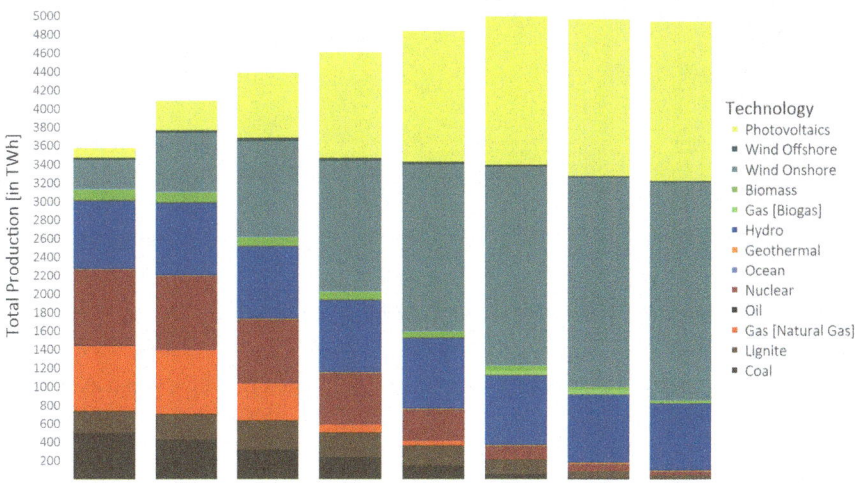

Fig. 13.13 Yearly power production in the BAU pathway. Source: Own illustration

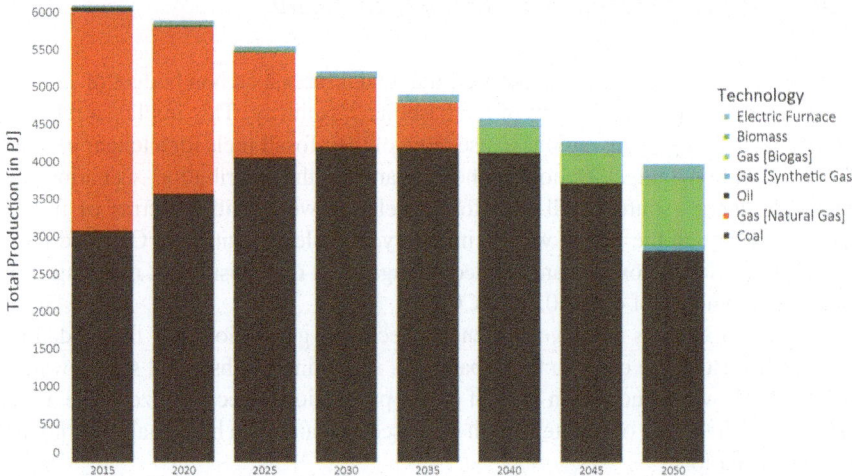

Fig. 13.14 Yearly high-temperature heat production in the BAU pathway. Source: Own illustration

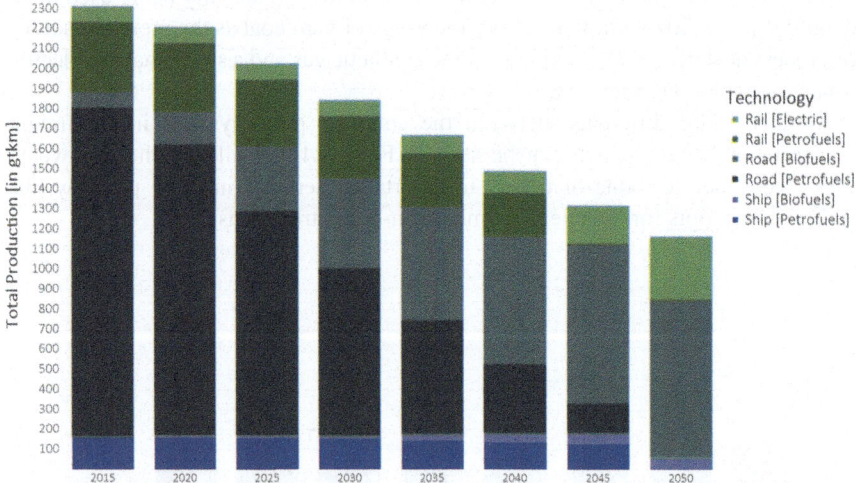

Fig. 13.15 Freight transportation services in the BAU pathway. Source: Own illustration

13.4.3 Emission Pathway: 1.5 Degrees

The 1.5°pathway has a total emission budget of 24.15 GtCO$_2$, which is not even half of the budget of the 2°pathway. Given today's emissions, this limit would be reached within the next years. Thus, a drastic reduction in emissions—and possibly—negative emission technologies are needed.

Figure 13.16 shows the development of electricity generation over the years. The stricter carbon budget leads to a major change in the development of power

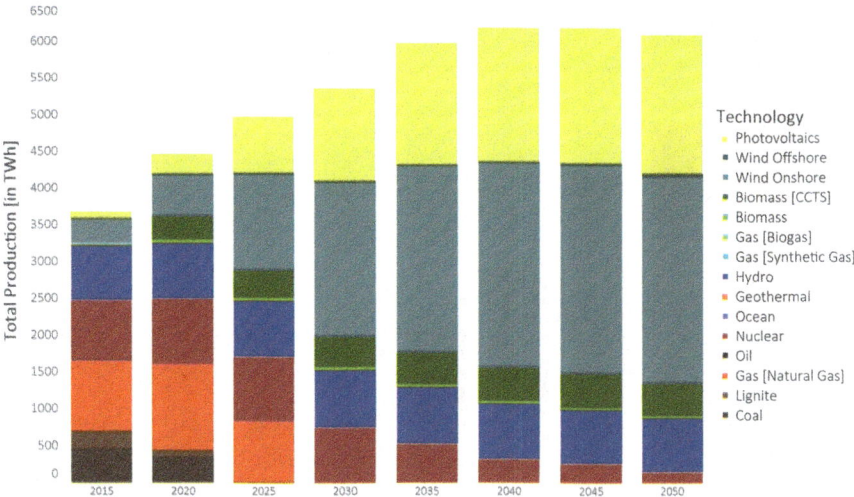

Fig. 13.16 Development of yearly power production in the 1.5°pathway. Source: Own illustration

generation. The sector is virtually decarbonized as early as 2030, with only nuclear power remaining in terms of conventional power sources. Bio-energy with carbon capture, transport, and storage (BECCTS) is used after 2020, as it is the only option for the model to achieve negative emissions and reduce the burden of current carbon emissions. Coal and lignite are phased out by 2025, with natural gas following in 2030.

High-temperature heat also shows great diversion from both the 2°, as well as the BAU pathways. While coal is the dominant fuel source for high-temperature heat in both other pathways, the model opts for natural gas instead, in order to save on emissions. Electric furnaces play a much more critical role than in the other scenarios, as it is one of the few options for emission-free process heat generation. Biomass, which would be the other option, is instead used in CCTS processes to generate negative emissions (Fig. 13.17).

Freight transport in 2050 is based solely on overhead-powered trucks for road-based transportation, contrary to the other emission pathways. As soon as the technology becomes available at low costs, a fast switch towards electricity-based freight transport can be observed. While parts of the existing fleet of trucks remain in the system, all new capacities from 2035 onward are electric trucks, powered by overhead lines.

13.4.4   Comparison of Emission Pathways

The model results show that for each emission pathway, a cost-optimal solution for keeping the set emission targets can be found. This also holds true for most CO_2

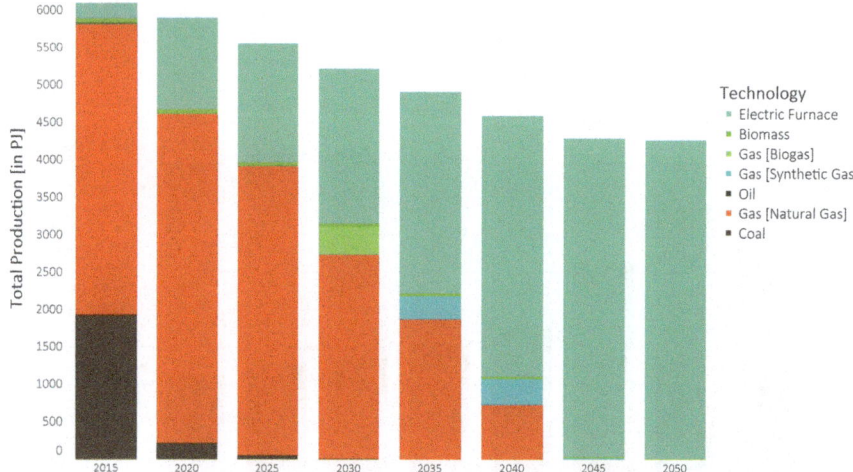

Fig. 13.17 Yearly high-temperature heat production in the 1.5°pathway. Source: Own illustration

Scenario	Free Distribution	Share by Current Emissions	Share by GDP	Share by Population	Free Distribution (CCTS disabled)
1.5° pathway	12,2%	22,4%	Infeasible	27,2%	133,4%
2° pathway	0,0%	0,3%	1,4%	0,8%	0,0%
BAU pathway	-1,2%	-1,2%	-1,1%	-1,2%	-1,2%

Fig. 13.18 Cost comparison of all emission pathways and distribution scenarios. Source: Own illustration

distribution scenarios, the only exception being the "Share by GDP" scenario in the 1.5°pathway, which does not yield a feasible solution. Figure 13.18 shows a comparison of total costs, relative to the base case (2°pathway, free distribution of emissions).

As expected, the 1.5°pathway generates the highest total costs, at least 12.2% higher than those of the 2°pathway. The BAU pathway is cheaper overall, albeit only by about 1% compared to the base scenario. When it comes to distribution scenarios, the planner-perspective "Free Distribution" scenario yields the lowest overall costs, since it distributes emissions solely on a cost optimization basis. When introducing region-specific limits of emissions, an overall increase in system costs can be observed, except for the BAU pathway, where the overall emission constraint is relaxed enough so that distribution only plays a minor role.

Whereas distribution only produces a cost difference of about 1% in the 2°pathway, for the strict emission targets of the 1.5°pathway, the difference is significant. A difference of around 15% between share scenarios can be observed, with the "Share by GDP" scenario being impossible to solve for the model, given the constraints.

13.4.5 Discussion

13.4.5.1 Fossil Fuel Prices

When taking a look at the possible transformation pathways towards renewable energies in an energy system, one has to pay close attention to the underlying prices for fossil fuels. Determining the future prices of fossil fuels is a difficult task, with only few reliable sources available. The IEA, for example, predicts fuel prices in line with their scenarios. The problem with this is that the model results are still based on large shares of fossil energy carriers, often in combination with CCTS. This is where the issue of the "green paradox" arises. Since we estimate large shares of renewables coming into the system, the demand for fossil fuels would fall drastically, and thus their price would have to decline as well. This would in turn lead to a slower transformation towards renewables, as cheap fuel prices could get fossil fuel based generation to become competitive once again. Current assumptions of fossil fuels priced as a finite resource (with thus constantly increasing costs) may have to be revised and updated. This task will become increasingly important in the future, as these price assumptions (together with potential carbon pricing) drive model-based results, and thus, decisions.

Although important to keep in mind, these issues are most adequately dealt with using scenario and sensitivity analyses. Multiple sensitivities for fossil fuel prices have been calculated and examined to test the robustness of the results, although there might be opportunities for future research to include such simulations into the scope of the model.

13.4.5.2 Solar PV Potentials

As with prices, the theoretical potentials for renewable supply technologies strongly drive model results. Assumptions concerning both the amount of available land and the definition of such are a heated topic in both science and policy (as seen in Clack et al. (2017) and Jacobson et al. (2017a, b)). The decision about these values directly influences the modeling constraints and can therefore steer results in certain directions.

The values chosen for our model runs (see Sect. 13.5.4) concerning solar PV potentials are quickly exhausted, with some regions reaching the maximal values as soon as 2030 (see Fig. 13.19). Given other results in literature (e.g., Ram et al. 2017), this seems rather early.

As mentioned in Sect. 13.2.2.4, an own assessment of solar potentials of all European regions has been conducted to provide ground for sensitivity analyses, as the actual potentials used in other models are usually unobtainable.

The results (depicted in Fig. 13.20) show that the available solar potential heavily influences the results for a transition towards renewables in Europe. Especially in the sunnier regions in the south of Europe, vastly larger amounts of solar capacities are

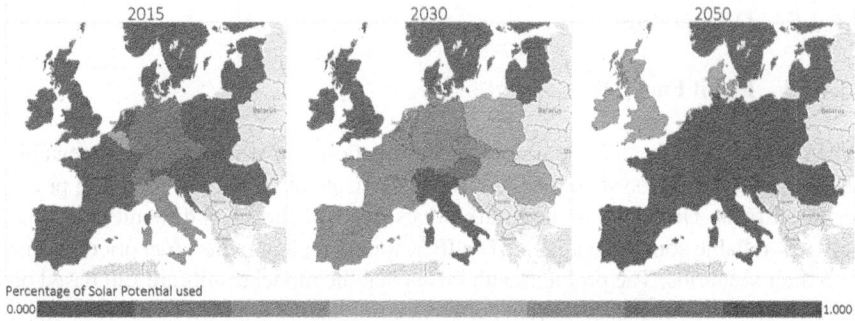

Fig. 13.19 Utilization of solar PV potential in Europe in the base scenario. Source: Own illustration

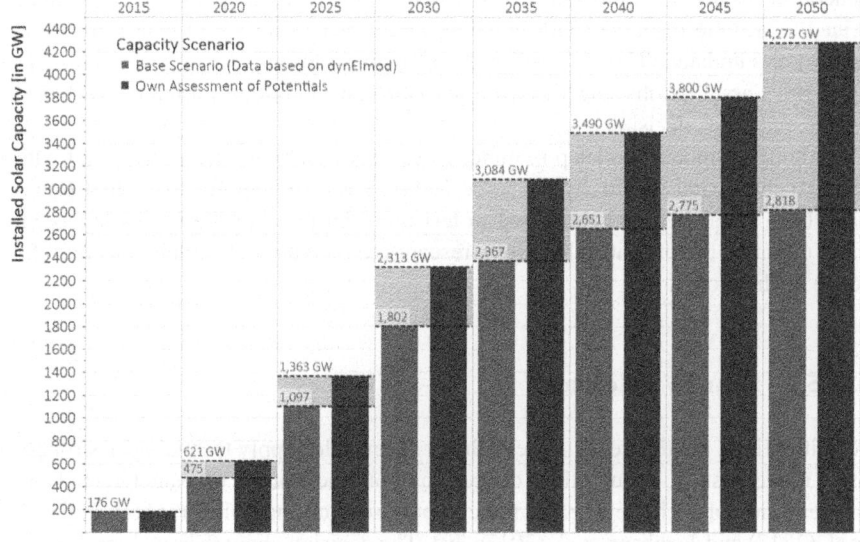

Fig. 13.20 Installed solar capacity between calculated scenarios in GW. Source: Own illustration

constructed and shift both the resulting production and capacity mixes, as well as the grid structure and expansion.

13.5   Conclusion

Over the last decades, climate warming has sparked a heated debate about the emission of greenhouse gases. If the concentration of these GHGs is not reduced significantly within the near future, irreversible and severe consequences for humans and natural systems are the to be expected (McMichael et al. 2006). One of the biggest contributors of GHG emissions is the energy sector, accounting for more than two thirds of the global emissions (IEA 2016). The most important greenhouse gas is CO_2, which is responsible for more than 80% of the emissions in the energy sector (Foster and Bedrosyan 2014). Therefore, various challenges arise for different countries when it comes to decarbonizing their energy systems. Especially highly developed countries and regions, such as the a leading role in the low-carbon transformation process.

In this chapter, possible decarbonization pathways were analyzed, using varying assumptions for carbon constraints and distributions among the chosen model regions. For the analysis, the Global Energy System Model (GENeSYS-MOD) has been used, a linear program, minimizing total system costs for the sectors power, heat, and transport, given external constraints, such as emission limits. The framework has been expanded with various new functionalities and improvements, such as an upgrade to the trading system with respect to power trade, or overall performance optimization. Additionally, a new and improved data set, introducing new technologies (especially in the transportation sector) and featuring 16 time slices,[9] has been added to GENeSYS-MOD v2.0. Europe is modeled in a total of 15 regions, with model calculations optimizing the pathway from 2015 to 2050 in five-year steps.

Three different pathways have been considered: a pathway that limits global warming to 2 °Celsius (C), a 1.5°pathway, and a business as usual (BAU) pathway. While the overall emission trajectory is relevant on a global scale, the distribution of these emission budgets onto the European countries is important, especially considering possible policy implications. Thus, a total of four distribution methods for the set carbon budget have been examined: free distribution, share by GDP, share by population, and share by current emissions. The results indicate that even ambitious climate targets can be met, both technically and economically. All modeled pathways and distribution scenarios were solvable, except for the share by GDP in the 1.5°pathway, which did not yield any feasible solution. It can be shown that reaching a climate target of 2 °C only implies a cost increase of about 1%, while reducing total emissions by almost 20% compared to the BAU case; the 1.5° target is achievable too.

---

[9]Instead of the previous six time slices.

No matter which distribution is chosen, the model results show that meeting ambitious climate targets requires widespread effort and strong policy instruments in the near future. While much of the renewable transformation is market-driven (as can be seen in the BAU scenario), goals of well below 2 °C can only be achieved if carbon constraints are set and maintained by policy.

The power sector sees a steady phase-out of fossil fuels across all pathways, usually starting out slowly from 2015 to 2020, with 2025 to 2030 usually representing large sums of fossil capacities going off-grid. This is due to old capacities growing obsolete, as well as renewables becoming more and more competitive. While overall electricity demands decline, the total power generation increases from about 3600 TWh to about 5100 TWh in 2050, as a coupling between the usually segregated sectors can be observed. Storage plays an important role of balancing grid infrastructure and demands, with about 739 gigawatt (GW) of installed storage capacities in 2050, most of which are Lithium-Ion (Li-Ion) batteries. The high-temperature industrial heat sector is the most difficult to decarbonize. Renewable alternatives for process heat are expensive and difficult to implement, which means that fossil fuels play a significant role for high-temperature heat generation, given the emission budget. Further research should take a closer look at real-life implications of the obtained model results, such as the stranded asset problem that could arise, given the fast phase-out of fossil generation capacities.

As always with quantitative, model-based research, certain aspects of the real world can only be included in a simplified version into the model. While the extension of the amount of time slices greatly improves the temporal setting of the model, there are still limitations to the amount of variability that can be observed with the model. Some effort should be placed into more model-improvements, such as adding more load-balancing options, for example in the form of reworked storages, or the implementation of BEV as electricity storage into the model.

References

Boden, T., R. Andres, and G. Marland. 2017. *Global, Regional, and National Fossil-Fuel CO_2 Emissions (1751–2014) (V. 2017)*. Oak Ridge, USA: Carbon Dioxide Information Analysis Center, Oak Ridge National Laboratory.

Bogdanov, D., and Christian Breyer. 2016. *Eurasian Super Grid for 100% Renewable Energy Power Supply: Generation and Storage Technologies in the Cost Optimal Mix*. Daegu, South Korea: International Solar Energy Society.

BP. 2017. *BP Statistical Review of World Energy – June 2017*. 66th ed. London, UK: British Petroleum.

Burandt, Thorsten, Karlo Hainsch, Konstantin Löffler, Heinrich Böing, Jenny Erbe, Ivo-Valentin Kafemann, Mario Kendziorski, et al. 2016. *Designing a Global Energy System Based on 100% Renewables for 2050. Student Research Project*. Berlin, Germany: Technische Universität Berlin.

Bussar, C., P. Stöcker, Z. Cai, L. Moraes, R. Alvarez, H. Chen, C. Breuer, A. Moser, M. Leuthold, and D. U. Sauer. 2015. Large-Scale Integration of Renewable Energies and Impact on Storage

Demand in a European Renewable Power System of 2050. *Energy Procedia*, 9th International renewable energy storage conference, IRES 2015, 73 (Supplement C): 145–153.

Clack, Christopher T.M., Staffan A. Qvist, Jay Apt, Morgan Bazilian, Adam R. Brandt, Ken Caldeira, Steven J. Davis, et al. 2017. Evaluation of a proposal for reliable low-cost grid power with 100% wind, water, and solar. *Proceedings of the National Academy of Sciences* 114 (26): 6722–6727.

Czisch, Gregor. 2007. *Joint Renewable Electricity Supply for Europe and Its Neighbours*. Institute for Electrical Engineering, University of Kassel.

ENTSO-E. 2013. *Scenario Outlook & Adequacy Forecast (SO&AF) 2013–2030*. Brussels, Belgium: ENTSO-E.

Farfan, Javier, and Christian Breyer. 2017. Structural changes of global power generation capacity towards sustainability and the risk of stranded investments supported by a sustainability indicator. *Journal of Cleaner Production* 141 (January): 370–384.

Foster, Vivien, and Daron Bedrosyan. 2014. *Understanding CO_2 Emissions from the Global Energy Sector*. Washington, DC: The World Bank.

Fraunhofer ISI, Fraunhofer ISE, Institute for Resource Efficiency and Energy Strategies (IREES), Observ'ER, Technical University Vienna - Energy Economics Group, TEP Energy. 2016. Mapping and analyses of the current and future (2020–2030) heating/cooling fuel deployment (fossil/renewables) (Report). Karlsruhe: European Commission Directorate - General for Energy.

Gerbaulet, Clemens, and Casimir Lorenz. 2017. dynELMOD: a dynamic investment and dispatch model for the future European electricity market. Data Documentation No. 88. DIW Berlin: Berlin, Germany.

Gerbaulet, Clemens, Christian von Hirschhausen, Claudia Kemfert, Casimir Lorenz, and Pao-Yu Oei. 2017. Scenarios for Decarbonizing the European Electricity Sector. In *2017 14th international conference on the European Energy Market (EEM)*.

Hainsch, Karlo, Thorsten Burandt, Claudia Kemfert, Konstantin Löffler, Pao-Yu Oei, and Christian von Hirschhausen. 2018. Emission pathways towards a low-carbon energy system for Europe – a model-based analysis of decarbonization scenarios. DIW Berlin Discussion Paper no. 1745.

Hosenfeld, Hans, Alexandra Krumm, Linus Lawrenz, Luise Lorenz, Benjamin Wechmann, Bobby Xiong, Thorsten Burandt, et al. 2017. *Designing an Energy Model for India and China for the Low-Carbon Transformation to 2050 – Assessing National Policies, Grids, and Modeling Regional Scenarios*. Berlin, Germany: Technische Universität Berlin.

Howells, Mark, Holger Rogner, Neil Strachan, Charles Heaps, Hillard Huntington, Socrates Kypreos, Alison Hughes, et al. 2011. OSeMOSYS: the open source energy modeling system: an introduction to its ethos, structure and development. *Energy Policy, Sustainability of Biofuels* 39 (10): 5850–5870.

IEA. 2015. *World Energy Outlook Special Report 2015: Energy and Climate Change*. Paris, France: OECD.

———. 2016. *World Energy Outlook 2016*. Paris, France: OECD.

IPCC. 2014. *Climate Change 2014: Synthesis Report. Contribution of Working Groups I, II and III to the Fifth Assessment Report of the Intergovernmental Panel on Climate Change*. Geneva, Switzerland: IPCC.

Jacobson, Mark Z., Mark A. Delucchi, Zack A.F. Bauer, Savannah C. Goodman, William E. Chapman, Mary A. Cameron, Cedric Bozonnat, et al. 2017a. 100% clean and renewable wind, water, and sunlight all-sector energy roadmaps for 139 Countries of the World. *Joule* 1 (1): 108–121.

Jacobson, Mark Z., Mark A. Delucchi, Mary A. Cameron, and Bethany A. Frew. 2017b. The United States can keep the grid stable at low cost with 100% clean, renewable energy in all sectors despite inaccurate claims. *Proceedings of the National Academy of Sciences* 114 (26): E5021–E5023.

Löffler, Konstantin, Karlo Hainsch, Thorsten Burandt, Pao-Yu Oei, Claudia Kemfert, and Christian von Hirschhausen. 2017a. Designing a model for the global energy system—GENeSYS-MOD: an application of the Open-Source Energy Modeling System (OSeMOSYS). *Energies* 10 (10): 1468.

Löffler, Konstantin, Karlo Hainsch, Thorsten Burandt, Pao-Yu Oei, and Christian von Hirschhausen. 2017b. Decarbonizing the Indian energy system until 2050: an application of the open source energy modeling system OSeMOSYS. *IAEE Energy Forum* (Singapore Issue 2017): 51–52.

McMichael, Anthony J., Rosalie E. Woodruff, and Simon Hales. 2006. Climate change and human health: present and future risks. *The Lancet* 367 (9513): 859–869.

Oei, Pao-Yu, Johannes Herold, and Roman Mendelevitch. 2014. Modeling a carbon capture, transport, and storage infrastructure for Europe. *Environmental Modeling & Assessment* 19 (6): 515–531.

Pfenninger, Stefan, and Iain Staffell. 2016. Long-term patterns of European PV output using 30 years of validated hourly reanalysis and satellite data. *Energy* 114 (November): 1251–1265.

Plessmann, G., and P. Blechinger. 2016. How to Meet EU GHG Emission Reduction Targets? A Model Based Decarbonization Pathway for Europe's Electricity Supply System until 2050. Energy Strategy Reviews. Berlin, Germany: Reiner Lemoine Institute.

PwC. 2011. *Moving Towards 100% Renewable Electricity in Europe and North Africa by 2050.* London, UK: PwC, PIK and IIASA.

Ram, Manish, Dmitrii Bogdanov, Arman Aghahosseini, Solomon Oyewo, Ashish Gulagi, Michael Child, and Christian Breyer. 2017. *Global Energy System Based on 100% Renewable Energy – Power Sector.* Lappeenranta, Finland and Berlin, Germany: Lappeenranta University of Technology (LUT) and Energy Watch Group.

Rasmussen, Morten Grud, Gorm Bruun Andresen, and Martin Greiner. 2012. Storage and balancing synergies in a fully or highly renewable pan-european power system. *Energy Policy, Renewable Energy in China* 51 (Supplement C): 642–651.

Scholz, Ivonne. 2012. *Renewable Energy Based Electricity Supply at Low Cost – Development of the REMix Model and Application for Europe.* Stuttgart, Germany: Universität Stuttgart.

The World Bank. 2017. World Bank Open Data. Free and Open Access to Global Development Data. 2017.

Tröndle, Tobias Wolfgang. 2015. Development of a Global Electricity Supply Model and Investigation of Electricity Supply by Renewable Energies with a Focus on Energy Storage Requirements for Europe. Dissertation, University of Heidelberg, Heidelberg, Germany.

UNFCC. 2017. *National Inventory Submissions 2017.* Bonn, Germany: United Nations Framework Convention on Climate Change.

WBGU. 2009. *Kassensturz für den Weltklimavertrag – der Budgetansatz: Sondergutachten.* Berlin: Wissenschaftlicher Beirat Globale Umweltveränderungen.

Welsch, M., M. Howells, M. Bazilian, J.F. DeCarolis, S. Hermann, and H.H. Rogner. 2012. Modelling elements of smart grids – Enhancing the OSeMOSYS (Open Source Energy Modelling System) Code. *Energy, Energy and Exergy Modelling of Advance Energy Systems* 46 (1): 337–350.

Zerrahn, Alexander, and Wolf-Peter Schill. 2016. *Long-Run Power Storage Requirements for High Shares of Renewables: Review and a New Model. Renewable and Sustainable Energy Reviews.* Berlin, Germany: DIW.

Part IV
Assessment, Perspectives, and Conclusions

Chapter 14
General Conclusions: 15 Lessons from the First Phase of the Energiewende

Claudia Kemfert, Pao-Yu Oei, and Christian von Hirschhausen

> *The thesis of this book is that a fundamental and radical change in the energy policy of the Federal Republic [of Germany] has become unavoidable. We want to introduce a strategy for the future energy supply, which—after due diligence—is technically feasible and economically and politically advantageous to avoid the wreckage of the current strategy.*
> *Florentin Krause (1980, Chapter 1): Why we need an energiewende. (p. 13) (Own translation from German original.)*

14.1 Introduction

Energiewende "Made in Germany": this is a relatively recent phenomenon, yet with a long germination period, going back to the 1970s, and it has attracted broad interest in many spheres, including academia, industry, and policy making. The previous

C. Kemfert (✉)
DIW Berlin, Berlin, Germany

Hertie School of Governance, Berlin, Germany

German Advisory Council on the Environment (SRU), Berlin, Germany
e-mail: ckemfert@diw.de

P.-Y. Oei
DIW Berlin, Berlin, Germany

Junior Research Group "CoalExit", Berlin, Germany

TU Berlin, Berlin, Germany

C. von Hirschhausen
DIW Berlin, Berlin, Germany

TU Berlin, Berlin, Germany

© Springer Nature Switzerland AG 2018
C. von Hirschhausen et al. (eds.), *Energiewende "Made in Germany"*,
https://doi.org/10.1007/978-3-319-95126-3_14

chapters have provided insights into specific aspects of the process, and have sketched out possible pathways for future developments. The chapters of this book share among them the conviction that, while many obstacles have yet to be over-come, the energiewende is well underway, e.g., increasing the share of renewables in the electricity sector, or taking nuclear power plants from the grid without adverse impacts; however, significant challenges remain, e.g., increasing energy efficiency, and reducing the carbon footprint of the energy system as a whole. From a public policy perspective, the energiewende is well justified because it enhances the welfare of society.

The objective of this concluding chapter is to draw some cross-cutting lessons from the first period of the energiewende. Until recently, the focus of the energiewende was on the electricity sector, but what is required is an energy system wide approach. There are at least three decades before us in which further reforms, technical innovations, and political consensus will be required to make the energiewende a true success. The empirical evidence from the recent past, together with a technical and political assessment of the feasibility of the next reform steps, allows us to formulate 15 lessons, both summarizing the previous chapters and opening up perspectives on the future. This will be done following the book's structure: the next section looks at lessons from the long-term analysis of energy and climate policies (Part I of the book). Section 14.3 focuses on the lessons from the ongoing energiewende in Germany (Part II), and Sect. 14.4 provides lessons on the interplay between the German setting and the low-carbon transformation at the European level (Part III). Section 14.5 discusses the findings, provides an outlook on the next phases, and concludes.

14.2   Lessons from the Long-term Trends: The Energiewende in the Context of Long-term Energy and Climate Policies in Germany

14.2.1   Lesson I: The Energiewende Has Challenged the Traditional Modus Operandi of Energy Policy in Germany

Historically, there has been a large degree of convergence between Germany's energy industry (in the larger sense, including equipment suppliers, traders, etc.) and the political establishment. Even in critical moments, when public opinion threatened the incumbent structures, the German energy industry was able to main-tain its power—for instance, after the First World War, when the sector simply ignored the law on nationalization. In the 1980s and up to the fall of 2010, German utilities were able to continue down the path of nuclear power, despite opposition from a growing majority of the population, by defining strategies and targets and by convincing policymakers to follow. Overall, one observes a pattern in German energy policy in which the corporate sector sets targets and roadmaps, and public energy policy is reduced to following suit.

The energiewende broke this pattern and led to an implosion of the four major incumbent utilities. In fact, the energiewende was a decidedly political maneuver that forced the incumbent fossil-nuclear industry to react. After pronounced political conflicts over nuclear energy, political leaders at the highest level sent a clear signal in the summer of 2011 that nuclear power plants would be closed down. The initial opposition to this decision abated rapidly. The phasing out of coal will take somewhat longer, perhaps into the 2030s, but it is structurally similar: In both cases, there was a political decision to favor public well-being over private profits, underpinned by a broad societal consensus.

14.2.2 Lesson II: The Energiewende Corresponds to a Certain Extent to the "Soft Path" Not Taken in the 1970s

Discussions about the orientation of energy policies date back to the social and environmental movements movements of the post-1968 period. The positions that emerged from these discussions were summarized concisely by Amory Lovins (1976) as the "soft" and the "hard" path of energy policy. The "hard path" represents large-scale technologies such as nuclear or big coal, and centralized corporate structures, generally not well controlled by public policy. The "soft path" refers to more widely distributed, mostly renewable generation, and democratic control of the sector and energy policy.

The core goal of the energiewende is to phase out nuclear and fossil fuels, to replace them by renewables in a much more energy-efficient economy, and to replace the monolithic governance structures of corporate monopolies with distributed generation, ownership, and political decision making, which corresponds to what Amory Lovins had coined the "soft path." The energiewende is in fact rooted in the idea of a soft path, whereas in the 1970s, the hard path was clearly predominant. A great deal of the terminology of that time—for example, the terms coined by Lovins (1976 for the U.S.) and those used by Krause et al. (1980 for Germany)—was used again by the proponents of the energiewende. In line with the "soft path," the objective of the energiewende is not only to modify the electricity mix, but also to abandon the incumbent power structures. Would anybody have anticipated, a decade ago, that RWE would focus its corporate strategy on becoming the third biggest renewables producer in Europe? Whether this will eventually succeed remains to be seen.

14.2.3   Lesson III: The Energiewende Is a Long-term Project, and It Is a Political and Societal Revolution as Much as It Is a Technological Revolution

What the energiewende is concretely, and when it started, is a topic of ongoing debate. Most observers have noted a continuous energiewende movement beginning with the energy policy struggles of the 1970s; they trace the strategies and partial successes of today's energiewende back to efforts in the 1970/80s to work towards a "soft path" of energy and climate policy. Yet the energiewende of the 1980s sought to abandon nuclear power and to make Germany independent of oil and nuclear power, but it accepted a large share of coal in primary energy supply. Today, there is consensus that the phasing out of coal, and perhaps also of other fossil fuels, is part of the energiewende as well. It is also important to note that the very nature of the Energy Concept 2050 (BMWi and BMU 2010) still corresponded to the "hard path," i.e., a low-carbon transformation with significant amounts of coal (assuming carbon capture) and nuclear energy. This path looked well on paper, but it was incompatible with an 80% or even higher share of renewables.

When thinking about the energiewende, technical change generally comes to mind first: the generation mix, efficiency, demand-side management, and so on. However, we have also seen that the technical objectives (share of renewables, nuclear phase-out, etc.) have been accompanied by efforts to make energy and climate policy more transparent, to foster public debate, and to increase the participation of citizens in the process (*Bürgerenergiewende*, or a citizens' energiewende), including large-scale ownership of electricity generation and other assets. Until recently, about two thirds of renewables investments were made outside the corporate sector. The broad success of distributed renewables since 2010 has been spurred by popular support which, in turn, was also driven by material benefits, i.e. lucrative investments by certain parts of the population. Policy will need to keep the "citizens' energiewende" on track and combat the lingering danger of a "counterrevolution".

14.2.4   Lesson IV: The Energiewende Has not Changed the Cross-Subsidization of Energy-Intensive Consumers by Small Businesses and Household Consumers

If there is one feature of German energy policy that the energiewende has not succeeded in changing, it is the lobbying power of the energy-intensive manufacturers for low electricity prices. This can be observed in the emerging period of electrification, during the late nineteenth century, throughout the twentieth century, and even in recent periods of the energiewende: In 2018, 50% of all industry electricity, that is, the share of electricity-intensive industries, pay only a minor share of the renewables surcharge, which was instead paid by regular industrial consumers and private households. Consequently, energy-intensive industries paid electricity prices in the range of 4–16 €cents/kWh, whereas the average household consumer paid around 30 €cents/kWh.

The unequal burden sharing of the costs of the energiewende between the energy-intensive industry and the remaining consumers confirms the political economy hypothesis expressed by Mancur Olson, and others: small, well-organized interest groups such as energy-intensive industrial sectors can lobby more effectively than large, difficult-to-coordinate interest groups or the general public. Any attempts to work against this trend may even backfire and undermine political support for the energiewende at large. In fact, one may consider the cross-subsidization of the electricity-intensive industry to be the political price that has to be paid for the successes of the energiewende, i.e., a political compromise. The government administrations that followed the initial phase of the energiewende were heavily criticized for leaving all the privileges of the incumbent, energy-intensive industries (that were exempt from paying the additional costs of renewables) intact, and for falling prey to the lobbying of a few well-organized interest groups. In hindsight, the distribution of costs to the general public may not have been fair, but it was effective.

14.2.5   Lesson V: Methodology Matters: Polycentric Approaches Are Superior to Monocentric ("One-Size-Fits-All") Approaches

From a methodological perspective, the energiewende proves the polycentric approach to public policy to be more appropriate than the monocentric approach still used by many conventional energy and climate economists. The latter often refer to the existence of one "optimal" or "first-best" policy instrument, supposedly relying on Jan Tinbergen's call to develop one (and only one) instrument for any given policy objective. The political economy realities of the energiewende (and other economic processes) are different: political action takes place at various levels from the global (e.g., climate negotiations at the UNFCCC), the national, and down to the local (e.g., climate legislation in the city of Berlin). It is risky to focus on one-size-fits-all" policy instruments, such as setting GHG emission reduction targets for specific sectors through a European Commission Trading System (ETS), as evidenced by the instrument's failure to provide guidance for technological innovation. A polycentric approach suggests that the policy instrument be opened up to debate at different levels—global, European, national, regional, and local—and that the scope of economic analysis be broadened to consider institutional and political economy arguments as well.

382 C. Kemfert et al.

14.3 The Ongoing Energiewende in Germany: Lessons from the "Engine Room"

14.3.1 Lesson VI: The Energiewende Is the German Version of Low-Carbon Transformation that Focuses Specifically on Renewables

The German energiewende is one of a range of different options for engaging in a low-carbon transformation of the energy sector. It focuses on renewables (>80% in the electricity sector, >60% in overall energy) and aims at ultimately eliminating coal (efforts to promote carbon capture, transport, and storage [CCTS] were abandoned in 2011) and nuclear power plants. Although the current objective is to reach above 80% renewables in the electricity mix by 2050, it may turn out that an even higher share is achieved, since the remaining technologies (e.g., natural gas) may be more expensive and no longer suit the system.

14.3.2 Lesson VII: Some Targets of the Energiewende Were "Low-Hanging Fruits" and Have Already Been Reached...

Even with its relatively high level of ambitions, the energiewende has already achieved many of its objectives. The adverse economic effects of the March 2011 nuclear moratorium were minor, and the planned closure of nuclear power plants is no longer controversial. The energiewende is also on track toward its renewables targets. Remarkably, the German economy seems not to have suffered, and the energiewende has served as a catalyst for technical change, innovation, and employment.

14.3.3 Lessons VIII:... But Carbon Emission Reduction and Energy Efficiency Have Not

An objective of the energiewende that has not been achieved is a significant reduction of carbon emissions in Germany. Although emissions decreased with the financial crisis (2008 and shortly thereafter), the overall level of emissions remains high, both in electricity, and even more so in the other sectors. Thus, a pillar of the energiewende, a 40% GHG emission reduction until 2020 (basis: 1990) has not been met, and the 2030 target (−55%) will only be reached with additional policies. Discussions are ongoing how to recover momentum and develop these policies, such as an accelerated coal phase-out, as well as an electrification and supply through lower-carbon fuels in other sectors.

Energy efficiency targets, too, are hard to reach. Electricity demand (2050 target: −25%) has decreased slightly but may pick up after the economic and financial crisis, when growth resumes. Final energy demand in transportation (2050 target: −40%) hinges on structural technical change in the sector, which is difficult to forecast. Reducing energy consumption for domestic heating (2050 target: −80%) requires up-front investments in what is currently an unfavorable economic and institutional climate. Finally, reducing primary energy demand (2050 target: −50%) also requires drastic measures in all sectors, which have not yet been undertaken. A similar danger arises at the European level, where the European Commission and the Member States have difficulties setting meaningful energy efficiency targets and translating them into operational objectives and corresponding legislation.

14.3.4 Lesson IX: Technologies Needed for the Energiewende in the Electricity Sector to Succeed by 2050 Are Already Available

Opponents to the energiewende have been very inventive in identifying potential obstacles: "Storage is a problem", "the speed of the energiewende hinges on the speed of network development", "without demand-side management technologies the system becomes unstable", and "offshore wind needs to become significantly cheaper" are just a few of the arguments used against the energiewende. However, the last decade shows that none of these issues pose a real technological threat to the long-term goals of the energiewende. In fact, all elements of a renewables-based electricity system are available, although not all of them have been tested at the appropriate scale:

- Generation-wise, technologies to reach a 80–100% renewable-based electricity system are already in place today; solar PV and solar thermal technologies are still seeing breakthroughs and learning rates of 15–20% (i.e., cost reductions from doubling capacity), onshore wind could supply the entire German electricity demand by itself, and offshore wind is readily available but still relatively expensive. Biomass and hydroelectricity provide additional renewable capacities.
- There are now storage technologies of all types that assure a functional electricity system, even with 100% renewables, including metal batteries, hydrogen, and power-to-gas. Cost degressions on some of them, e.g., lithium-ion batteries, have been so breathtaking that "variable" renewables, previously considered as inter-mittent sources, are becoming competitive as regular baseload.
- Similarly, electricity transmission and distribution infrastructure is available in sufficient quantity and quality to support the transformation of the electricity system. The closure of nuclear and coal power plants will open up extensive network capacity, and more flexible use of the system will have the same effect

(re-dispatch). The idea that the expansion of renewable capacities would need to be constrained by network expansion has been proven wrong.

Thus, while there is uncertainty about how these technologies will play out, there is no doubt about the technical feasibility of a renewables-based electricity system.

14.3.5   Lesson X: The Electricity Sector Is Relatively Easy to Decarbonize, Whereas the Transportation, Industry, and Household Sectors Are Not

The energiewende is proceeding rapidly in the electricity sector, where there is a clear roadmap to a decarbonized, renewables-based sector. However, the path is less clear for other sectors that are more reliant on fossil fuels. This holds for the transport sector, where alternative fuels (e.g., electric cars, biofuels, hydrogen) are difficult to implement against a large "installed base" and interest groups representing the German car manufacturing industry. The household sector also encounters difficulties changing its energy mix and carbon footprint, mainly regarding the persistence of significant fossil-fuel based heating systems. Decreasing the energy demand and increasing sector coupling by switching the transport and heating sector to (renewable) electricity is a challenge for the next years. Some technical obstacles also remain in the industry sector, such as steel, cement, and petrochemicals, as well as in the agricultural sector.

14.3.6   Lesson XI: The Economic Benefits of the Energiewende Surpass the Costs, So the Energiewende Is Economically Efficient

A major criticism of the energiewende is its supposedly high cost. However, a dynamic analysis suggests a comparative advantage of renewables, with continuously falling costs. Approaches that consider social costs—externalities such as environmental damage or risks of nuclear accidents—conclude that the energiewende is socially efficient. Nuclear energy has by far the highest costs, not only high and generally increasing capital costs, but also the social costs of accidents, the costs of insurance, and the unknown costs of decommissioning of power plants and storing nuclear waste. Likewise, the social costs of fossil fuels, including climate change and other negative environmental externalities, surpass the market value of the electricity produced by far. Initial back-of-the-envelope calculations of the overall costs and benefits indicate that the energiewende is a costly investment but that it will have even higher returns, and that it is therefore economically efficient from a social and public policy perspective.

14.4   Lessons from a European Perspective: The Energiewende in Germany and Low-Carbon Transformation in Europe

14.4.1   Lesson XII: The European Low-Carbon Transformation Is a Mosaic of National Strategies, Many of Which Include Substantial Shares of Coal and Nuclear Energy...

The choice of the national energy mix is a national prerogative according to Article 194 of the Amsterdam Treaty. So even though there is a common European climate policy and some coordination of national energy policy instruments at the European level, the "low-carbon transformation" of the European energy system is essentially a mosaic of national strategies. It would therefore be neither fair nor useful to transpose the objectives of the German energiewende to the EU level. Some of the other EU Member States have also set ambitious GHG emission reduction targets, but they use different energy mixes, including nuclear power and "clean coal". The UK and France, for instance, may still be counting on new nuclear power to reach their respective climate targets into the 2020s and beyond. Some traditional coal countries, such as Poland, Hungary, and Romania, are still clinging to hopes for carbon capture, transport, and storage (CCTS) to justify their extensive coal electrification.

There are no signs of political consensus being reached on the superiority of particular energy carriers such as renewables. The European Union has therefore pursued a less specific "low-carbon transformation," allowing countries to determine their own fuel mixes. There appears to be tacit agreement among the Member States not to step on each other's toes. This applies to the enormous subsidies planned by the UK for new nuclear power plants at Hinkley Point C, but also to the ongoing support for renewables in Germany. Although the German energiewende and the European low-carbon transformation interact, the two can also be pursued separately.

14.4.2   Lesson XIII: ... Yet the German Energiewende Can Benefit from Regional and Europe-wide Cooperation

The objectives of the energiewende in Germany are fully compatible with the creation of a single European energy market based on the principles of sustainability, supply security, affordability, and public acceptance. By definition, and also due to geographic realities, reforms in one country have to be coordinated with neighboring countries (regional approach) and with the overall European internal energy market trends (pan-European approach). This does not imply normalization of political instruments or of national energy mixes, however, which will continue to be affected by national preferences in the European context. Even though Member States may pursue very different objectives in their low-carbon transformation, the energiewende in Germany (and similar processes in other Member States) can benefit from regional as well as pan-European energy policy coordination, such as capacity sharing, joint balancing markets, redispatch, and congestion management. While it has been relatively easy to reap the operational benefits, this has proven more difficult when investments are involved.

Pan-European regulation, such as the infrastructure network codes, generally helps to strengthen the national markets, but in the context of the energiewende, regional cooperation has also proven crucial. Thus, Germany has negotiated bilateral agreements on sharing hydro storage with Switzerland, has created cooperation schemes with its Eastern neighbors for loop flows, and has negotiated bilateral sea cables with Norway and Sweden. The German government has also contributed to a joint declaration with all 12 of its electrical neighbors on market design.

14.4.3   Lesson XIV: Infrastructure Is an Important Element of the European Low-Carbon Transformation, But Does Not Constitute a Bottleneck for Decarbonization

Infrastructure is a pillar of any transformation process, but has not been a binding constraint for either the German energiewende or European low-carbon transformation so far. There is ample infrastructure capacity available across Europe, both for electricity transmission and for natural gas transport. More efficient use of existing capacities, regionally focused, well-designed local expansion schemes (such as reverse pumping of natural gas), and some congestion management can relieve the situation considerably. Complex incentive schemes have not helped the sector, which should be regulated as efficiently and transparently as possible. Thus, while some overinvestment is useful, the high rates of return on equity provided to network operators in Germany and elsewhere (e.g., 9% guaranteed return on capital employed, leading to a weighted cost of capital [WACC] of ~5–7%) have distorted the incentives to substantially overinvest.

14.4.4   Lesson XV: The Low-Carbon Energy Transformation Needs Strong European Energy and Climate Policies

The low-carbon energy transformation needs strong European energy and climate policies. We have identified very diverse European policies over the last decades, first betting on coal, later on nuclear, and then a gradual introduction of renewables. Despite disagreement about other policies, e.g., monetary policy, social policy, or immigration, Europe can remain strong in energy and climate policy, to the benefit of national transformation processes, such as the German energiewende, but also to the benefit of sustainable development in Europe and elsewhere around the world.

14.5   Conclusions

The energiewende in Germany is an ambitious, complex undertaking that needs to be analyzed in the historic, technical, and political-societal context. It has repercussions and interdependencies well beyond the German energy system, foremost in the European context but also worldwide. In this concluding chapter, we have drawn 15 lessons from the process to summarize the essence of the preceding chapters. Like the book as a whole, the conclusions focus on the electricity sector, but provide lessons for the overall energy sector as well, where similar challenges lie ahead in the areas of transport and heat.

After the first phase of the energiewende, there is no longer any doubt about its political or technical feasibility in light of the progress achieved so far. Public support remains strong, with all major democratic political parties continuing to back the energiewende goals. Differences of opinion on specific legislation, e.g., further updates of the renewables law ("EEG 4.0") exist, but there is no fundamental disagreement on the way to move forward. The renewables-based low-carbon energy transformation has given rise to new political lobbying groups that assure continuity of the process above and beyond the day-to-day political bargaining process. The energiewende is here to stay, and more research is needed to explore the most efficient pathways for its future (global) development.

The manufacturer's authorised representative in the EU is Springer
Nature Customer Service Centre GmbH, Europaplatz 3, 69115 Heidelberg,
Germany. If you have any concerns regarding our products, please
contact ProductSafety@springernature.com

Printed and bound by CPI Group (UK) Ltd, Croydon, CR0 4YY

23/04/2026

02095644-0001